P9-DMS-432

WIN AT ALL COSTS

WIN AT ALL COSTS

INSIDE NIKE RUNNING AND ITS CULTURE OF DECEPTION

MATT HART

DEY ST.
An Imprint of WILLIAM MORROW

 For information, address HarperCollins Publishers, 195 Broadway, New York, NY 10007.

HarperCollins books may be purchased for educational, business, or sales promotional use. For information, please email the Special Markets Department at SPsales@harpercollins.com.

FIRST EDITION

Designed by Angela Boutin

Library of Congress Cataloging-in-Publication Data has been applied for.

ISBN 978-0-06-291777-5

20 21 22 23 24 LSC 10 9 8 7 6 5 4 3 2 1

TO TESSA

CONTENTS

Sport is no longer a release from
the harsh everyday American business world
but its continuation and apotheosis.

—ROGER ANGELL, *THE NEW YORKER*, 1975

WIN AT ALL COSTS

PROLOGUE
HONEY, THE FBI IS AT THE DOOR

ON A SUNDAY AFTERNOON IN JULY 2017, TWO INCONSPICUOUSLY DRESSED MEN AP-proached the front gate to Adam and Kara Goucher's yard in Boulder, Colorado. The family mongrel, Freya, started barking, which alerted Kara to them. She was in no mood for visitors. Earlier that day she had bombed out of a sixteen-mile race-pace workout on the road and was now both irritable and hungry. She turned over the salmon she was cooking for dinner and said to Adam, "Someone's coming to the door. Do not let them in."

The special agents arrived in Boulder earlier that day and decided to change out of their suits after an hour of driving around town. Though Boulder's reputation as a city for hippies is steadily being supplanted by one of opulent wealth, it's rare to see a suit within city limits, and they were worried they would draw unwanted attention.

Now in jeans and polo shirts, the men fussed with the front gate. Adam went out to meet them and after a brief introduction, led them inside. Kara glared at her husband, but he gave her a warning look.

"These guys are FBI agents," he said.

Kara set the stove to simmer as their six-year-old son, Colton, peeked up from behind the couch where he was playing. The investigators told the couple they'd like to ask a few questions about their former mentor, Alberto Salazar, and their onetime employer, Nike, Inc. Until now, the Gouchers had no idea the Federal Bureau of Investigation was looking into the controversy the couple had been mired in ever since they turned whistleblowers on their coach in 2013, when they reported his alleged misdeeds to the United States Anti-Doping Agency (USADA). No one involved had yet been sanctioned, but the visit that day from the FBI was a soft assurance that they hadn't been whistleblowing into the wind and that there might, at some point, be retribution. USADA's power lies in its ability to ban and suspend athletes, coaches, and doctors from involvement in competition, but they are not a law enforcement agency. The FBI, however, has the power to incarcerate.

Still, Kara wasn't ready to discuss the details of an investigation that had upturned her life in more ways than one. Turning to the agents she said, "Can you come back after dinner?"

THE SUMMER OF 2017 HAD BEEN A PECULIAR ONE, HOT WITH ANGER OVER POLITICS and a growing feeling that life in America was manifestly unfair. Truths about rich and powerful men getting away with whatever they pleased, from fraud in the case of Donald Trump to Harvey Weinstein's multiple rape allegations, were all over the news, and the country seemed more divided than ever on the basic facts of our collective reality as the internet gave every conspiracy theory fertile ground, and an audience to grow, regardless of how spurious. It was the same summer that convicted doper and Nike athlete Justin Gatlin denied Usain Bolt, in his last solo race, a fairy-tale ending to a storied career. It was the summer the running world was still trying to digest Nike's controversial multimillion-dollar attempt at the marathon world record. Pursuing the sub-two-hour milestone on a Formula 1 track, the Autodromo Nazionale Monza outside Milan, in northern Italy, Nike broke from International Association of Athletics Federations (IAAF) rules by having pacers swapped in and out of the race, running ahead of the athletes in perfect flying-V formation to block the wind. The competitors, who had footwear technology designed to lessen energy expenditure—a shoe called the Nike Zoom Vaporfly Elite—chased a green laser that beamed out from the back of an exhaust-free Tesla. Thirty-two-year-old Kenyan Eliud Kipchoge came within twenty-five seconds of the coveted sub-two-hour target, but many saw the event as little more than a publicity stunt and raised skepticism about the legitimacy of the results. Meanwhile, Kara Goucher had recently learned that her bronze medal from the 2007 IAAF World Championships in Osaka a decade earlier would be upgraded to silver. She had found out on social media that the Turkish athlete Elvan Abeylegesse, who finished ahead of Kara in the 10,000-meter race, had tested positive for a banned substance. Using new technology, the IAAF had retested urine from both the 2005 and 2007 World Championships; the reanalysis uncovered thirty-two adverse findings from twenty-eight athletes, and Abeylegesse's name was the first to be leaked to the world. Abeylegesse tested positive for stanozolol, the same artificial steroid that caused Canadian sprinter Ben Johnson to be stripped of his Olympic gold medal in 1988.

Two thousand seventeen also saw the news break that the federal government's $100 million civil lawsuit against Lance Armstrong—instigated by his former teammate and admitted doper Floyd Landis—would be moving to trial. But perhaps the biggest splash in doping news that year had been the release of the documentary film *Icarus* by playwright-turned-filmmaker Bryan Fogel. A passionate amateur cyclist, Fogel originally set out to expose the inefficacy of the anti-doping system, but his efforts led him to Dr. Grigory Rodchenkov—then-director of Moscow's anti-doping agency—and Fogel found himself embroiled in an unfolding international scandal centered on the 2014 Winter Olympics in Sochi. Russia, as the film depicted, had never had an anti-doping program. Instead, the Russian Anti-Doping Agency (RUSADA) had been serving its athletes, likely for the past forty years, with both the drugs and the means to avoid detection. (Despite more than 1,500 documents made public by the World Anti-Doping Agency (WADA), Russian officials, and their president, Vladimir V. Putin, denied the existence of a state-sponsored doping program.)

Icarus would garner much acclaim, even winning an Oscar for Best Documentary Feature. Lost in these new revelations, however, was the Russian athlete who first exposed her country's national doping program in 2014—a young track star named Yuliya Stepanova.

Yuliya had grown up on a farm with what she described as an alcoholic and abusive father in the city of Kursk, in western Russia. While watching the Sydney 2000 Summer Olympics, she daydreamed of what it'd be like to be an Olympian who competed on television and was admired by her entire country. At a running camp years later she overheard girls in the dining hall talking about taking pills and receiving injections. Her coaches eventually confided to her that without performance-enhancing drugs, the fastest a female athlete could expect to run the 800 meters, Yuliya's best event, was about 2 minutes and 5 seconds. Below two minutes would be world class, and to achieve that time she'd have to start doping, they told her. Yuliya asked her coach Vladimir Mokhnev to put her on a drug program (Mokhnev has denied her account). Yuliya claims he started her out with testosterone and later added the powerful drug erythropoietin (commonly known as EPO) to her regimen. For a time, Stepanova received an EPO shot every other day, and began taking Turinabol (the anabolic steroid given to East German athletes in the 1970s and '80s).

The effects of the drugs, for Yuliya, were dramatic. Her personal record dropped from 2:13 to 2:02, and after winning an age-group national championship, she earned a coveted spot on her country's international running team. Performance-enhancing drugs were a normal part of athletic training for many Russian athletes. Yuliya knew the risks—being caught cheating meant she could be banned from competition—but as she later told the *New York Times*, she was committed to doing everything her coaches asked of her: "I train like they say, I take drugs like they say." And with her athletic dreams beginning to come true, she had little incentive to stop. She was a professional athlete with a modicum of fame in her home country, and the government paid her about forty-eight thousand dollars a year (a considerable sum in Russia) to do nothing other than to train and race.

In early February 2013, as Yuliya was preparing for the IAAF World Championships in Moscow, everything came crashing down. A drug test revealed a banned substance in her blood, and the IAAF suspended her from competition for two years. "When I just found out about being sanctioned, the world that I imagined to myself collapsed in front of my eyes," she later wrote in a statement to WADA. The forced time away from the sport proved to be clarifying for Yuliya. "Now they tell me to keep doing everything they say. Take my punishment, my two-year ban, and don't say anything. Rest, get healthy, and I will still get paid my regular salary. After two years . . . I can start training again for the Rio Games." Despite her coaches' assurances, it was now obvious to Yuliya that she was a sacrificial lamb. She wrote a ten-page confession to WADA, detailing the coordinated administration of performance-enhancing drugs by the very organizations tasked with policing it in Russia, but nothing came of it.

In an even more courageous effort to expose the truth of her country's drug corruption, Yuliya returned to training and began secretly recording her daily interactions at Russian training facilities. She and her husband, Vitaly, a former educator and doping-control officer with the RUSADA, shared the recordings with German investigative journalist Hajo Seppelt, who then broke her story in a documentary, *The Doping Secret: How Russia Makes Its Winners*, which aired in Germany in December 2014. Russia's track-and-field team was subsequently banned from the 2016 Summer Olympics—a stunning and humiliating turn of events for the prideful nation.

Yuliya is now considered one of sport's most important whistleblowers, but the sentiment in her home country is one of contempt. A spokesperson for Russian President Vladimir Putin called her a "Judas," and Russian newspapers featured cartoons and headlines portraying Yuliya and her husband lying for fame and fortune. On the internet, there are even calls for her murder, or "liquidation." For fear of reprisal, her family has been forced into exile in the United States. By the time I contacted them in 2018, for an assignment from the *New Yorker* about what it's like to hide from Russian assassins, they had already been forced to move to more than six different secret locations and were leery of participating.

The outrage directed at the Russian athletic program in the wake of the scandal seemingly ignited the Russian hacking organization Fancy Bear. The group, who is also believed to be behind the breach of the 2016 Democratic National Committee, began a hacking spree to unearth evidence of doping among athletes around the world and lessen the criticism directed at their own country.

In September 2016, Fancy Bear began publishing information from a WADA database showing the widespread use of the therapeutic use exemption (TUE) system, which allows athletes with valid medical needs to take pharmaceuticals with ingredients on the banned-substances list. Though prominent athletes were named, including tennis champions Serena and Venus Williams, Rafael Nadal, gymnast Simone Biles, runner Mohamed "Mo" Farah, and cyclist Bradley Wiggins, the athletes shrugged off the incident, arguing that this use of banned drugs was entirely within the system provided and that, therefore, they did nothing wrong.

Fancy Bear was also blamed for unearthing an internal USADA document that laid bare years' worth of information, including damning details of possible cheating, prescription drug misuse, and medical malpractice in the United States at Nike's premier professional running program, the Nike Oregon Project (NOP), led by legendary coach Alberto Salazar. For years, stretching back to Salazar's own illustrious running career, rumors swirled of his use of performance-enhancing drugs, but until his assistant coach Steve Magness blew the whistle on him in 2012 by taking his concerns directly to USADA, these allegations had amounted to little more than chatter.

Details of the hacked report first surfaced in late February of 2017 in the *Sunday Times* of London, who published four articles in two days reporting that USADA had assembled a compelling case against the Nike Oregon Project. They focused their coverage on UK track star Mo Farah, while choosing not to publish the stolen document in its entirety. The news, at least in the United States, barely made an impression.

IN MAY 2017, A USB DRIVE FULL OF EXPLOSIVE ONES AND ZEROS CAME INTO MY POSsession through a confidential source. When it arrived at my house in Boulder, Colorado, I plugged the drive into my computer, which revealed a single file, a 4.7-megabyte PDF titled "Tic Toc, Tic Toc . . ." Awash in adrenaline, I read the pages in front of me. The document I had on my screen—the "Interim Report of the US Anti-Doping Agency to the Texas Medical Board"—laid out USADA's case against NOP's team coach Alberto Salazar, team physician Dr. Jeffrey Brown, and some of Nike's highest-profile professional runners. In 269 illuminating pages, it depicted a culture of secrecy, coercion, cheating, and possible medical malpractice at Nike.

The *New York Times* assigned me to unpack the damning report, focusing on the evidence of coercion. It was a complicated case, involving alleged abuse of prescription drugs and gray-area tactics, but since no one had been caught red-handed with a syringe in their arm it would require careful reporting to convey exactly what had gone wrong and why it mattered. For this, I needed to speak to people who'd been there—who had trained under Salazar and would be willing to take me behind the scenes of the Nike Oregon Project. I attempted to contact people named in the document to see if they would comment on the record. Nike's PR representatives dragged their feet in responding to my inquiries, even as I stressed that the piece was slated to be published in the coming days. Alberto Salazar and his protégé, Galen Rupp, another key figure of the report, did not respond to calls or emails.

The USADA report partially focused on two athletes, Olympian Dathan Ritzenhein and former Loyola Marymount University track athlete Tara Welling (née Erdmann). Neither was eager to speak on the record. Ritzenhein and I eventually did a phone interview. He was gracious but didn't want to talk about any of the substances allegedly given to him by Brown or Salazar.

Welling would not speak to me and instead passed me off to her husband, Jordan Welling, who works at Nike (though he has no relationship with the Oregon Project or the sports marketing department). I stressed to him how important it was to get an actual interview with Tara, even offering to fly out to Portland to sit with her in person, but they demurred.

Over email, she said that when she joined the team she "had nothing but the utmost respect for Alberto and the staff" and that "I personally never had concerns regarding long-term health while with the Oregon Project." I found these comments hard to reconcile with what Welling had told USADA.

As the report laid out, she had not been truthful with their investigator during their first interview over the telephone, when she denied ever seeing Dr. Brown. The anti-doping agency, in their further review of the evidence, found an email that undercut her claim, and a second, in-person interview was scheduled, which took place in Portland on February 5, 2016. During the questioning, Welling recalled Salazar "abusing prescription medications," and that he carried two large bags of medications that he would give to athletes; he would give her prescription Celebrex whenever she needed it. Dissolving into tears, she told officials, "I don't know if Alberto did something to me."

I FIRST MET OLYMPIANS ADAM AND KARA GOUCHER AT THE TRANSROCKIES RUN STAGE race in Colorado in the summer of 2016. Adam was participating in the multiday trail race, and Kara was there to cheer him on and serve as a celebrity speaker for the event. Both former Nike Oregon Project athletes, Adam and Kara had been silently sitting on their own story of disillusionment with their erstwhile coach and sponsor, until 2013, when they sat down with USADA and revealed what they'd seen and experienced on the team, providing key testimony for its eventual report.

When I approached Adam and Kara a year later about the story I was writing for the *New York Times*, they weren't interested in being quoted in the article, but they were willing to help me understand the background on what had taken place during their time at Nike. They confirmed elements of the report that strained credulity—such as the claim that Salazar frequently gave Rupp private massages. *If NOP employed some of the country's best sports-massage therapists, then why would Salazar need to personally do the job?* I wondered. But Kara assured me, "Oh yeah, that absolutely happened, and it happened a lot." (This was corroborated in interviews with Dathan Ritzenhein, Steve Magness, and others.) The Gouchers also told me—as Kara had testified during USADA hearings—that on a couple of occasions, Salazar left tubes of testosterone cream around the condo at training camps where he acted as Rupp's personal masseur. (Salazar has denied administering banned substances to any of his athletes.)

The story—titled "'This Doesn't Sound Legal': Inside Nike's Oregon Project"—ran on the front page of the *New York Times* on May 20, 2017, and immediately my inbox filled with messages from fellow journalists, athletes, and crackpots alike. One person, who inundated my inbox with attachments and was insistent on speaking to me, was Robert Lyden, a running coach, footwear designer, and outspoken former Nike employee from Oregon. He had sued Nike (and Adidas) for patent infringement and was embroiled in a lawsuit with the megacorporation (both cases were voluntarily dismissed in early 2018). He presented me with what he thought was enough circumstantial evidence to damn Nike once and for all in the court of public opinion. He relayed a possibly apocryphal story I had heard once before about Nike athlete Joan Benoit, the 1984 Olympic marathon champion and running world sweetheart, being offered performance-enhancing drugs by Nike coaches in the 1980s, with Phil Knight present in the room. (Benoit hasn't responded to requests for comment.) He also shared a story about a red dot from a gun scope appearing on his stairwell's white baluster at 1:30 in the morning. He thought someone was trying to send him a message. "Be careful," he told me.

I was beginning to realize just how extensive the tale of coercion and cheating might have been at Nike. The next time I met with the Gouchers, July 6, 2017, we sat down for lunch to talk about a possible book. I told them I wanted to expose the unsettling details of this premier running program and tell the wider story of cheating and misconduct at Nike and at the highest levels of endurance sports. Adam, who has a preternatural sense of fairness, was uncompromising in his attitude toward Nike. When I asked about their relationship with Kara's brother-in-law, who works at Nike and helped design the latest Nike shoe to hit the headlines, the Vaporfly 4%, he said, "Fuck 'em. They're dishonest cheaters." Kara was far more empathetic. Her gaze dropped and her mind drifted as she told me, "I don't want this to be all negative, we had a lot of good times."

Leaving Adam and walking toward our cars, Kara's eyes welled with tears. "They were our family," she told me. "Galen [Rupp] was like my little brother and Alberto [Salazar] was like a father to me, for years. I loved them." Although this was only the third time we'd ever met, I wondered if I should have tried to console her. But what we had started that day—what we had pried open—could not be put neatly back.

AFTER LEAVING THE NOP IN 2013, THE GOUCHERS RETURNED TO BOULDER, A TOWN that calls itself The Bubble because there doesn't seem to be anyone within its city limits who isn't astoundingly fit, accomplished, and wealthy. Here, your neighbor is likely to be a former Olympian, and you might recognize the guy next to you in line at Whole Foods from the latest Patagonia catalog. Because of Boulder's altitude, climate, and topography, it's a place where athletes move to focus their lives on training for endurance sports. Take, for instance, the road up Flagstaff Mountain, just a few blocks from the Gouchers' home. It's a route on which both professional and recreational cyclists often test themselves against the more than 2,300 feet of uphill, and the clock. A ride on which known-dopers Tom Danielson and Levi Leipheimer still hold record times.

Neither of the Gouchers has ever ridden up Flagstaff on a bike. They don't rock climb the Flatiron peaks that jut out of the ground on Boulder's western flank. They don't fight the I-70 traffic to ski on the weekends at one of the nearby world-class resorts. The Gouchers are runners. They have dedicated their lives to the simple pursuit of being as fast as humanly possible in the most basic and primal of sports.

Today, their familial tides are dictated by the dual moons of Kara's training requirements and parenting. She is in the twilight of her running career, a critical point in which an athlete either goes back to the private sector or floats up from earth, like a child's balloon, and somehow maintains a life in the limelight, propelled by the winds of past achievements. There is little doubt which one Kara will do. At age thirty-nine, she still looks like she could win a major marathon, and the fresh face that endeared her to running fans all over the world has barely aged.

Adam retired from running in 2011, after a string of injuries that left him with a hitch in his stride and compromised mobility. Now he more closely resembles a CrossFit athlete than a waiflike distance runner. He is affable but intense, and easily animated by talk of performance-enhancing drugs. "I think that ninety percent of US sports are dirty," he told me. "I'm talking baseball, football, basketball, tennis, running, all of them. And they're stealing money from clean athletes. It's not fair."

When the Gouchers welcomed the FBI agents back into their home promptly at 7:30 p.m. (after they had heeded Kara's demand), their yard sat bathed in fading light. Beyond them, the Flatiron mountain range bustled with rock climbers, cyclists, and trail runners. The couple sat at the dinner table and for the next few hours began to tell the story of their athletic careers to the two agents seated across from them.

"Okay, now tell us exactly what happened at Nike."

01

THE BEST FEELING I'VE EVER HAD IN MY LIFE

ON AUGUST 4, 2012, TWENTY-NINE MEN FROM EIGHTEEN DIFFERENT COUNTRIES LINED up in northeast London to find out who could run 6.2 miles—the 10,000-meter Olympic event—the fastest. Among them was Ethiopian legend Kenenisa Bekele, who owned the Olympic record for the distance at 27 minutes and 1.17 seconds, the world record of 26 minutes and 17.53 seconds, and had won the previous two Olympic gold medals in the event. Add this to his four World Championships in the 10,000 meters and it was clear to see why the British television broadcasters, in their preamble to the event, called Kenenisa "the world's greatest distance runner." The race, however, was anything but a foregone conclusion: it was a stacked field of the fastest men in the world. Kenenisa's younger brother, Tariku Bekele, was a favorite, as was the world half-marathon champion, Kenyan Wilson Kiprop. Kiprop had run the fastest 10,000 meters of the year in front of Oregon track fanatics at the University's famed Hayward field during the 2012 Prefontaine Classic, which hosted the East African Countries Trials.

The XXX Olympiad of 2012 was London's third time hosting the modern Olympic Games since the ancient games were revived by Pierre de Coubertin in 1896. The United Kingdom spent an estimated $15 billion to stage these Games including $775 million on the new stadium the men were now lined up in. London's first time hosting an Olympics, in 1908, saw approximately two thousand athletes—though only thirty-seven women—compete for 110 gold medals. The Games of 2012 would involve some 10,500 athletes—with women now representing almost half of the field—competing in 302 events.

Nike had several athletes racing, but the distribution of their largesse had focused mostly on their two most prominent Oregon Project runners, Galen Rupp and Mo Farah, from whom they were expecting great things. The men had been training together under the diligent eye of Alberto Salazar, a former Nike athlete and American record holder in the 10,000 meters. Salazar had been working with Rupp, who was now twenty-six, since before he could legally drive and had grown to love him like a son. After a somewhat lackluster beginning to his professional running career, Farah and his family moved to Oregon early in 2011 to join the secretive Nike program based at the Beaverton world headquarters. He was still under contract with Adidas when he arrived, contractually obligated to wear the three stripes while training—an unheard-of breach of etiquette on the insular campus and a sin for which Farah was forced to endure stares of disdain from campus employees, until Salazar got word out that it was okayed at the highest levels. Farah had great reverence for his new coach and the pair enjoyed an almost instant leveling up in the athlete's performance.

Eleven years into its tenure and untold millions invested, the Oregon Project had not produced to a commensurate level of expectation, or expense. An American athlete hadn't earned a medal of any color in the 10,000 meters since the 1964 Olympics in Tokyo, Japan—twenty-two years before Rupp was born. A European runner hadn't won the event since 1984, and an Englishman never had. Today would change all of that.

The race plan for Farah and Rupp was simple. *Stay near the front of the pack*, Salazar had explained to them before the start.

If anyone makes a move to pull away, cover the distance and put yourself in a position for a final-burst victory. "We felt they could out-sprint anyone in the race, we didn't care if it was a fast pace, a slow pace, whatever," said Salazar, after the event. "They wouldn't be trying to win it until the last four hundred, maybe even the last two hundred meters."

As the network cameras panned past the athletes on the start line, Rupp, wearing a long gold chain, a red Nike USA singlet, and shorts pulled up high, kissed his wedding ring. As they stopped on Farah—who, despite now being a Nike athlete, was required by United Kingdom Athletics to wear the Adidas sponsored jersey of Great Britain—held his palms up behind his ears, as a way of saying *Let's hear it!* to the crowd. The Somali-born athlete moved to England as an eight-year-old and considers himself British. The London crowd responded with deafening enthusiasm. Buoyed, Farah blew the audience a kiss, raised his arms, and started jumping up and down in place to keep the crowd going and warm up his legs. Then he brought his palms down over his face and, with a big exhale, cleared his expression back to all-business, as an actor might do to prepare himself for the drama ahead.

In professional running there are two events that matter more than all the others combined. The Olympics is the premier affair. A distant second, but still the third-largest sporting event in the world, is the International Association of Athletics Federations World Championships. The Olympics take place every four years, with the World Championship alternating every other year, on the odd year, since 1991.

Winning an Olympic medal, then, as now, comes with no prize money from race organizers; instead many countries' national sports federations provide the financial incentives. And they vary widely among countries. An athlete representing the United States of America in the 2018 Winter Olympics, for instance, would have received $37,500 for a gold medal, $22,500 for silver, and $15,000 for bronze. Singaporean athletes are paid the most for ascending the Olympic podium, receiving $1 million for a gold medal, $500,000 for silver, and $250,000 for bronze. However, for most athletes in most countries this bonus is not life-changing money. The *real* financial windfall for a successful runner is made through sponsorships, where finishing on the podium can gain an athlete millions of dollars in corporate backing. Signing an endorsement contract with Nike or Adidas or Puma can bring fame and fortune, provided the athlete continues to perform at the top level.

When the starting gun fired, eighty thousand fans watched Kenenisa propel through the herd to take the lead and set the pace, a tactic he hadn't employed in many of his preparatory races so far this year. The field of runners spread out into a long line, with everyone clamoring to stay with the front pack. Farah and Rupp were patient, at times dropping back out of the top ten, but always in the mix of runners in the lead group. More than half the athletes racing were in Nike spikes—their allegiance visible from the stands thanks to the neon yellow of their shoes.

There are two "long distance" Olympic track races: the 5,000 meters and 10,000 meters. The "ten" or "10K" as it's called, is the longest of the Olympic running events held on a track. Though the 26.2-mile marathon begins and ends on the Olympic stadium oval, it usually sends the athletes out to the streets of the host city, rather than forcing them to run 105.5 laps around the track. The 10,000-meter event was added to the Olympic program in 1912, though just for men. Women runners had to wait an additional seventy-six years to compete in their own 10K. The 6.2 miles is exactly twenty-five laps of a standard four-hundred-meter track.

Six minutes into the event, Eritrean Zersenay Tadese took the lead. Comparatively short and compact with a slight forward bob, the five-time World Half Marathon Champion was known for pressing the pace in past competitions. He pulled the group through the halfway point. After eighteen minutes, Kiprop jogged off into the center of the track and dropped out. Twenty minutes in, Rupp found himself in fourth place, sandwiched in between the two Bekele brothers, with Kenenisa in front of him in third and his brother, Tariku, behind. Farah waited patiently as his coach had instructed, in sixth place. The front pack was now twelve athletes that had pulled away from the others, with fellow NOP runner American Dathan Ritzenhein meters off the back in no-man's-land struggling to regain contact. Farah made a move to the front but didn't attempt to drop the group. By the time the bell sounded for the last lap around the track, the race had seen no significant power plays.

With increasing urgency, Farah and Kenenisa battled around the final turn before the last lap with Tariku just off their heels—but Farah would not be denied.

As Farah accelerated ahead, the crowd drowned out any chance of hearing the stadium announcers. With his graceful long legs gobbling up the track, Farah's feet flashed Nike yellow as he ran the final 400 meters in about 53 seconds. Rupp began his move into third at 27 minutes and 12 seconds and passed Tariku, and then, with just meters to go, overtook Kenenisa for silver.

Crossing the finish line first, Farah, arms outstretched, wore an expression of disbelief. As the men decelerated, Rupp, less than half a second behind Farah, screamed "Moooo!" The two men hugged. Then Farah, who is a practicing Muslim, got down on his knees, put his head on the polyurethane track, and bowed three times toward Mecca to the south. The cameras cut away until he was done with his prayers. He then rolled around like a joyful child before Rupp helped him up into an embrace. With arms slung around each other, both men flashed the number one sign with their index fingers before Rupp remembered he was second and added the additional finger to make two.

The day would come to be known as Super Saturday in England, for the three gold medals won by their Olympic team in quick succession, ending with Farah's.

"Look at the scalps of Africa taken by Mo Farah, and of course, Galen Rupp," a broadcaster blared.

"It was the greatest feeling, perhaps, I've ever had," said Salazar of the race. "It was better than anything I ever did in my own running career. Other than marrying my wife, and my kids' births, this was the best feeling I've ever had in my life."

His two stars had ascended to the top of the sport, just as he'd envisioned way back in 2001 at the inception of his running program. The stunning and decisive vindication for Salazar and the NOP would soon be immortalized in a banner featuring a photo of the duo's triumphant finish that hung in the Lance Armstrong fitness center on the Nike campus. The decade-plus effort to help top American distance runners not only compete with but defeat the Kenyans and Ethiopians had finally paid off.

After the event, Salazar received more than sixty text messages, but the most crucial one came from Tom Clarke, the Nike vice president who had originally green-lit the Oregon Project. Watching from a hotel room in Atlanta he couldn't hold back the tears of joy. His text message to Salazar read, "I'm so proud. Congrats. TC."

AS FARAH AND RUPP CIRCLED THE TRACK ON THEIR RESPECTIVE VICTORY LAPS IN LON-don, back in Houston, Steve Magness's cell phone lit up with texts of his own. As former assistant coach and scientific adviser to the prestigious program and Salazar's former right-hand man, Magness had helped coach both athletes in their preparation for this day. However, since quitting his job two months earlier, the entire fiasco made him sick to his stomach as he considered what his former athletes had done to achieve greatness.

"It's supposed to be this grand moment where you played a role in helping someone do something that no one thought was possible, and it's the complete opposite," Magness told ProPublica and the BBC. Instead, it was "one of the most disheartening moments of my life."

He had moved back home to Texas crestfallen after leaving the palatial Nike campus and was trying to put his career and his life back together. Though he accepted a position as an assistant coach to the University of Houston's track-and-field and cross-country programs, he ruminated on the many unsettling things he'd witnessed while employed at Nike world headquarters, which dampened his excitement for the sport that had become his life's work.

Magness felt an obligation to tell anti-doping authorities what he'd seen, but this was no trivial decision for an exercise scientist deeply involved in the sport. He had a tough decision to make: keep his mouth shut and pretend he didn't see anything untoward or contact the sporting authorities and let them decide what to do. Whistleblowing is typically a thankless act of sacrifice, Magness understood. Shining lights into dark and secret worlds and speaking truth to power comes with great risk, and here it was an even greater risk—because it directly implicated him. There was a very real chance coming forward could destroy any prospect of a career as a running coach, get him sued, and possibly land him with the scarlet letter of a USADA doping suspension.

"I've never experienced war or anything like that, but I experienced PTSD from this," Magness said when we met in Houston for interviews in the fall of 2018. He's of average height, slight build, and looks every bit a scientist with dark un-moussed hair, glasses, and an introverted demeanor.

Growing up in the North Houston suburbs in the academic shadow of his older brother, Magness turned to sports as a way to differentiate himself. He loved soccer, but was the progeny of track men (his father ran and his grandfather was a state champion).

In 1966, President Lyndon B. Johnson started a program called the Presidential Physical Fitness Test, as a way to encourage exercise among school-aged kids through a variety of challenges. One of which was the timed mile. This was where Magness caught the first glimpse of his future.

By high school his coaches encouraged him to stop playing soccer altogether, then his teammates coerced him to take training seriously. He would run before and after school and started putting in fifteen miles every day. "My entire senior year of high school, I stayed up past ten p.m. a total of six times," Magness later said of his dedication. His mile time improved from 4 minutes and 21 seconds his freshman year to an astonishing 4 minutes and 1 second as a senior, the fastest mile run by any high schooler that year.

"I thought, straight up, I was going to college and then I was going to run professionally," said Magness, "and I was going to go to the Olympics." That year, the *Houston Chronicle* listed him as a "Legend in the Making," in an article about the city's next generation of sports superstars.

He chose to attend Rice University to study exercise science and ran well at cross-country as a freshman, but his track season went poorly. "I didn't care about school work, I didn't care about grades. I didn't care about any of it," said Magness. "I was just all running." His entire identity was wrapped up in being fast, so when running wasn't going well, he questioned his value as a person. Looking for a change he transferred to the University of Houston just five miles away. His performance improved on the new team, but not to the level of expectation he had garnered coming out of high school.

After graduation Magness moved into a running commune of sorts with a few friends in Los Angeles. The two-bedroom condo was subsidized by the local running store so the six post-collegiate athletes, who all slept in bunkbeds, could have an opportunity and some breathing room financially to train and race. Magness focused on road 5K events, but only lasted six months in the house before he left for graduate school in Virginia.

As it became painfully clear that his dream of running for his country in the Olympics wasn't going to happen, Magness gradually turned his attention to sport science. He began training with Alan Webb and his coach Scott Raczko while studying physiology at George Mason University. The time spent with America's best miler was formative for Magness. He witnessed how hard Webb worked, day in and day out, with Raczko. "And he was clean," said Magness.

In December 2010, he visited Webb, who was now living in Oregon and training under Salazar. Magness tagged along for Webb's daily training at the Nike campus, where he unintentionally impressed the coach. Crowded around a small screen, watching a video of Paula Radcliffe's running form, Magness made an astute comment that no one had asked him for.

Still, when Salazar called to offer him employment, he thought it was a practical joke. In 2011 Magness accepted what he can only describe as his dream job, working with the most powerful coach in track and field and some of the fastest distance runners on the planet at Nike in Beaverton, Oregon. He ignored red flags—like Salazar's rumored testing of testosterone or the flippancy with which he passed out prescription drugs—with the typical gusto of a new Nike employee. A wide-eyed Magness worked in the Mia Hamm office building and met daily with his athletes in the Lance Armstrong fitness center. He ate in a cafeteria named Boston, after the marathon. He saw NBA star Manu Ginóbili one day while working with an athlete on an AlterG treadmill and LeBron James on another day in the Bo Jackson gym. His friends and even his parents would frequently remind him of his amazing luck and privileged position.

"I'm thinking my future is set," said Magness. "I'm going to do this and if Salazar retires, maybe I'll get his position. I don't have to do anything else for my life."

Over time, however, the veneer wore off, and the reality of the high-pressure environment began to seem at odds with Magness's moral compass. He had witnessed Salazar acting in dishonorable ways, but, at least at first, nothing overtly illegal. He'd seen him bully doctors into giving both himself and his athletes prescription drugs they didn't have a medical need for. He had watched as Salazar celebrated injuries to Nike athletes trained by rival coaches. Salazar was obsessed with beating Jerry Schumacher, a fellow Nike running coach he had hired to succeed him, who eventually broke off to run his own, rival program within the brand's ecosystem. And the most cringeworthy of all, Salazar had a habit of making inappropriate comments about female athletes' bodies, especially his star athlete Kara Goucher.

Though Salazar tended to act as his athletes' doctor, his undergraduate degree was in business and marketing. By the time he hired Magness, he had a reputation as a trainer who chased one scientifically suspect trend after another, often changing direction and making decisions on gut feeling instead of solid data, at great expense to Nike. In an interview with *Sports Illustrated* years earlier, Salazar appeared cavalier about his own drug use. He admitted to experimenting with the corticosteroid prednisone in an attempt to revive his deficient adrenal system, and noted that he was on Prozac when he won his last race, the 55-mile Comrades Marathon in South Africa, in 1994.

According to the USADA report on the Nike Oregon Project, Salazar admitted to also secretly taking testosterone around this time—a drug which is unambiguously illegal in all sports—a fact the world wouldn't find out about until the report went public in 2017.

Magness, on the other hand, had both a bachelor's and a master's degree in exercise science, having graduated summa cum laude. The day he arrived at the leafy Nike campus he instantly imbued Salazar's running program with scientific credibility.

Early in his tenure at Nike, a number of disconcerting things began happening. While Magness was on a trip with Galen Rupp in Germany, Salazar mailed the runner drugs, which he'd placed in a secret hole cut into a Clive Cussler novel. Shortly thereafter, Salazar asked Magness to investigate a substance called L-carnitine—a drug that is not on the banned list—after researchers in England discovered that it could possibly improve running endurance.

The substance, though easy to find on the open market, had a notoriously long load time, which meant that to get L-carnitine into the athletes at the levels required to realize the performance-enhancement they'd have to employ experimental techniques.

Salazar feared other coaches would soon begin exploring the supplement, too. With the 2012 Olympic Trials approaching he ratcheted up efforts with Dr. Jeffrey Stuart Brown to use an infusion method that the supplement researchers had used as a proof of concept. The method bumped up against WADA rules around infusions, but, Magness told me, Salazar said he'd cleared everything with USADA.

"The next one was the testosterone," said Magness. Later the same year, while reviewing some historical medical data on the Oregon Project athletes, he noticed a line item for Rupp that read, "Presently on prednisone and testosterone medication."

He took it to Salazar, who reviewed it and then told him it was a mistake. Salazar blamed Dr. Loren Myhre, the Nike physiologist who had entered the early notes. "He said Loren was crazy, that's why it was wrong, he never said anything about it being a supplement," Magness explained.

Initially, Magness managed to convince himself the notation was probably a mistake, though he logged it in the back of his mind as yet another red flag before going back to work. From that day on, he and Salazar would never see eye to eye again.

About a year in, Magness said Nike stopped paying him his salary, a turn of events he is convinced was orchestrated by Salazar. "Any leverage Alberto had to make you afraid of your job or just putting you on edge," he explained, "he took advantage of." So Magness quit, and after watching Rupp, who he described as "not even on the same sphere of talent level as Alan Webb," take silver at the 2012 London Olympics, right behind Farah, decided to send the most consequential email of his life.

On December 10, 2012, Magness sent the following message to USADA's "Play Clean" email address:

> Look into the Nike Oregon Project athletes.
>
> I'm strongly suspicious of using testosterone cream as I saw it labeled in test results for Galen Rupp before. Along with the fact their head coach has a prescription himself for testosterone cream.
>
> Also, Hemoglobin levels are regularly in the 17–17.8 levels for athletes with total red blood cell mass as high as 1100g for an athlete like Galen Rupp. Those are pretty high levels, even with the use of altitude or simulated altitude. Unfortunately I was not privy to blood volume and therefore hematocrit levels.
>
> They're also pretty good at doing legal injections under 50ml instead of infusions. I know it's permitted, but they've done this with L-carnitine, magnesium, and iron, plus a few others probably. L-carnitine they took an infusion protocol and instead went with 3–4 small injections while drinking a high-glucose drink instead of the glucose+carnitine infusion that was done in the medical journals.

What began as a cathartic truth-telling exercise for Magness ignited a full-blown investigation by USADA, which typically follows up on all credible tip-offs. They interviewed Magness multiple times in the proceeding months and commandeered his laptop and phone, which they scoured for evidence of further wrongdoing. They told him that they were going to ban him first, for the L-carnitine infusion he received as NOP's guinea pig, and then work on getting Salazar and Dr. Brown suspensions.

Whistleblowers have protection for a reason: in exchange for their valuable and reliable information, they are usually given a lighter sentence, or none at all. In our interviews, Magness said he told USADA, "Look, if you think I deserve a ban, well okay, but this sucks because along with the Gouchers, this whole shit starts with me coming forward and risking everything.

"What if you ban me and then Alberto goes through his arbitration and he gets off scot-free? So then the one person who is going to be banned is me. I'm not painting myself as some righteous person who's one hundred percent in the right all the time. I'm flawed and I'm human and I did some stuff that I deeply regret. And if I went back, I wouldn't do it again, but it's not like you caught me, I'm giving you all this stuff because, for the greater good, there is some shit that needs to be fixed here."

02

FORT KNOX WEST

TO GET TO THE HEART OF THE NIKE CORPORATION, YOU TRAVEL NORTH ALONG THE famous green berm in Beaverton, Oregon, then take a left on Bowerman Drive and pass under an elevated white bridge with a giant orange checkmark on it. The initial buildings were cast around Lake Nike, a man-made feature at the center of campus. The elevated earthen berm that surrounds the campus buildings was built with the excavated dirt removed to create the lake, and the bridge connects the nearly two miles of running trail that surrounds much of the four-hundred-acre corporate headquarters. It's a fitting threshold for the world's largest athletic shoe and apparel manufacturer, whose humble roots began with the then-nascent sport of recreational running in 1964 and now makes its presence known around the globe.

Phil Knight once called it "a topographical map of Nike's history and growth," and driving onto campus has the feeling of entering an amusement park, museum, embassy, and college campus all in one. An ongoing $1 billion expansion project has brought the total number of primarily white, gray, and black buildings to more than seventy-five, and interspersed amid bucolic streams, glens, and waterfalls are tennis and basketball courts, soccer fields, and gyms. Black Mercedes vans with orange swooshes on them and orange bikes with white swooshes provide transportation through the sprawling campus. Like most places in the Northwest, Nike's headquarters is idyllic when the sun comes out—blooming green and full of waterfalls, bamboo forests, and statues—but there are rain-covered sidewalks and bridges between buildings for the reality of the weather in this region of the country.

Bowerman Drive is named for Bill Bowerman, the force of nature who coached University of Oregon runners from 1948 to 1972 and cofounded Nike with Knight. Bowerman was an Oregon original of true pioneer stock (his grandmother traveled the Oregon trail "in utero" as his father liked to say) who not only coached Phil Knight, but National Collegiate Athletic Association (NCAA) championship teams, and an Olympic gold medalist during his twenty-four-year career. He played an integral role in ushering in the running boom itself, which helped create the very market that Nike so masterfully took advantage of. The company that began selling Japanese shoes out the back of a green Plymouth Valiant at high school track meets now enjoys a market value of about $160 billion and boasted of a yearly revenue of more than $39 billion in 2019. Nike and its swoosh are possibly the most recognizable brand on the planet, and its cofounder Phil Knight is one of the richest men to have ever lived, with a net worth estimated by *Forbes* at some $35 billion.

Many factors contributed to the rise of running in the early 1970s, including Frank Shorter's Olympic gold medal in 1972, an increasing public interest in fitness, and the birth of some of America's major endurance races. But Bowerman's success on the track, his community outreach running classes, and his groundbreaking book, *Jogging*, are considered essential ingredients to the movement, at a time when, by one estimate, 48 percent of Americans had at least tried running.

As a coach, Bowerman was a tinkerer who pondered every aspect of his discipline for ways to make his athletes faster. He conceived of a sports drink that would replenish what the runners lost through sweat, ten years before Gatorade (though he abandoned the idea after not being able to improve the taste past "sheep's piss") and would work tirelessly on improving the surfaces that running tracks were made of. His athletes loved and revered him, but some of them have also called him a tyrant and considered him a sadistic prankster. He was as likely to recite a biblical verse as he was to take his penis out and urinate on his young athletes in the showers, or brand them with a hot key in the sauna as a perverse test of their pain tolerance.

He was convinced that shoes were central to athletic performance, and believed that the lower the weight, the less encumbered athletes would be, and the faster they'd run. The track shoes that were available in post–World War II America, however, were either expensive German shoes or "Crap. Crap. Crap. Crappy, hard leather, crappy ten ounces per shoe," as Bowerman put it. He had fought the Germans in World War II as a member of the famed US Army Tenth Mountain Division and resented paying Adidas, a German company, for his team's track spikes.

LONG BEFORE NIKE GOT IN THE SHOE GAME, ADIDAS WAS THE FIRST PREEMINENT global athletic brand. Its founder, Adolf "Adi" Dassler, began his life's work amid the economic depression of World War I, in Herzogenaurach, Germany, where he and his brother founded Gebrüder Dassler Sportschuhfabrik, the Dassler Brothers Sport Shoe Factory, in 1924, specializing in athletic shoes. Adi was generous with samples, sending them out to sports clubs, players, and team managers. By 1926 the brothers were producing one hundred pairs of shoes a day, including football (soccer) shoes and leather track shoes with hand-forged spikes. Sales were brisk, and the Dasslers rapidly expanded their operation.

As Adolf Hitler rose to power in the 1930s, the brothers had little choice but to pledge allegiance to the Nazi Party. To refuse would have put everything they had built, and the livelihood of their more than one hundred employees, in jeopardy. Though he preferred athletic competition to politics, Adi became a supervisor of the Hitler Youth sports league in Herzogenaurach with the goal to support athletes with sporting goods equipment without concern for political affiliation, religious faith, or ethnicity. Regardless, the brothers' shoe factory was employed to make marching boots for the Nazi troops.

Hitler viewed sport as a way to develop and display the "genetically pure" Aryan race as physically superior to the rest of the world. A Nazi Germany that dominated international sports would, by necessity, be fit and hardened when needed for the battlefield. The training of soldiers therefore included the systemized practice and playing of sports.

When Berlin was selected as the location of the XI Olympic Games in 1936 the Reich chancellor seized the opportunity to show international visitors an open-minded, cosmopolitan, and peaceful regime. Never in their history had the Games become such a political chess match. Though the German Jewish population was living in ever-increasing fear, the international community hadn't yet realized how bad it had become, and certainly not how bad it was going to get. The Nazis hid their true intentions, and as the Games approached they took great pains to disguise the systematic disenfranchisement of the Jews (even suspending distribution of *Der Stürmer*, Berlin's weekly anti-Semitic newspaper).

For sixteen days in August the Nazi Party enchanted the world with their clean streets, efficient public transit, and beautiful grand hotels. "Foreigners are spoiled, indulged, flattered and fooled," the journalist Bella Fromm wrote in her diary. "The propaganda machinery is trying to give visitors a positive impression of the Third Reich using the Olympics as camouflage."

The XI Olympiad of the modern era proved to be the most dramatic and intriguing spectacle in all of Olympic history, and became the apogee of Hitler's popularity—and his hypocrisy. The "peace-loving statesman" had already begun building his concentration camps and had set in motion a plan to grow the army for a forthcoming war.

It was the biggest and the most impressive Games ever, with

more countries participating, more athletes running, and more competitions than ever before (just shy of four thousand athletes from forty-nine countries traveled to Berlin to compete). It was a media event of unprecedented scope and scale. This was the first time the Olympics were televised, and eighteen hundred journalists from fifty-nine countries reported on the competitions. Radio broadcasts of the events reached forty-one countries.

Fifteen world records were set in August 1936, as well as forty-one new Olympic bests. Germany won thirty-three gold, twenty-six silver, and thirty bronze medals. They were far and away the most successful country at the Games with eighty-nine total medals, followed by the United States, which won fifty-six.

The undisputed hero and fan favorite of the Games was an American track-and-field athlete from Ohio State named Jesse Owens. Much to the displeasure of Nazi leadership, the African-American athlete became the first to win four gold medals in one Olympiad . . . and he did so in Dassler shoes. His performance in front of the Führer was an affront to the entire idea of white superiority. This would prove an extremely risky boon for the Dassler brothers' business in the narrowing band of acceptable behavior of Nazi Germany. But it established their worldwide reputation as a true sports performance brand, and sales took off.

When Germany invaded Poland in September 1939, the Nazi troops fought for days on end without rest. The speed and efficiency of the invasion shocked the Allied Forces. Recent research by author Norman Ohler for his book *Blitzed: Drugs in the Third Reich* has unearthed a possible reason. Ohler has claimed that the Nazi ranks were awash in drugs: cocaine, opiates, and most of all, methamphetamine.

Many countries repurposed their shoe manufacturing facilities for their respective war efforts. The US army wore flying boots made by the Converse shoe company out of Massachusetts, while one of the oldest cleat makers in England, a company called Gola, went from creating soccer "boots" to marching boots. Because of industry shortages, the Gebrüder Dassler Sportschuhfabrik factory evaded shutdown, but was employed to build *panzers* (tank armor) and bazookas, as well as footwear for the troops. They even managed to expand their running line to include shoes with new names such as “Blitz” and “Kampf.”

As the war came to an end and Soviet forces closed in on the Berlin chancellery in April 1945, Hitler committed suicide by swallowing cyanide and shooting himself in the head.

The Dassler Brothers Sport Shoe Factory survived the war, but in 1948, after twenty-four years of working together, brothers Adolf and Rudolf went their separate ways due to irreconcilable differences. Each man continued the family profession and built their respective brands into West Germany’s showpiece sports companies. Adi combined his first and last name to arrive at Adidas. Rudolf, who went by “Rudi” among friends, named his business Ruda and set up shop just across the Aurach River in Herzogenaurach (though he would later rename his company Puma).

By the mid-1950s, Bowerman was fiddling with his own creations in the United States. As the head coach at the University of Oregon, he was obsessed with making a better performing and lighter-weight running shoe for his athletes. He spent time with a local cobbler to learn the basics of how footwear was put together, then systematically experimented with new materials, including nylon, mesh, lightweight leather, rattlesnake, and even kangaroo skin on the upper part of the shoe. He very quickly managed to cut the ounces in half and began gluing the shoes together instead of using the traditional sewing method employed by industry leaders. He reached out to American footwear manufacturers of the time, like Spalding and Rawlings, with his ideas and track shoe improvements. “He tried to send pictures to all the shoemakers and nobody would have anything to do with it,” Bowerman’s wife, Barbara, told PBS in 2007. “He decided, ‘Well, I’m going to make my shoes myself.’ ”

Knight, who went by Buck as a young man, joined Bowerman’s Ducks in 1955, and became one of the coach’s pet projects. The gruff sensei set to toughening the young man—who was viewed as having more enthusiasm than talent—with some good-natured hazing, reportedly subjecting him to the shower urination ruse at least three separate times. He also questioned out loud if the shy, scrawny Knight really wanted to be on the team. He once forced the towheaded Oregonian to run a long, hard workout with the team when he was clearly ill and should have been in bed. "He was contradicting everything he'd taught us about not running sick and making it worse. I knew it was just to make me acknowledge that his word was law, no matter how arbitrary. But that didn't make it any easier to take. I was so mad I ran a personal record," said Knight. "When I hung in, he saw character, and that was the end of it. Later, it was funny to watch him hazing other guys the same way—funny after you'd been through the cycle."

Though Knight ran a respectable personal best of 4 minutes and 13 seconds in the mile while attending the University of Oregon, his résumé of achievements has but one line from his time as a Duck. In 1959 he was part of a four-man, four-mile relay team that set an American record at the Drake Relays, beating the previous mark by seven-tenths of a second. Bowerman referred to Knight as "a good squad man."

After running for Bowerman at the University of Oregon, Knight was accepted into Stanford University's MBA program. He graduated in 1962 and moved back into his parents' spacious white house on Claybourne Street in the Eastmoreland neighborhood of Portland. The twenty-five-year-old then convinced his father to fund a vacation-cum–business trip to Japan, armed with an idea he'd developed in graduate school to import cheap Japanese shoes and sell them at considerable markup in America. He left on Thanksgiving Day of 1962 (a time, he points out in his 2016 biography, when "90 percent of Americans still had never been on an airplane").

Once in Japan, Knight paid a visit to the track at the University of Tokyo, where he noted the shoes on the athletes' feet. After touring sporting goods stores as well, he concluded Onitsuka Co., Ltd., was the best hope for US exports. He cold-called the company and managed to set up a meeting with their executives. Though virtually unknown in America, Onitsuka dominated the Japanese basketball market and sold more than $8 million worth of sports shoes each year; however, many in the running community considered Onitsuka track spikes, which also featured a three-stripe emblem, a poor Adidas knockoff. Their owner, a former military commander named Kihachiro Onitsuka, started the company in 1949, and as legend has it, created his original shoe last—the basic three-dimensional mold upon which a shoe is constructed—by melting candles from a Buddhist shrine and shaping them to his own feet.

Knight arrived at the Onitsuka headquarters confident but admittedly clueless, and proceeded to begin his business life with a creative lie. Fresh out of an MBA program, he wasn't even employed, let alone a business owner, but when asked who he represented in America, he sputtered "Blue Ribbon Sports," a name he came up with on the spot. Knight has floated two different explanations for the name throughout the years: One involves the fact that he had been drinking the night before the meeting and recalled the beer company Pabst Blue Ribbon. The other—more polished but no less unimaginative—explanation was that he thought of the first-place ribbons given to the winner of a foot race. Knight also told the Onitsuka executives that he had run for the famed track coach Bill Bowerman. The gruff, old veteran held cachet to anyone who knew track and field, and the executives were summarily impressed.

Knight didn't leave the meeting with a business deal, but he did receive a handshake from Onitsuka confirming that they would mail him twelve pairs of their Tiger brand shoes as soon as they received his payment. In his travel diary that day he wrote that he had "faked out Tiger Shoe Co." He then sent a letter to his father asking him to wire the $50 to the company.

While Knight awaited samples in Oregon, halfway across the world in Herzogenaurach, Adi Dassler reigned over the world's largest athletic shoe company. Despite being entangled in a family feud with Puma, Adidas had seven hundred employees in four factories, and sold their wares in fifty-eight different countries—far and away the biggest brand in the industry.

The samples arrived in Portland three weeks after the assassination of the thirty-fifth president of the United States, John F. Kennedy, on November 22, 1963. By then, the pragmatic Knight had already taken another job at an accounting firm that would become Coopers & Lybrand, which paid him $500 a month. He immediately sent two of the five Onitsuka Tiger Limber Ups to Bowerman with a note saying he'd be willing to sell his old coach shoes for his athletes at cost, $4.50 for flats and $7.00 for the track spikes. On January 22, Bowerman wrote back that he liked the looks of the shoes and that he wanted to be cut in on the deal.

Neither of the men had much extra cash to use as seed money in their new business venture, however. Bowerman's teacher salary allowed him to support his wife and three sons, but not much else, and Knight was still a junior accountant. Regardless, later in the week, they shook hands on a partnership that gave Knight 51 percent and Bowerman 49 percent ownership of their new entity, Blue Ribbon Sports. With no business plan to speak of, they signed a contract a few days later, with both men investing $500 into the venture. The business model was simple: import Japanese shoes, mark them up, and sell them to Americans. All of this sounded agreeable and opportune to Knight, but a week later it struck him that he better add an addendum to the contract. In a note under his signature in legible longhand he wrote: "P.S. I forgot to make it part of the original agreement, but I think it ought to be made explicit: There will be no pissing on partners in the shower."

THOUGH PORTLAND IS OFTEN DEPICTED IN POPULAR CULTURE AS A TOWN OF BOHEmian eccentrics bumming around among the artisanal cafés, knitting circles, independent bookstores, and marijuana dispensaries, the area has thrived from the historic agreement Bowerman and Knight made back in 1964.

The transformation into the Silicon Valley of the athletic footwear industry began as soon as the campus opened in October 1990, just seven miles west of downtown Portland. Over the years, as Nike blossomed, so too did the local economy, as the company brought a reported sixteen thousand jobs and nearly $2 billion to the state.

There are now more than eight hundred athletic and outdoor industry companies in the area, employing more than twenty thousand people. German juggernaut Adidas Group moved some of its operations just outside the City of Roses in 1993 after admitting it was having trouble recruiting talent to live in the Bavarian village of Herzogenaurach. Under Armour, which is based in Baltimore, recently did the same. Smaller but equally competent brands, such as Columbia, Mizuno, and KEEN, also call Portland home.

Every year, America imports an average of seven and a half pairs of shoes per person, but the market remains difficult to break into, requiring a large initial investment and a constant stream of new products.

Nike is the largest company in Oregon's history and for many years it was the only *Fortune* 500 company headquartered in the state (there are now two). Their influence runs deep in many facets of the community, where they donate, volunteer, and sponsor events. The University of Oregon now offers a master's degree in sports product management, and Portland State University added an athletic and outdoor certificate program for undergraduates in 2013.

Back on Bowerman Drive, now on foot, past the charging station for electric cars, the first building one comes to is Prefontaine Hall, named for the running icon who ran for Bowerman and was Nike's first professional runner (though they were still called Blue Ribbon Sports back then). The building sits atop a peninsula jutting out into Lake Nike and houses Knight's office and the corporate museum. Inside is memorabilia integral to the mythology of the brand. There is the long-lost waffle iron that Bowerman used to create the outsole of the first Nike Waffle Trainers and the Volkswagen bus that was used to sell his sneakers in the 1960s.

Those were simpler times, when the infamous boys' club would call their offsite upper-management gatherings "Buttface Meetings." For years the rambunctious, arrogant crew would cause sensible diners to flee restaurants and bars they would take over. Those early years have left an unflattering legacy, one that the "*Saturday Night Live* of the *Fortune* 500" has been unable to shed.

Though Nike is now well known for their sports endorsements, there were strict rules around amateur athletics when the brand first launched. Olympic and collegiate athletes, for instance, could not be paid for their sport in any way and if caught doing so would risk losing their eligibility; Nike worked around this by paying athletes under the table, a tactic that they had been using since at least the mid-1970s to lock in the best athletes as early as possible.

J. B. Strasser and Laurie Becklund wrote in their book, *Swoosh: The Unauthorized Story of Nike and the Men Who Played There*, about the early years at the company, that aside from Prefontaine, "the endorsement of Henry Rono, a Kenyan distance runner recruited by Washington State University who had long worn Nike, was probably the company's first significant NCAA rule violation."

Though Knight was initially opposed to the idea of marketing in general, paying celebrity athletes to wear his shoes was a tactic that proved so effective the company started to feel as if they *had* to do it. As the brand was transitioning from Blue Ribbon Sports to Nike, the winged goddess of victory, they launched new basketball and tennis shoes. Realizing the wears would die of obscurity if they weren't attached to an athlete, Nike made its first legal endorsement deal with Romanian tennis star Ilie Năstase. From the outside this must have seemed like an odd choice for the all-American brand. He was a petulant child on the court, a fact that earned him the nickname "Nasty" on tour. But he was talented. Really talented. And he was dominating the competition in a brand-new pair of Nike Match Point shoes. Nike offered him a $5,000 contract to wear their shoes with "Nasty" emblazoned on the heel. Năstase countered at $15,000. Knight, who hated to negotiate, agreed to $10,000, but later said of the ordeal, "I felt that I was being robbed."

In hindsight, it's now easy to see that this was the beginning of an overall marketing strategy that has served the brand brilliantly throughout: take a brash young athlete with the possibility of superstardom and turn them into a countercultural icon who moves shoes. So-called bad boys are interesting for brands like Nike. Bad boys get press. Bad boys move shoes. "Every time he hit his patented overhead smash, every time he went up on his toes and stroked another unreturnable serve, the world was seeing our swoosh," Knight wrote in his biography. "We'd known for some time that athlete endorsements were important. If we were going to compete with Adidas—not to mention Puma and Gola, and Diadora and Head, and Wilson and Spalding, and Karhu and Etonic and New Balance and all the other brands popping up in the 1970s—we'd need top athletes wearing and talking about our brand."

In their battle against Adidas, Knight said his company took on "a total David versus Goliath mentality," and the brand rallied around being the underdog. Brazen backdoor maneuvers and brilliant marketing allowed Nike to surpass Adidas in worldwide sales for the first time in 1983, a long-standing goal in an often-bitter rivalry with the German corporation.

The following year, with the help of basketball scout Sonny Vaccaro, Nike had the brilliant insight to sign a young player named Michael Jordan, fresh off the game-winning shot in the NCAA Division I National Basketball Championship for the University of North Carolina (a game in which he was wearing Converse shoes). Jordan, who was leaving college early to enter the NBA draft, would go on to become, arguably, the best basketball player of all time.

Jordan's favorite brand, however, was Adidas, and he was dead set on signing his shoe contract with the German brand. Though Nike had numerous NBA players on the roster by 1984, most notably George "the Iceman" Gervin, many athletes, including Jordan, still thought of them as the running guys. Adidas offered him $100,000, essentially the same deal they had with basketball star Kareem Abdul-Jabbar at the time, but, at the urging of his mother, Jordan traveled to Oregon to hear Nike out.

Vaccaro, who was also a Nike sales representative at this time, was emerging as a valuable backroom dealmaker for the brand, and urged Nike to "give the kid everything you got." J. B. Strasser's husband, Rob Strasser, an aggressive and rotund lawyer that Knight had poached from his law firm, wowed Jordan in their meeting. Strasser unveiled the Air Jordan concept, then offered the twenty-one-year-old a contract worth $2.5 million over five years, with an all-in price tag of more than $7 million once the stock options, annuity, performance guarantees, signing bonuses, and advertising commitments from Nike were taken into account. He would get his own shoe and more say in the design, a unique prospect at the time, but unheard-of in the deal was the royalty structure they offered. He would not just receive a percent of the sale on Air Jordan–branded shoes and apparel, but also on each Nike Air basketball shoe sold once they exceeded the previous year's benchmark. All Jordan had to do was appear in photo shoots, film commercials, make appearances for the brand, and, of course—win.

Regardless, Jordan had never even worn a pair of Nikes and remained insistent on signing with Adidas. He took the deal to the Germans and pleaded, "This is the Nike contract, if you come anywhere close, I'll sign with you guys." Adidas turned him down. Jordan signed with Nike, in a deal that would come to fundamentally reshape the American endorsement business. By September 1985, Nike had sold more than 2.3 million pairs of Air Jordans and within a year the shoes had generated $100 million in sales.

The Air Jordan brand would go on to become so popular (and expensive) that they would be blamed for physical assaults and even murders, causing some high schools to ban them before the 1980s were over. Jordan would dominate the NBA and win six championships with the Chicago Bulls, all while wearing Nikes, transforming both the NBA's and Nike's fortunes forever.

Jordan wasn't enough, at least initially, to help Nike avoid the ceding of industry dominance to Reebok International Ltd., however. From on-high Nike had lost touch with ever-changing fitness trends and were outmaneuvered by Reebok in the aerobics marketplace. The incident proved to be a wake-up call for the brand, who had so obviously forgotten about the other 51 percent of the population—women.

"We saw in the late 1970s what we thought was the running

revolution, but it wasn't," Knight said in 1987. "It was the first shot of a fitness revolution."

It was Mark Parker, a former Penn State distance runner who had been hired a decade earlier and mentored by Nike's first employee, Jeff Johnson, who helped lead the company back to the top through the mid-eighties. (Knight, once asked what had turned his Stanford thesis into a company, replied, "Jeff Johnson.") Parker had started as a shoe designer after he graduated from college in 1975, before becoming an "all-purpose product guy," with extraordinary advertising instincts. "It was obvious from the get-go that he was an imaginative guy," Johnson told me.

The next great opportunity, as they saw it, was the NCAA. As college basketball and football became national TV pastimes for Americans, Nike was keen to get their shoes on every collegiate star athlete. But there was much consternation internally around paying college athletes under the table. Nike found an innovative way in, one that would change the economics of sports forever: pay the coaches.

"We were the first corporate entity to be involved with a coach or university," Sonny Vaccaro has said of his days with Nike. Once a college coach was on Nike's team, the corporation would ply them with free apparel—a significant expense incurred by the universities before this time—and watch collegiate stars wear their products during the championship games on national TV.

Just two years after Nike's first all-school deal with the University of Miami in 1987, revenue doubled, profits quadrupled.

Once the NCAA door was open, Nike simply had to offer more money to get the teams to bend to their will. Despite the objections of his team, Penn State's coach Joe Paterno was the first to agree to a more prominent placement of the swoosh, after Nike made the program an offer they couldn't refuse. When the players ran onto the field October 15, 1994, in an important televised game against Michigan, the Nike logo had migrated from the bottom of the sleeve to the front of the jersey on the left shoulder pad. The swoosh now showed up hundreds of times during each football game. Not with a bang, but with a whimper (and the signing of a large check), amateur sports had been co-opted to commercial interests.

Working closely with the advertising firm Wieden+Kennedy, Nike would move forward and create some of the most innovative, groundbreaking, and memorable marketing campaigns, including "Just Do It," "Bo Knows," and "It's Gotta Be The Shoes." It took them about three years, but by 1989 Nike officially regained the so-called czar of athletic footwear title from Reebok, who were left with a bad taste in their mouths.

"Nike always think of whoever is their number one opponent as a warlike enemy," said Paul Fireman, Reebok's CEO through this turbulent time. "They're insane, sick, disgusting, I think."

Orchestrations of grand marketing plans were becoming a Nike signature move—the best illustration of which may have taken place, even if by accident, at the 1989 Major League Baseball All-Star game. As a Nike athlete, Bo Jackson was about to step to the plate from the on-deck circle when he noticed a plane flying overhead. Behind the plane flapped a massive banner that read "Bo Knows." He looked to the stands and suddenly noticed "Bo Knows" signs and fans wearing "Bo Knows" hats.

He launched a home run over the fence. The nationally televised program then cut to commercial whereby the audience was treated to the premiere of his ad campaign's famous "Bo Knows" television commercial, in which Jackson, who already played two professional sports, was a tongue-in-cheek expert in many others, including cycling, hockey, and golf. The goal of the campaign was to sell Jackson's new cross-training shoe, one that was versatile enough to be worn for many different sports. Though it looked as if Nike had bent time and space to their will, in truth, they had simply put all the pieces into place. It was just plain good luck launched by a home run that caused them to fire off in seemingly perfect order.

"ATHLETES ARE EVERYTHING AT NIKE," ITS WEBSITE PROCLAIMS, BUT NOWHERE IS THIS more evident than on campus, where banners featuring top athletes in action line the walkways and where there are 281 bronze statues of key figures in their history sprinkled throughout. In addition, many of the buildings are named after prominent sporting stars like Mike Schmidt, John McEnroe, Michael Jordan, Joan Benoit Samuelson, Alberto Salazar, and Mia Hamm. The Tiger Woods Conference Center overlooks the Ronaldo Field's two international-sized soccer patches (named for Nike athlete and professional soccer player Cristiano Ronaldo).

As is the case sometimes with erecting buildings in someone's namesake, irony struck when Nike was forced to remove Penn State coach Joe Paterno's name from the campus childcare center after it came to light that he knew his assistant, Jerry Sandusky, was a serial rapist and had been assaulting young boys in their charge. Paterno not only failed to report the incident to police, but seemed to look the other way. (The Penn State Nittany Lions football program brings in tens of millions of dollars a year and the school is still sponsored by Nike.)

In interviews with current and former contractors and employees, Nike's corporate culture has been described to me as insular and cultish, a place where very few, with the possible exception of the public relations department, seem to be affected by the steady drip of incendiary news about the company.

The worst of which, and the one that left an indelible mark, was a firestorm of "sweatshop" criticism that heated up in August of 1992 when *Harper's* magazine published a story about the working conditions in the foreign factories where Nike products were made. Though Nike had started in Japan's low-cost factories, over the years they chased cheap labor to Taiwan, South Korea, China, and Vietnam (this was a key tactical decision that provided the production cost benefits that allowed Nike to surpass Adidas in profits). Nike was paying some of its workers in Indonesia fourteen cents an hour, taking advantage of a mostly female workforce from farming families. Activists found dangerous working environments in unventilated factories where toxic chemicals were being used.

What the public found hard to reconcile, once all the numbers were laid out, was the fact that a $3 billion company sold $140 sneakers made by people earning around $1.50 a day. To highlight the disparity, *Harper's* also published a worker's pay stub with the report and compared it to Michael Jordan's endorsement contract, noting that it would take this employee 44,492 years to earn equivalent compensation.

For years, Nike largely remained silent on the topic. Newspapers and labor activists filled the void, reporting on below-minimum wages and allegations of physical abuse on the factory floors. Knight told a reporter at the time, "Our culture and our style is to be a rebel, and we sort of enjoy doing that. But there's a fine line between being a rebel and being a bully. And yeah, we have to walk that line."

Despite the criticism over the years, Nike stock purchased in 1998 for around $5 is worth more than $100 today. As I write this, the brand is coming off a social media call-to-arms that had people posting videos of themselves burning their Nike products after the company released advertisements featuring blackballed football player Colin Kaepernick, who had taken a knee during the national anthem in protest of police brutality against African-Americans. The company is also currently fighting accusations that their sports contracts discriminate against women, allowing Nike to reduce or stop paying female athletes while pregnant or in the postpartum period. (Following the public outcry, Nike announced a new maternity policy for all sponsored athletes.)

A 2018 lawsuit brought by four former Nike employees described the company hierarchy as "an unclimbable pyramid—the more senior the job title, the smaller the percentage of women." A lawyer for the women said, "The way Nike marginalizes women at its headquarters is completely contrary to how it portrays itself to its customers as valuing women in sports and the importance of providing equal opportunity to play." In early 2019, a federal judge denied Nike's attempt to dismiss the women's class action claims. The case is still making its way through the courts.

While reporting this book, I visited Nike's Beaverton headquarters for the first time on a sunny September day in 2018. Driving under the bridge, I passed the foreign flags on Bowerman Drive that represent the first countries Nike did business with in the early years. Their products are now sold in more than 170 countries, but there is a special connection to Japan, and the campus's Nissho Iwai Japanese Garden pays homage to the country where it all started for Bowerman, Knight, and Nike.

The campus was bustling and beautiful—an ideal playground for the sports-obsessed minions that populate Nike's roster. There was a Ferrari parked in Ray Allen's reserved spot by the Bo Jackson Fitness Center and a Lexus LC500 in the one reserved for Michael Jordan in front of the Mia Hamm building. My professional attire included a sports jacket and some leather Oxford shoes. Every one of the more than twelve thousand employees on campus wears sneakers with a swoosh, though there was the odd Converse All Star mixed in (Nike purchased the company in 2003). In front of the Seb Coe building, and right on cue, a tall athletic blond woman looked at my shoes, then glared up at me. They weren't Nikes.

On your average workday in Beaverton, employees and professional athletes test product and take meetings amid the ample open space between buildings—weather permitting. Nike's many intramural teams and training camps give employees a chance to stay fit and remain connected to sports. Running clubs gather at my last stop, the Michael Johnson track in the southeast corner of the campus, to warm up and run laps. This is where Nike's flagship professional running team, the Nike Oregon Project, once gathered to do their daily workouts. The halcyon track is partially covered with trees, which gives it the feeling of running into a protective green tunnel after you round the northern corner. Had I arrived at the track in October 2004, I would have seen two of America's best runners arriving for their first workout on their sponsor's campus with their new coach, Alberto Salazar.

03
WHAT ARE YOU ON?

ON AN APRIL MONDAY IN 2001, ALBERTO SALAZAR SAUNTERED INTO THE BOSTON DELI on Nike's campus eager to watch the 105th running of the Boston Marathon that was being broadcast on the television in the corporate cafeteria bearing its name. He sat down with Tom Clarke, a former competitive runner and a Nike vice president, to enjoy the race.

As a maniacally focused twenty-three-year-old professional Nike sponsored athlete, Salazar had won the event in 1982. His stride for stride battle of wills with Dick Beardsley is likely the most thrilling Boston Marathon ever run, a day that came to be known as the "Duel in the Sun." (It is also the subject of John Brant's book by the same name.)

As the 2001 race proceeded, the two men grew increasingly frustrated with the poor showing of the US athletes. They weren't competitive. They weren't impressive. And they weren't even in the top five.

It hadn't always been this way. In the 1970s, America produced some of the best distance runners in the world. Bill Rodgers won both the Boston and the New York City marathons four times each. At the Munich Olympics in 1972, the United States claimed first, fourth, and ninth in the marathon event, the best finish by three male runners from a single nation since 1908. One of Steve Prefontaine's close friends, former track star Frank Shorter, won the gold medal that year, and he was on his way to a repeat in the following Olympiad when he was robbed of the victory by an East German athlete named Waldemar Cierpinski, who was later found to be part of his country's infamous doping program, though Shorter still managed to finish second in the world.

US athletes, wearing Nike shoes and swooshes, had once dominated the global running scene, setting world records, medaling at the Olympics, and winning the sport's marquee events. Salazar himself had personally ruled the early 1980s, winning three New York City marathons and one Boston. But it was now painfully obvious that the national performance drought—which arguably began when an injured and overtrained Salazar begrudgingly stepped out of the spotlight—was starting to bleed into the new century.

Salazar and Clarke knew the state of affairs. From 1983 to 2001, American distance runners had won only 4 percent of World Championship and Olympic gold medals, and only one male American-born runner had won a top-tier marathon since Salazar won the 1982 New York City Marathon.

East African countries like Kenya and Ethiopia had superseded the American athletes in the elite ranks. But it wasn't just that US athletes were simply losing to the Africans; they were also, somehow, getting slower.

As the televised event proceeded, Clarke and Salazar watched as Korean athlete Bong-Ju Lee outpaced an Ecuadorean runner and three Kenyans to win the Boston Marathon in the time of 2 hours, 9 minutes, and 43 seconds. When the first American—thirty-four-year-old Rod DeHaven—crossed the finish line the announcers gushed about him finishing in the top ten. "Tom and I just looked at each other," Salazar told the *New Yorker*. "It was, like, wow. We've sunk so low that sixth place is considered an accomplishment."

Almost two decades of lackluster American performances had made the US sports establishment desperate for a homegrown hero to challenge the African dominance. "I told Tom that I could coach Americans to do better," said Salazar. "Tom got out a pen and started writing down some ideas: coaching, talent, sports science." The two men hatched a plan to bring American running back to prominence on the international stage—if the state wasn't going to support American athletes, then Nike would.

The idea was simple, incubate the next great American world beater—a Tiger Woods or Lance Armstrong for the running world—and watch sales take off. Clarke convincingly pitched the idea to Knight, who quickly signed off on it. Salazar, who had been a Nike stalwart for years—as both a sponsored athlete and more recently an employee in the marketing department—would lead a program that would lean on new technologies and methods. Throughout his running career, unorthodox procedures, chemicals, techniques, supplements, and tinctures had been his stock-in-trade. He suggested a marathon-training approach that would harness obscure technology as a way to equalize the African dominance. Clarke, a veteran of thirty-five marathons who has a doctorate in biomechanics, was sold.

The timing seemed to be right, too, due to the crop of young high school boys that had the nation talking about a possible American competitive resurgence. Deemed the "Big Three," Alan Webb, Dathan Ritzenhein, and Ryan Hall had each dominated their high-school competitions. The previous December, they had met at the national crucible of high school running, the Foot Locker Cross Country Championships, where Ritzenhein won, Webb placed second, and Hall came in third. In May, Webb ran the mile in 3 minutes and 53.43 seconds, breaking Jim Ryun's high school record for the distance that had stood for thirty-six years. The future looked bright, and if Nike could get the program up and running it would be primed to receive the next great wave of American talent.

With Salazar at the helm, a new team—which would later be called the Nike Oregon Project—would take advantage of the corporation's generous financial reserves to create a closely monitored, scientifically advanced, and technologically sophisticated training milieu like nothing ever before attempted with US athletes.

Salazar's focus on coaching began at home, when he became obsessed with making his sons—Tony, who played football, and Alex, who played soccer—into superstar athletes. Tony was working with a personal trainer before he was even in junior high. Both sons have proven loyal to their father, and refuse media interviews, but Tony once told *The New Yorker* that his father used to wake him up at three in the morning to give him protein shakes. "Back then, I was trying to gain weight for football, and sleep was the longest period that I'd go without eating," Tony said. "He'd go downstairs, turn on the blender, and then bring the drink to me in bed." Salazar's focus on the details of athletic training lent itself to coaching, where athletes at the top level are often separated by one-hundredths of a percentage point and taking care of the little things can cumulatively make a big difference.

By the mid-nineties, Salazar was coaching Nike athlete Mary Decker Slaney in her bid to compete in the 1996 Olympic 10,000 at age thirty-seven. She had been a running phenomenon and a precocious talent, who was once considered the most remarkable female runner of her time, but was now in the later stages of a stellar career that saw her win almost everything, with the notable exception of an Olympic medal. Decker Slaney had been featured on the cover of *Sports Illustrated* four times, and was once nominated their "Sportswoman of the Year," following her double at the first IAAF World Championships in 1983, when she took on the Eastern Bloc and won gold in the 1,500 and 3,000 meters.

As she approached forty, Decker Slaney asked Salazar to help her train, and hopefully fill in the gap in her résumé with an Olympic berth to the Games in Atlanta, Georgia. The popular press was rightly excited about a former marathon great helping coach an aged and road-weary veteran who had never quite reached her Olympic dreams. In 1994, *People* magazine featured Decker Slaney in a profile where she acknowledged she "turned to her friend Alberto Salazar . . . to help devise an injury-free program for her," while a May 1996 article in the *New York Times* refers to Salazar as "her coach of two years."

Just a month before the Games, a dual interview with Decker Slaney "and her coach, Alberto Salazar" also appeared in the June 1996 issue of *Runner's World*. It began, "If Joan Samuelson is the soul of US women's distance running, Mary Slaney is the heart. The broken heart," then went on to describe how, through all her success, Decker Slaney never medaled in the Olympics. Beside the text is a photo of coach and athlete running together on a wet, cloudy day and features the caption, "Alberto Salazar has been coaching Mary Slaney as she prepares for her latest run at an Olympic Medal."

At the US Olympic Trials in June 1996 Decker Slaney provided urine to the United States Olympic Committee (USOC; USADA didn't yet exist) that showed an impermissible ratio of testosterone to epitestosterone. She contended the result and argued that the test was unreliable for women on birth control pills. Her team took the case to the USA Track and Field (USATF) Doping Hearing Board, which found in her favor, before the IAAF took the case to arbitration, where the panel found that Decker Slaney had committed a doping offense. She received a two-year retroactive ban from sports.

NIKE'S HIGHEST-PROFILE RUNNER AT THIS TIME, IN 2001, WAS SPRINTER MARION Jones, a two-sport athlete who had been competing internationally since she was twelve years old. Jones won the California 100-meter state championship in all four of her years in high school and then turned down a chance to join the Olympic team for a basketball scholarship at the University of North Carolina, where, as the team's freshman point guard, she helped them take home the 1994 NCAA National Championship.

After graduation, however, Jones turned her attention to professional running. She won three golds and two bronze medals in Nike shoes at the Sydney Olympics in 2000, which made her a household name, and helped her accomplish a rare feat in running: turning her speed into a multimillion-dollar empire. Jones was strikingly beautiful and charismatic, which aided in her marketability, and in addition to Nike was also sponsored by Oakley sunglasses and TAG Heuer watches. Not since three-Olympic-gold-medal winner Florence Griffith Joyner had a female athlete been able to turn running fast into a platform for global fame and fortune to such a degree. (It should be noted that Joyner's own career was plagued with allegations of doping, even though she never failed a test. She died in her sleep after suffering a seizure at just thirty-eight years old.)

Under the tutelage of Nike coach Trevor Graham, Jones didn't lose a race in 1998, solidifying herself as the fastest woman on earth at both the 100- and 200-meter distances. Jones was soon earning a few million dollars a year.

As the Sydney 2000 Olympics approached, she doubled down on her ambition. "I'm not looking to win silver or bronze," Jones told reporters. She said she wanted to win five gold medals. In the Nike commercials preceding the games she spoke into a microphone like a radio DJ and admonished athletes listening to be better role models. "The drug use, the spousal abuse, the violence—it's got to stop," she said. "We need more role models, the more the better. Can you dig it?" she added, as a Nike swoosh and a new URL, Nike.com/MrsJones, appeared in the corner before a fade to black.

While publicly admonishing others to stop drug use, Jones was injecting and ingesting performance-enhancing substances in private. Jones and her husband, Nike sponsored shot-putter C. J. Hunter, were both on various illicit performance-enhancing drugs (PEDs), which Graham helped Jones procure through Victor Conte and Bay Area Laboratory Co-Operative (BALCO).

During the Sydney Olympic Games, it was announced that Hunter had failed an earlier drug test. He then gave a press conference after Jones's first event of these Games in an ill-advised attempt to explain how the failed test was all a big mistake. By his side were Jones, Conte, and Johnnie Cochran, the famous lawyer who managed to set O. J. Simpson free from a murder charge in 1997. (Nine years earlier he'd helped Jones, then a high-schooler, ensure a missed, out-of-competition drug test didn't become a full-blown scandal.)

Jones performed well in the Games, but ultimately came up two shy of the five-gold-medal goal. She did, however, win the prestigious 100- and 200-meter double in shocking fashion, crossing the 100-meter line ahead of the next racer by the largest margin recorded since a hand-timed race in 1952. She graced the covers of *Sports Illustrated* and *Vogue* magazine at the apex of her commercial fame, but PED rumors would haunt her for the rest of her career after Sydney.

Aside from the drug rumors, the corporate nightmare of Nike's factory conditions reemerged during this time in 2001. The company's original sin, exploiting cheap labor in foreign countries, was brought back into the spotlight by an MIT graduate student named Jonah Peretti. Procrastinating on a paper he had to write, he started playing around online with a Nike promotion, called Nike iD, that allowed customization of the materials and colors used to construct the shoes customers ordered. The service also allowed users to have a motto embroidered onto them, provided none of the words used were on Nike's banned phrases list. After trying a few four-letter words and getting them instantly rejected, Peretti chose the word "sweatshop" hoping it'd be placed under the Nike logo on his Zoom XC's. The order went through, but as soon as it was reviewed by Nike it was canceled. Peretti received an email notification that said his personal iD violated one of their corporate policies—his now famous volley with Nike corporate is below:

From: "Jonah H. Peretti" peretti@media.mit.edu
To: "Personalize, NIKE iD" nikeid_personalize@nike.com
Subject: RE: Your NIKE iD order o16468000

Greetings,

My order was canceled but my personal NIKE iD does not violate any of the criteria outlined in your message. The Personal iD on my custom ZOOM XC USA running shoes was the word "sweatshop."

Sweatshop is not: 1) another's party's trademark, 2) the name of an athlete, 3) blank, or 4) profanity.

I chose the iD because I wanted to remember the toil and labor of the children that made my shoes. Could you please ship them to me immediately.

Thanks and Happy New Year, Jonah Peretti

From: "Personalize, NIKE iD" nikeid_personalize@nike.com
To: "Jonah H. Peretti" peretti@media.mit.edu
Subject: RE: Your NIKE iD order o16468000

Dear NIKE iD Customer,

Your NIKE iD order was canceled because the iD you have chosen contains, as stated in the previous e-mail correspondence, "inappropriate slang".

If you wish to reorder your NIKE iD product with a new personalization, please visit us again at www.nike.com

Thank you, NIKE iD

From: "Jonah H. Peretti" peretti@media.mit.edu
To: "Personalize, NIKE iD" nikeid_personalize@nike.com
Subject: RE: Your NIKE iD order o16468000

Dear NIKE iD,

Thank you for your quick response to my inquiry about my custom ZOOM XC USA running shoes. Although I commend you for your prompt customer service, I disagree with the claim that my personal iD was inappropriate slang. After consulting Webster's Dictionary, I discovered that "sweatshop" is in fact part of standard English, and not slang. The word means: "a shop or factory in which workers are employed for long hours at low wages and under unhealthy conditions" and its origin dates from 1892. So my personal iD does meet the criteria detailed in your first email.

Your website advertises that the NIKE iD program is "about freedom to choose and freedom to express who you are." I share Nike's love of freedom and personal expression. The site also says that "If you want it done right . . . build it yourself." I was thrilled to be able to build my own shoes, and my personal iD was offered as a small token of appreciation for the sweatshop workers poised to help me realize my vision. I hope that you will value my freedom of expression and reconsider your decision to reject my order.

Thank you, Jonah Peretti

From: "Personalize, NIKE iD" nikeid_personalize@nike.com
To: "Jonah H. Peretti" peretti@media.mit.edu
Subject: RE: Your NIKE iD order o16468000

Dear NIKE iD Customer,

Regarding the rules for personalization it also states on the NIKE iD web site that "Nike reserves the right to cancel any Personal iD up to 24 hours after it has been submitted".

In addition it further explains: "While we honor most personal iDs, we cannot honor every one. Some may be (or contain) others trademarks, or the names of certain professional sports teams, athletes or celebrities that Nike does not have the right to use. Others may contain material that we consider inappropriate or simply do not want to place on our products.

Unfortunately, at times this obliges us to decline personal iDs that may otherwise seem unobjectionable. In any event, we will let you know if we decline your personal iD, and we will offer you the chance to submit another."

With these rules in mind we cannot accept your order as submitted. If you wish to reorder your NIKE iD product with a new personalization please visit us again at www.nike.com

Thank you, NIKE iD

From: "Jonah H. Peretti" peretti@media.mit.edu
To: "Personalize, NIKE iD" nikeid_personalize@nike.com
Subject: RE: Your NIKE iD order o16468000

Dear NIKE iD,

Thank you for the time and energy you have spent on my request. I have decided to order the shoes with a different iD, but I would like to make one small request. Could you please send me a color snapshot of the ten-year-old Vietnamese girl who makes my shoes?

Thanks, Jonah Peretti

Nike never responded to Peretti's final email. Peretti, showing the savvy that would later make him millions of dollars as a co-founder of the *Huffington Post* and the viral news website *BuzzFeed*, contacted *Harper*'s magazine, who had done some of the earliest reporting on factory conditions, asking them if they'd be interested in printing the email exchange. They turned him down, so in January 2001 he forwarded the email string to a dozen friends. It traveled from person to person and group to group like a virus until it reached millions of inboxes, reminding everyone of Nike's shady past.

The *San Jose Mercury News* was the first outlet to publish the emails, with most of the largest news organizations soon following suit. Peretti was flown out to New York and appeared on NBC's *Today Show*. Nike sent their director of Global Issues Management, Vada Manager, to debate the grad student. Katie Couric served as moderator on the television program.

The email string is now widely considered one of the first pieces of viral content on the internet and began a firestorm of shared outrage. Peretti was inundated with emails, some of which were berating, but most of which were positive in nature. One in particular excoriated him for his "rather worthless" opinion and claimed to be from someone who wasn't "a Nike apologist or the like." It included no name or identifying details apart from a Hotmail address, but the author, likely proud of his work, cc'd the website *Salon*.com, among others, with hopes that they would publish the admonishing screed. Struck by the audacity, one of Peretti's colleagues at MIT decided to look up the sender's IP address—a number that essentially acts as a street address for every computer on the internet—and found that the message was sent from the Nike campus.

Nike spokeswoman Beth Gorny told *ESPN* she wished the fuming employee had kept his opinion to himself, and added, "This has sparked a lot of interest in the iD [product] . . . Sales are up." Indeed, the Nike iD site enjoyed its third-highest single day of sales after the *Today Show* aired.

THROUGH ALL THIS CONTROVERSY, NIKE WAS STEADILY ACTING ON THE OREGON PROJ-ect idea with hopes that the champions that it would create would help get the brand back in its country's good graces and return it to its institutional core. Although America's most popular sports—football, basketball, and baseball—had proven to be lucrative for Nike, the company was eager to regain its foundational market. This was part of a broader move to corner endurance sports in general, which it had been steadily gobbling up with the rise of the cyclist Lance Armstrong.

Cheating rumors had been swirling around Armstrong since his first Tour de France stage win in 1993. He was the perfect Nike athlete, an outspoken cycling powerhouse like none before him who brought droves of consumers into the sport. Every one of them would need bike kits and bike shoes that Nike planned to provide in *maillot jaune*, the yellow jersey color of the Tour de France leader. Armstrong had always been a brash and confident athlete, but as his success grew in tandem with his large corporate sponsors, he morphed into a litigious and vindictive bully with an unchecked ego and bottomless resources.

Just over a month before Salazar and Clarke conceived of the

NOP team, the Nike campus was abuzz with the ceremony to christen their headquarters' employee gymnasium the Lance Armstrong Fitness Center. (A month later, the Tiger Woods Conference Center made its debut.) Armstrong's story—the proletariat kid raised in a single-parent home who beat cancer and all the world's professional cyclists on bread and water alone—was in many ways closer to fiction than reality, but it moved Nike-branded gear.

Before Armstrong and his family arrived for the ceremony, Salazar, who had become a big fan of the cyclist, emailed Howard Slusher, who was in charge of organizing the event.

"Howard I've never asked for a favor like this. I've never asked to meet Tiger Woods. I've never asked to meet Mia Hamm. I really want to meet Lance Armstrong. I think he's the toughest athlete in the world, that I've ever seen. You can just see it in his eyes, the guy is just tough as nails. I just want to meet him."

During the presentation, Howard read Salazar's email to the crowd. "This comes from someone who is supposed to be pretty tough himself," he said, before he brought Salazar out to meet Armstrong in person.

"Having Nike name a building after me is big," Armstrong said to the crowd. "I've often said it was a dream of mine even as a young boy to become a Nike athlete. To have that dream come true, but then also be the recipient of this honor is beyond exciting for me. I am proud to have my name associated with something so solid and strong as the Nike brand, the people at Nike, and this amazing fitness center. It represents Nike's belief in me—and I'll be thinking of that as I climb the mountains ahead in my life."

Knight told the crowd that "had an aspiring writer concocted a screenplay three years ago that essentially mirrored what Lance Armstrong actually accomplished, movie studios would have rejected it as too 'out there,' even for Hollywood. I've been fortunate to know a lot of truly remarkable athletes, but there are no more inspiring stories than Lance's, and Nike is extremely pleased to honor him with a building in his name."

Inside the fitness center, parts of the facility were named after people and places that were important to the two-time Tour de France champion, including his wife, Kristin; mother, Linda; son Luke; and even his childhood swimming coach from Plano, Texas, Chris MacCurdy.

Throughout the year, Nike ran the "What are you on?" commercial on national television, which jabbed a finger in the eye of Armstrong's many critics and accusers. It showed him being photographed and drug tested, then riding his bike in the rain. This was the "clean" process by which he ostensibly beat the best cyclists in the world. A story that was so compelling that Armstrong—with Nike's sizable marketing efforts behind him—became one of the most famous people alive and one that caused sports fans the world over to abandon reason and any semblance of healthy skepticism.

Nike rode Armstrong's popularity to record-level profits and brand recognition in a new sport. And the money seemed to blind most people to what everyone in the industry knew—that Armstrong was a cheater. In addition to his illicit drug use, he allegedly paid cyclists on other teams to let him win.

Though Nike has always maintained that they knew nothing of Armstrong's misdeeds, Kathy LeMond, the wife of cycling legend Greg LeMond, testified under oath that the mega-corp paid the then-head of the Union Cycliste Internationale, Hein Verbruggen, $500,000 to cover up Armstrong's positive test for corticosteroids in 1999. (Nike and Verbruggen have denied this claim.)

In the Nike commercial Armstrong's voice-over is now shocking in its temerity: "This is my body and I can do whatever I want to it. I can push it, and study it, tweak it, listen to it. Everybody wants to know what I'm on. What am I on? I'm on my bike, busting my ass six hours a day. What are you on?"

SALAZAR, MEANWHILE, WAS BUSY TRYING TO FIND RUNNERS WITH THE RIGHT MIX OF talent, work ethic, and willingness to experiment. They sought athletes who had impressive race results, but who also had the potential to run 10,000 meters in under 28 minutes and 30 seconds, the accepted benchmark for athletes who hoped to break 2 hours and 8 minutes in the marathon.

Initial recruits included Dan Browne, Karl Keska, Mike Donnelly, Dave Davis, and Chad Johnson. Donnelly told journalist Andrew Tilin, who was writing a story for *Wired* magazine, that he had been training by himself while working part-time at a local bank when Salazar called with an offer to join the new team in Oregon. Davis said he was completely broke and about to take a job on a landscaping crew when he received his call.

Recruits moved into a five-bedroom Portland bungalow that Salazar filled with several scientifically suspect devices. In the rec room was a vibration platform that supposedly increased leg power simply by standing on it. There was a coffin-like hyperbaric tube that the athletes could lie in to get the recovery boost of highly pressurized oxygen. And there was the $35,000 worth of Russian software, called Omegawave Sports Technology System, that measured heart rate variability through electrodes stuck on the athletes' bodies, which supposedly predicted whether they were ready for hard training from day to day. Add an electrode to the forehead and the software purported to scan their liver, kidneys, and central nervous system to produce an overall health score.

Another software package Salazar used was a $3,000 program called DartTrainer, which provided side-by-side comparison of footage of the athletes' running form. The workout video, slowed down to thirty frames per second, allowed coaches and athletes to analyze their gait for inefficiencies. Though many top-level running coaches believe that an athlete's form shouldn't be tinkered with, Salazar said, "You don't have a chance to compete against the very best unless you run like them." And he estimated the DartTrainer to help take about ten seconds, or approximately 1 percent, off each runner's 5,000-meter time.

"The plan remains doing whatever is necessary to create winners. We'll think out of the box, not just for the sake of being different but to find what really works," Salazar told *Wired*. "I know this is the way to go."

East Africa's rise on the international running scene forced an increased scrutiny of what, exactly, the African countries were doing that made them so good. Of all the countries, Kenya was disproportionately dominating world athletics. Journalists and researchers were dispatched by their respective outlets and faculties to attempt to figure out what made them so good. David Epstein, author of *The Sports Gene*, has called one tribe of Kenyans, living mostly in the highlands of western Kenya, "the greatest concentration of elite athletic talent, ever, in any sport, anywhere in the world."

Conclusions were drawn about Kenyan athletes' slight lower legs and distal swing weight, as well as their proximity to altitude. They seemed to be born for forward propulsion, and most of them had been doing just that, running to and from school much of their lives.

But there is an economic reality that can't be ignored when discussing the East African dominance. A Kenyan athlete who wins a major marathon, for instance, will change his family's lot for generations to come. There is also the inconvenient truth that until very recently, the East African countries have had no WADA-approved anti-doping agency and were therefore rarely tested when compared to the efforts of other prominent Olympic countries, like America. When analyzed, Kenyan athletes' specific run training methods, nutrition, and sleep weren't all that different from their contemporaries in other countries. And they didn't have access to any of the latest sports technology that Salazar had placed much of his hopes on. Two points did stand out in the analysis: they trained in groups, rather than the siloed efforts of most American athletes, and they lived and trained at altitudes of 7,000 to 8,000 feet above sea level.

Research has shown that spending time in a low oxygen, high altitude environment is a hormetic stress on the body, which it responds to by producing more red blood cells over time. Red blood cells transport oxygen from working lungs to working muscles during physical activity. The higher your hematocrit levels, a measure of the percentage of your blood made up of red blood cells, the faster and better you will feel while running.

Sports scientists didn't begin seriously considering the effects of altitude on human performance until the lead-up to the Mexico City 1968 Olympics, which took place at around 7,350 feet above sea level. Knowing how important oxygen is to athletic performance, there was a fear that forcing athletes to push their limits to win gold medals in Mexico's thin air could damage them in some way.

What their tests proved was that blood-boosting, or increasing an athlete's red blood cell count by re-injecting their own blood, worked wonderfully as a performance enhancer. It was also during this time period, at an annual meeting before the 1968 Games, that the International Olympic Committee defined "doping" for the first time as "the use of substances or techniques in any form or quantity alien or unnatural to the body with the exclusive aim of obtaining an artificial or unfair increase of performance in competition."

Beaverton, Oregon, sits around 100 feet above sea level, where the air contains its maximum amount of oxygen. This is ideal for training and racing but doesn't provide the stress that causes the body to produce an advantageous boost of those cherished red blood cells. Simply moving to live full-time at high altitude poses its own problems—mainly that an athlete can't reach their speed potential in a low oxygen environment, thus making it unfeasible to do the bulk of their daily training at altitude.

With all this information swirling inside his mind, Salazar spent approximately $110,000 of Nike's money on air-thinning technology, and more still to seal the house and rooms in which the altitude would be manipulated. There were special doors, light fixtures, and windows to keep the ambient outside sea-level air from leaking in. Athletes ate, slept, rode bike trainers, watched TV, and played video games in an environment that their bodies believed to be thousands of feet above sea level. Salazar told the runners to stay in the house for twelve hours each day and to keep the room above 12,000 feet.

At one point, some of the athletes began to struggle, and soon discovered the house had accidentally been set to mimic an altitude of 14,000 feet. Johnson complained of being winded after rearranging his room while Keska had trouble sleeping, a common occurrence at high altitudes, and staggered through subsequent workouts after restless nights.

As the *Wired* magazine story details, it took just a few months for the athletes to revolt. They delivered the Nemes vibration platform to the garage to collect dust, canceled their Pilates sessions, and had Salazar set the bedrooms to a more manageable 7,000 feet. "We run ten miles in the morning, five miles in the afternoon, and the altitude rooms don't allow us to recover so quickly. That doesn't leave a lot of energy to frivolously waste on whatever else," Dan Browne told Tilin. "We can only do so much in a day."

In 2006, an Executive Committee of the World Anti-Doping Agency gathered to discuss the legality of hypoxic chambers—a category that would include Nike's costly altitude house. To be placed on WADA's List of Prohibited Substances and Methods, a chemical or procedure has to meet two of three determining criteria:

1. It has the potential to enhance sport performance;
2. It represents a health risk to the athletes; and
3. It violates the spirit of sport.

Though WADA determined that artificial low-oxygen environments were potentially performance enhancing and against the spirit of sport, the Executive Committee ultimately decided not to include them in the list.

In the summer of 2002, Keska told Tilin, "I'd like to do well in an Olympic marathon. As frightening as it sometimes sounds, this seems like a very natural and normal environment for helping me reach my goal."

Browne, whose red blood cell count had increased 11 percent since he joined the team, won the US marathon championship in 2002, and the program seemed to be on its way to putting American athletes back on international podiums. But it became increasingly apparent, race after race, that although the Nike Oregon Project was competitive on the national level, none of its athletes were on track to win a major marathon or a gold medal. And none of the first group ever broke the 2-hour-and-10-minute barrier, let alone a sub-2-hours-and-8-minutes.

Phil Knight remained patient through the Oregon Project's colicky infancy and allowed his coach to continue to pour millions of dollars into the program even if the first two years fell far from expectations. In the media, Salazar suggested the athletes themselves were responsible for the lack of success, referring to the initial recruits—Olympians and national champions among them—as "B-plus" competitors. The implication was that there was only so much he could do with them. He needed more competitive athletes.

He needed Adam Goucher.

04

TAKING RUNNING OFF THE BACK PAGE

ADAM GOUCHER WAS BORN IN HOLLYWOOD, FLORIDA, IN FEBRUARY 1975, THREE months before Steve Prefontaine died tragically, and three years after Nike signed their first legal athlete endorsement contract with Ilie Năstase. As the Nike Waffle Trainer was becoming popular, Adam's father, Richard Goucher, took a job in Colorado Springs, and moved the family to Colorado when Adam was five years old.

The friendly, midsized town sits in the shadow of Pikes Peak, one of Colorado's fifty-four mountains that rise more than 14,000 feet above sea level. Founded in 1871 by Civil War general and railroad builder William Jackson Palmer, the frontier town grew in population due to its warm, dry climate that was thought to be advantageous to curing tuberculosis. It's now home to almost half a million people, as well as a handful of military facilities, and a long list of religious institutions. It also lays claim to some of sports' most important organizations, including the Olympic Training Center, USA Cycling, and the United States Anti-Doping Agency.

Growing up, Adam's father was largely absent from family life, traveling for months at a time maintaining betting machines and scoreboards at dog and horse racing tracks. "There was always a slight void in not having him around," Adam told *Runner's World* in 2001. "Even before my parents got divorced, a lot of my friends had dads around to go fishing or hunting, neat stuff fathers do with their sons. He was as involved as he could be, but his job took him away. That was hard."

Looking for ways to win his father's attention, Adam channeled his prodigious energy and familial angst into sports and found that he was a natural-born athlete. Running came easy to him, and he soon discovered he was one of the fastest kids in town. He played nearly every sport on offer, but he was particularly good at basketball, largely due to his running speed. He had a competitive nature that shined in whatever sport he set his attention to (he once broke an intermediate-hurdles city record and cleared six feet in the high jump). In ninth grade he was voted the MVP of his basketball team, even though he didn't have the latest pair of Nike Air Jordan shoes. "We didn't have the money for that kind of stuff," he said. He won the city championship in the mile on his first attempt in seventh grade, then placed second in eighth grade, and won again in ninth grade.

By the time he was a teenager his father had left the family, their house was on the verge of foreclosure, and Adam's mother, Lois, spent most of her time at her boyfriend's house. One night, Adam's older sister, Debbie, decided to take advantage of the lack of parental oversight and invited some new acquaintances over to drink alcohol. (Adam, who was fourteen years old at the time, was spending the night at a friend's house.) One of the young men attending the party had brought a gun. Another man began playing with it, and assumed it was empty as he joked about Russian roulette. He then put the gun to his head and pulled the trigger, accidentally killing himself.

Adam was kept busy with his friend's family, running errands around town while his mother mopped up pools of blood. The tragedy was formative for Adam, who watched in disbelief as the police treated his family as if they were responsible. As they finished the investigation, one of the officers intentionally drove over a skateboard ramp that Adam had built himself. The officer then backed over it a second time, to be sure it was destroyed. The message, as Adam internalized it, was that his family was trash—that he was trash.

This experience changed things for Adam. He started working with the local Drug Abuse Resistance Education (D.A.R.E.) program, speaking to elementary school students about the dangers of underage drinking. And he mentally began to set himself against those who doubted him and dedicated his life to a pursuit of excellence that has been his North Star ever since. He threw himself at disparate disciplines like jazz band and student council and was elected student body president of his class.

During Adam's sophomore year at Doherty High School he committed fully to distance running, and by his junior year, he was unbeatable. He faltered at Nationals, however, and finished in 15th place. All things considered he'd raced well against the best high school athletes in America, but Adam was devastated, and unable to contextualize his defeat.

His senior year he hung a handmade poster on the wall across from his bed that he could see every morning when he arose, one that laid out his goals clearly and prominently—it read:

Win Footlocker National Championship, San Diego, California.
Be the first Colorado runner to break 15:00 for 5K—Goal: 14:53.
Long Term—Collegiate National Champion.
Make Olympic Team in 1996, 2000, and beyond.

Adam was open and resolute about the 5,000-meter goal, and the team would raise a banner at the meets that read "14:53." His high school coach, Olympic Trials marathoner Judy Fellhauer, remembers his public attempt to become the first Coloradan in state history to run under fifteen minutes for a 5K in high school cross-country. "There were articles in papers all over the state every time he competed. That had never happened until then," Fellhauer told

Runner's World. "He took running off the back page." Running at elevation, Adam broke the 5,000-meter record with a time of 14 minutes and 43 seconds.

At the national finals of the Foot Locker Cross Country Championships—still six years before the "Big Three" came to prominence—Adam finally proved he was the best high school runner in the country by winning the race and beating future running star Mebrahtom "Meb" Keflezighi. It was at the Foot Locker Cross Country Championships where Adam briefly met a young runner named Kara Grgas-Wheeler. In the bustle around the event, he stopped to ask his future wife where a mutual friend of theirs had gone off to, then quickly trotted away to go find her (Kara doesn't think he actually remembers their first meeting).

Adam was recruited by almost every prestigious college running program, but says he flipped a coin when deciding between the University of Colorado Buffaloes and University of Wisconsin Badgers. When pressed, however, he admits that if the coin had come up Wisconsin, he would have likely just flipped again.

Adam enrolled at the University of Colorado (CU) a year before the team's assistant coach Mark Wetmore took over as the head coach of their running programs in 1995. Wetmore began his coaching career guiding a municipal children's team in his hometown of Bernardsville, New Jersey, in the 1970s, before assuming the assistant coaching position at his alma mater, Bernards High School. He then worked at Seton Hall University from 1988 to 1992 before moving to Boulder to become the assistant coach at CU in 1993. Wetmore is known for his no-nonsense approach, his stoicism, and his aversion to the limelight.

Adam burst onto the Division I NCAA scene not just ready to compete, but ready to win. As a freshman racing in the NCAA cross-country championship, he stunned the elders of Division I when he almost won the race outright, finishing just behind Martin Keino, a Kenyan immigrant and the son of Kenyan Olympic gold medalist Kipchoge "Kip" Keino. His performance went down as the second-best freshman finish ever, behind Bob Kennedy's 1988 win.

"He's the best young runner to come along since Craig Virgin," Frank Shorter told the *New York Times* when asked about Goucher. (Virgin was a two-time world cross-country champion and former American record holder in the 10,000 meters.)

The following September, a runner from Minnesota moved to Boulder to join the CU Buffaloes.

KARA GRGAS WAS BORN IN NEW YORK CITY, NEW YORK, IN JULY 1978, THE SECOND daughter to Patty and Mirko Grgas. Her grandfather on her father's side had immigrated to America from Croatia in search of a better life after World War II.

Kara's father, Mirko, excelled at soccer, which helped him get into college at Ottawa University, where he met Kara's mother, Patty. By the time Kara and her older sister came along Mirko was playing as a semiprofessional, and would eventually transition into coaching. "He really encouraged females to get involved in sport," Kara told me, "and was one of the first people to coach an all-girls team."

The budding family moved from the Queens borough of New York City to New Jersey in 1981 so they could have a yard and more space for the kids. Mirko and his brother ran a successful building insulation company, appropriately named Brothers Insulation, working on the Twin Towers and some of Donald Trump's eponymous buildings in the city.

On his way in to work early on July 2, 1982, a car in the opposite lane on Harlem River Drive, piloted by a drunk driver, launched over the divide and went airborne. If Mirko had left seconds earlier that morning he might have unwittingly continued on about his day, but in a deadly case of horrible timing the car speared his vehicle and killed him instantly. Kara was just a week from turning four years old and her sister Kendall had just been born six weeks prior.

Kara can't decipher which memories are hers and which have been told to her so many times that she has made them her own at this point, but she says about her father, "I know that he was super gentle. My mom is more rules-oriented and type-A, so they balanced each other out."

Patty, realizing that she couldn't raise the family by herself, moved them to Duluth, Minnesota, to be near her parents, Calvin and Ola Jean Haworth. Her grandparents loved the girls, cherished having them close, and tried to instill the importance of hard work. "The only thing we told them was that you've got to get your education," said Calvin. "We believe in education."

Patty married Tom Wheeler, who also had three children that had survived a tragedy (their mother had been brutally murdered). "We became the Brady Bunch," Kara said. "We were called the Brady Bunch my entire life."

The family committed to one another and rarely spoke of their past tragedies. "We did not talk about my dad growing up at all," Kara said. "It was almost as if it didn't happen. My mom would give me photos and I would hide them in my room." Kara bristled against her stepfather's verbal harshness at times, but created lasting bonds with all of her stepsiblings. "We were just this super weird family living on the edge of the woods," she told me.

Family life for the Wheelers was all about sports. "It was really important to the family, especially to my stepfather, who had tried to make it as a hockey player," said Kara. "Although he was a financial adviser, his passion was sports."

It was her grandfather Calvin who took Kara to her first road race, a small, local event at the Hermantown Summerfest. In the confusion of her first mass start, Kara set off with less grace than she'd hoped and immediately fell down. But to her grandfather's delight, she got back up and powered on, leaving him behind as her competitive nature took over.

She was an overachiever who got straight As and fancied herself a strong athlete.

In the sixth grade, following her older sister's lead, Kara told Tom she wanted to try hockey. "He laughed and said, 'Oh, Kara, you're way too delicate. You're graceful, but not a real athlete.' This was horrifying to me because we come from such an athletic family and I had done all these sports, but no one had ever told me that I wasn't athletic," said Kara. His comments were devastating and the self-doubt they gave rise to would echo in Kara's head for years to come.

In middle school, Kara decided that she simply had to win the school's Triple-A award, which was given to anyone who excelled at academics, athletics, and the arts. She had straight As already and had been playing the French horn since the fifth grade, but she needed to find her sport. "I went to cross-country and ran at a couple of the meets," she said, "then won the city championship." This is the race her grandfather loves to retell. Kara was so far out ahead of everyone that she kept missing turns, and had to double back to retrace parts of the course. And though she covered more distance than her competitors, she still won the event. In Minnesota, you can start running for high school in seventh grade, and Kara got the call up.

Attending Woodland Middle School, while running for Duluth East High School, Kara experienced her first, of what would become many, stress fractures, which took her out of contention for her next logical goal, the state championship. The following year, either Kara or her friend and teammate Amy Hill won every race they entered, five each, as she remembers it. Then Kara beat the defending state champion to qualify for the state meet. Recruiting letters from some of the most distinguished running programs in the country began appearing in her mailbox.

"That's when I started to feel, 'Oh my God, everyone thinks I'm going to win,' and that's the first time I ever experienced anxiety about running, like big-time anxiety, crippling anxiety. I couldn't eat that morning I was so nervous." To win, however, she'd have to contend with the athlete who would become her high school running foil—future Olympian Carrie Tollefson, who was the heavy favorite. Kara finished third in the state, while Tollefson added to her long list of championships.

At the state meet her freshman year, she was again thwarted by anxiety, overcome with emotions, and placed fifth. "I was vomiting," said Kara, "so I was taken to the hospital and given an IV."

The fact that she didn't quit, however, helped Duluth East High win the overall team championship, which began a string of wins that would last her entire high school campaign.

After running the fastest time in the state for the mile and the two-mile distance, Kara was then expected to dominate the track season. But instead, at the state championships a nervous Kara false-started off the line in the mile, and was disqualified from the race. She was devastated but also motivated to take action. "It was horrific, for me, but this was the very beginning of my activism," said Kara. "I wrote a letter to the Minnesota State High School League saying 'This is crazy, everybody gets a false start in the Olympics. We deserve a false start here.' I was denied."

Unlike a lot of high school running stars, Kara says her high school career was well balanced, and that her world wasn't built around running, at least not yet.

Kara sat on the "exec board," which planned the school's social events, and to her family's surprise she was even nominated as a finalist in the "Miss East" competition, which is the equivalent to being on the homecoming court. One year, after the hockey team qualified for the state championship, the school threw a send-off assembly, canceled school, and bused the students into the city to watch the game. They didn't win but received a three-page spread in the yearbook that trumpeted the team's minor success.

What about the state championship running team? she thought. Everyone knew hockey was king in Minnesota, but she felt as though the school was ignoring a championship team in their midst. "I just couldn't believe it. It was the first time I realized that the world isn't fair," she said.

Sophomore year, she won all her races and finally won the cross-country state meet. In track she again won the two-mile, but was running slower times in training than she had before.

She was invited to the Olympic Training Center and roomed with her nemesis, Tollefson. "It was good to get to know her and like her because she was really intimidating to me," said Kara. "I really liked her. She was so funny and friendly."

The following year, Tollefson was undefeated, and going for her fifth state cross-country title. As the race approached Kara's anxiety and dread grew. To make matters worse, both of their local papers were running articles about the inevitable face-off between the state's two best runners. Tollefson did in fact make it a record number five championship.

For Kara, it marked a confusing time for the young Minnesotan, who was growing through puberty, gaining weight, and running slower. As her performances dipped, the full-ride scholarship offers began drying up. At the regional Foot Locker event her senior year, she dropped out of the race. "We always thought I'd get a full-ride and I really wanted to go to Stanford, but after I dropped out of Foot Locker, that just ended, that was over."

She went on recruiting trips, largely alone, and simply couldn't picture herself at colleges like the University of Alabama and Georgetown University. As a graduation gift, Patty paid for Kara to see one last school. She chose the University of Colorado after reading an article about running prodigy Melody Fairchild, a homegrown Boulder running star.

This trip was also "a total disaster," as Kara was accidentally abandoned at the Denver Airport for hours, but for the first time Kara could see herself on this campus. They offered her a 20 percent scholarship, but Boulder was almost an impossible distance away from home for her, as a self-professed homebody, to conceptualize. She clung to the idea that even if she couldn't get back to her previous level of athletic success, at least she could be on a good team and contribute to something in a fulfilling way. She committed to CU.

At her final state championship Kara placed third in the two-mile and expected to get second in the mile event the following day, but said a little prayer, *God, whoever needs this victory, give it to them.* As the finish line approached she realized she had erased a huge deficit and was gaining on the leader. Newly determined, she pushed as hard as she could. "I beat her by a hundredth of a second," said Kara, "so I got to go out on top. I got to go out as a state champion."

At CU Kara would study psychology, and envisioned a career as a therapist. "I was in therapy as a child after losing my father. I was in therapy in high school for my anxiety," she said, "and I just thought, *I'm going to help people that have problems like I've had.*"

Adam was in full bloom by the time Kara arrived at the beautiful Boulder campus of the University of Colorado—everyone was talking about his performances and his Spartan attitude toward hard work and suffering. He won the cross-country championship his senior year, beating a stacked field, and solidifying himself as one of the best runners in NCAA history. University of Colorado Athletics now considers Adam "arguably the best runner to wear CU's black and gold."

After graduating in December 1998, Adam could have run for any shoe brand of his choosing after his stellar college career. "I was at the top of my game," he told me. "I thought I was going to go to Nike. They had the best offer, cash up front, no doubt about it."

Mark Wetmore encouraged Adam to consider other potential sponsors. Adam remembers Nike's global athletics marketing department head, John Capriotti, as being aggressive and presumptuous during recruiting. Cap, as he's called, controls Nike's exceedingly large purse strings and has likely dispensed more money to more runners than anyone else in the history of running.

Capriotti found his way to Nike after abruptly leaving his job as the head track-and-field coach at Kansas State University in November 1992. The university had discovered he had made illegal payments to amateur athletes in direct violation of collegiate rules. "I knew what I was doing the whole time," Capriotti told the local paper during the controversy, "and I knew what I was doing was against NCAA rules."

His former high school and college coach, Steve Miller, had just taken a job with Nike as the director of global track and field before heading the sports marketing department. And for this, Miller reached out to Capriotti to join him in his new venture. Although the Kansas State scandal gave him pause, Miller believed in Capriotti and after explaining the situation to Phil Knight, got the go-ahead to hire the firebrand.

"Phil and I talked about it," Miller told the *Oregonian*'s investigative journalist, Jeff Manning. "He (Capriotti) was the right guy. He said the infractions were genuine and real. I don't think Cap ever tried to hide from it."

Nike was a place where raw ambition was celebrated, and Capriotti found himself at home among even the most truculent executives. Unbound by the restrictive NCAA rules, he quickly rose through the corporate hierarchy. As Capriotti was on the ascent, American track was in trouble, and many now credit him with expertly piloting the company through troubled times for the sport and putting money behind events and federations with no clear return on investment . . . other than saving track and field.

When Nike offered Adam a $90,000 base salary with their standard bonuses, Wetmore advised Adam to go with someone else. The Italian brand Fila, which was founded in 1911 by brothers, Ettore and Giansevero Fila, had made more honorable inroads. The brand's origins were in manufacturing knitwear and sportswear for the people of the Italian Alps, but they were now making a push into the US running market—and they were eager to have an American superstar. Although they offered a base salary of only $50,000, they impressed Adam with their pitch. "They were the underdog," he told me. "They were excited about getting into the running space and wanted a flagship athlete." To sweeten the pot, Fila offered Adam sticky bonuses that would roll over for each subsequent contract year, adding to his base salary. He liked that his earning potential was tied more closely to his performance. He signed his first pro contract with the Italian brand and was lost to Nike for the next six years.

Early in 1999, Adam set a personal record time in the mile of 3 minutes and 54 seconds. This was an important milestone for the American runner. Even casual fans of the sport know how hard the four-minute barrier was to break. And they might even know the name of British middle-distance runner Roger Bannister, who broke it in May 1954. Either way, the distance has become an important barometer for runners, serving as shorthand for how fast they are and a way to compare them to generations past.

Of all the distance events, the mile is one that American audiences seem to understand. Olympic and Championship events use the imperial meter standard for their distances, from the 100-meter dash through the 10,000-meter slog. The metric push of the late 1970s wiped out all the American-centric events, and since 1976 the IAAF has allowed one nonmetric distance to be recognized for record purposes—the mile. Times over distances don't lie. In this way track lends itself to real and valid historical comparisons more than any other sport.

Coming off a senior year in which he had won three NCAA national championships (cross-country, outdoor 5,000 meters, and the indoor 3,000 meters) vastly improved Adam's confidence, but that was college, he was a pro now, and Bob Kennedy was the reigning king of American *professional* distance running.

Kennedy was the only non-African to ever run the 5,000 meters in less than thirteen minutes, and he had the commensurate air of invincibility about him. "Internally, I was in awe of Bob Kennedy," said Adam, "but I wanted a shot at him."

Kennedy had two signature model shoes with his name on them—the Zoom Kennedy and Zoom Kennedy XC. Even among the world's best athletes, a "marquee" sneaker was an exalted honor and a sign of ultimate commitment from a brand. To put an athlete's name, initials, or logo on footwear that the athlete also often had a hand in designing showed that the company had the utmost confidence in the athlete to perform, and move shoes.

It wasn't just the fact that Kennedy stepped to the line wearing Nike shoes with his initials on them that gave him confidence—he also hadn't lost a 5K to another American in six years. With the 1999 national championship looming just a few weeks away, Kennedy and Goucher faced off in a stacked international field on Long Island for a 3,000-meter race. African speedsters took first and second, leaving Kennedy and Goucher to battle for third. Kennedy, lunging at the finish line, barely bumped Adam off the podium. Adam had lost, but not by much, and there was suddenly a distinct shift in his perception of the competitive hierarchy—*Kennedy was beatable.*

Back home, Adam studied the Long Island race film over and over, trying to figure out what he could do differently when he met Kennedy on the track again for the 1999 national championship in the 5,000 meters. It appeared to him that in the final straightaway he had been straining so much that his form deteriorated into an overstriding mess that surely slowed him down. As a result, Adam altered his training to practice sprinting whenever he was tired and focused on shortening his stride while leaning slightly forward to get off his heels.

Weeks later, at nationals, Kennedy made a move just before the halfway mark. Adam expected him to throw in a couple of suicidally fast laps to create a gap on the field with hopes of "breaking the collective will" of the group, as he put it. Surging ahead of the pack, Kennedy did a double take over his shoulder, surprised to see Adam still with him. He then slowed, essentially forcing Adam to take the lead. "He was setting me up to outkick me down the final stretch like he did in that three-thousand-meter race," said Adam, "but I was ready."

As soon as Kennedy's body contracted in the first millisecond of an aggressive final push, Adam responded, bursting toward the tape, and finally outkicking America's best runner. "The elation of winning that race was greater than any other race up to that point in my career," said Adam. "I felt like I had finally arrived."

Back in Boulder, Adam continued to work with Wetmore as his coach and train at his alma mater. He joined the CU staff as a volunteer assistant coach as a way to give back to the team that helped elevate him; he was already on campus all the time anyway, working with Wetmore. Kara, who was in her senior year of college, admitted to not initially being happy about the new addition to the coaching staff. "I was like, 'What? He's always been good, he's never had to work through anything major, and what does he know about female athletes?'" To this point, the pair had been friendly, hanging out together in big groups, though not individually, but as Kara steadily improved, Adam's experience began to come in handy.

Suddenly, Kara was the odds-on favorite to win. "Up to that point I had never done anything big. I had gotten second at nationals in the 3K, but it was unexpected," Kara said. "I had never had attention like that before and Adam was awesome. He talked me through things, encouraged me to really stretch myself at practice. It was just cool, and well, I started to want to hang out with him outside of practice. He was kind of like a mentor at first, but only for a couple of weeks because then I started to be attracted to him."

With his relationship blossoming and his fitness peaking, Adam headed to North Carolina in February for the 2000 USATF National Cross Country Championship in the 4,000 and 12,000 meters. After winning the 4K race, he says, "I was a little sore in general, but felt pretty good and I blew everyone away again the next day to win the double," he admitted.

But trouble came during a trip with the CU team to the big twelve championships when, while running a workout on an outdoor track on a cold day in a driving wind, he strained his Achilles tendon.

Now that Adam had beaten America's best, it was time to beat the world's best. Typically, he would train in the morning, then join the team for practice, where he'd run with them or help Wetmore time laps and call out splits.

As a business adventure, Adam bought a newly constructed house with his best friend, former CU teammate Tim Catalano, twelve miles east, in Erie, one of the adjacent commuter suburbs that crowd Boulder. Their plan was to increase its value by adding a deck and improving the landscaping, then sell it at a profit. Hearing of the endless manual labor Adam had signed up for worried Wetmore. One day after practice he told Adam, "Do me a favor and just don't do the yard work stuff. Just hire someone to do it."

Adam, however, has more than a dash of mutineer in him. He and Tim got to work, moving tons of gravel from the driveway to the backyard, laying thousands of square feet of sod, installing a sprinkler system, and building a deck. This was in addition to the Olympic hopeful's 100-mile weeks of running, and on top of the ancillary weightlifting and plyometric exercises on his training schedule.

One morning after a particularly grueling day of manual labor, Adam awoke with a minor pain in his back. He ran his assigned twenty-mile training run regardless, and was scheduled to join his sister for a long car drive home to Colorado Springs, but the next day he could barely get out of bed, or bend over. Regardless, with his soldier's mentality, Adam again went out for his run. "Gotta go run," he said, ironically, "Gotta get my run in. So I went out for ten miles easy and had to stop halfway through." A concerned citizen who happened to be driving by stopped their car and asked if he was okay.

"Can I get a ride home?" Adam said.

Panicked, with just over two months until the Olympic Trials, Adam worked with Ed Ryan, a respected athletic trainer at the Olympic Training Center, and consulted with different specialists. One side of his body was spastic and tight, while the other was hypermobile and weak. He didn't risk running until ten days before the Olympic Trials preliminary event, and though his SI joint still bothered him, he had to start training again if he was to have any chance at Olympic glory. Muscle relaxers helped, as did the work of his massage therapist, Al Kupczak, who would set his table up at the track so that Adam could go back and forth from the table to the oval, and back again, in an effort to work the tissue into compliance.

Kennedy called Adam before the race. Someone had rear-ended his car, he explained, and the accident had caused his own back injury that he was struggling to work through. With this less than ideal news, Kennedy had a suggestion in mind. "You and I are the only two people that belong on that team," he told Adam. "If we take it out hard from the gun and we trade laps and help each other out, people will try to come with us and then die off."

"Bob, any other year I'm your man, but this year I can't do

that," Adam said. "If I did, I wouldn't make it. I have to run my own race."

On July 22, 2000, in front of twenty thousand people Bob Kennedy did exactly what he said he was going to do in a valiant effort to get out so far ahead of America's top runners that they would simply resolve to race for the second and third spot on the Olympic team. As the group passed through the first mile, Adam was in last place. "I let the race unfold and gradually moved my way up," said Adam. "Then Bob started to come back to us."

The group eventually blew past Kennedy, who couldn't maintain the aggressive effort, and Adam focused for a time on Marc Davis, who was running a solid pace. With 800 meters to go, Nick Rogers took the lead. Adam used Davis to bridge into the gap and closed in on Rogers, who had a four-second lead on Adam and Brad Hauser.

Then Adam unleashed everything he had. "One thing that Mark [Wetmore] always used to say was 'turn it off.' He uses the analogy of turning it off, like a switch, like a light. Relax. When people go all out, they fight and waste a lot of extra strength and energy. If you focus on turning that off, you can actually run faster."

Adam passed Davis and took the lead on the final stretch as the stadium went ballistic with applause. He ran the 12.5 laps around the oval in 13 minutes and 27 seconds, winning the race, the national championship, and an Olympic Team berth. As he staggered around in oxygen debt, the broadcast television crew corralled him and stuck a camera and microphone in his face. He babbled nonsense. "He literally can't form words, it's amazing video," said Kara. "It's the biggest moment in his career to that point and he can't even talk, he can't even capitalize on it."

By sponsor, the results read: Fila first place, Nike second, Nike third, William and Mary fourth, Nike fifth, and Nike sixth, with the Reebok Enclave team athletes rounding out the bottom half. It was the biggest victory of Adam's career, but one for which he would come to pay a high price.

05

JUST A COACH DOING THE RIGHT THING

BACK IN BOULDER, SIX WEEKS AFTER THE TRIALS, ADAM STILL DIDN'T FEEL QUITE right. Even while holding a pedestrian jog on the flat and fast Bobolink Trail his heart rate skyrocketed. "I'd do an easy six-mile run and I would be so exhausted. My heart rate would be like one hundred and eighty-five while running nine-minute miles," he said. "I fell into massive, chronic fatigue."

During a workout with fellow CU track-star-turned-professional-runner Alan Culpepper, his anxiety forced him to run off the track in tears after Culpepper blew him away in a sprint. Adam managed to pull himself together enough to drive straight to the closest emergency room, but at check-in found it hard to describe why he was there. "Something's wrong with me," he told the nurse. "I'm so exhausted."

"So, you're tired?" the nurse asked incredulously.

"Listen, I know I look like I'm really fit and healthy. I made the Olympic Team and I'm just a few weeks from going over to Sydney, but I'm jacked up. I don't know what is wrong with me, but you need to figure it out." The medical team took blood and told him he just needed time to recover from pushing his body too hard.

We know more about overtraining syndrome now, but it remains an insidious condition without a clear clinical diagnosis. It can present as a cascade of symptoms, including lethargy, depression, and insomnia, coupled with hormonal imbalances. There is no one test for it, and some physicians don't even believe such a condition exists; the cure prescribed is simply additional rest and recovery. For the same reasons it is hard to discern if an athlete has overtraining syndrome, it's additionally difficult to declare them cleared of it.

Adam spent considerable time resting, more than normal for an athlete of his caliber with the Olympics approaching, and it worked, to a point. He was able to have a few good workouts and resume some semblance of normal training in time to line up at the Olympics, though he was lacking in top-end speed.

Adam describes the 5,000-meter race at the 2000 Olympics as "a joke," and the one type of race—called a sit-and-kick—that he didn't have the sharpness to be competitive in. Every time he moved to the front to push the pace "a couple of Africans would come up past me, move in front of me and slow the pace down again," funneling Adam out the back, repeatedly. With one thousand meters to go, the real race began and Adam simply didn't have the explosive kick-power to win the final sprint. He finished in thirteenth place.

In a three-day span in July 2001, Adam ran his fastest-ever times, PRs, at both the 1,500 meters (3:36.64) and 3,000 meters (7:34.96). But his performance at certain benchmark distances seemed to be slipping, ever so slightly, due to recurring injuries and health concerns that he now attributes to the overtraining bout that began in 2000. He placed third in the 5,000 at the US Outdoor Championship, but only managed tenth place at the World Championship in Edmonton, Canada, in August (the top three spots went to Kenya, Ethiopia, and Kenya).

Kara graduated from the University of Colorado as arguably the most decorated and celebrated women's athlete in its history. She was spectacular in her senior year, winning the NCAA Championship in 3,000 meters, 5,000 meters, and cross-country while leading her team to their first overall national title. She remains the only two-time recipient of the CU Female Athlete of the Year award. What she didn't realize at the time was that an accident during a trail run in Eugene, Oregon, had cracked her kneecap, an injury she was able to run through, but one that became increasingly worse as time went on.

Shortly after Kara graduated the couple was married at a quintessential Colorado wedding in Lyons with dramatic red rock cliffs as their backdrop. "After Kara and I got together, we both knew this was it," Adam later told *Runner's World*. "Given Kara's and my family histories, it was pretty clear our own marriage was either going to be a train wreck or something incredibly strong."

Nike signed Kara for $35,000 a year, based mostly on potential, and the prospect of one day getting her husband to sign with the brand. They treated her well in the time before Adam was also a Nike athlete, even though she spent the next three running seasons mostly injured. Through this time, Nike never took advantage of contract clauses that would have allowed them to reduce her salary. Nike didn't hide the fact that they wanted to sign Adam, and they knew that after seeing how well they treated his wife that when his contract was up with Fila, the only reasonable decision would be to sign on with the Beaverton brand himself.

After the elation of the wedding and the Nike endorsement began to fade, Kara began a dark period of injuries and disillusionment. "Part of my patellar tendon ended up dying," she remembered. "I had to go on bed rest. I ended up sitting on the couch watching soap operas all day. It was a completely depressing time." She wouldn't compete at all in 2002.

Over time, Kara's dedication to rehabilitation began paying off and she slowly came back to running. She staged her comeback at a track in Walnut, California, deciding to try her hand at a race distance twice as long as she'd ever attempted previously—the Mt. SAC Relays 10K. She was rusty, having not raced in almost two years, and felt terrible within the first few minutes.

"I wasn't fit. I've run a lot of races, and I'd never walked before," said Kara. "But at Mt. SAC, I walked."

Adam had been extremely productive during his years as a Fila-sponsored athlete, winning championships and raising his profile. And as he'd hoped, the gamble to rely on the sticky performance bonuses paid off big. By the end of his first contract year they had pushed his salary over the $180,000 mark. Regardless, once Kara was in the Nike stable, it just didn't make sense for him to hold out anymore. He signed with Nike in 2003. And pundits and fans alike predicted Olympic berths for them both in the upcoming Trials.

THERE WERE 201 COUNTRIES COMPETING IN THE 2004 OLYMPICS, AND THERE ARE, OF course, more Olympic hopefuls than there are Olympic spots. To winnow potential competitors down to the truly elite, each country that competes in the Games holds a big winners-take-all event, called the Trials. Runners who finish in the top three spots in each distance's final make the Olympic team and will go on to race for their countries provided they have also run an Olympic standard time within the allotted qualification period. In the United States, during Olympic years, the Trials also serve as the de facto Outdoor Track and Field Championship event, which crowns the winner the best in the nation at their distance.

Bowerman, who coached the 1972 US Olympic Team, used to call the Trials "ordeals." The 2004 ordeals would be even more fraught for the athletes since federal agents had raided BALCO headquarters the previous summer in search of evidence of widespread cheating by athletes. Agents, led by Jeff Novitzky, interviewed Bay Area Laboratory Co-Operative (BALCO) founder Victor Conte and found troves of documents that seemed to implicate some of the sporting world's biggest stars, Marion Jones, Barry Bonds, Tim Montgomery, Bill Romanowski, and Gary Sheffield among them. In August of 2003, things began to unravel for Conte and BALCO when Nike-sponsored runner Kelli White became the first American athlete to ever test positive during a world or Olympic championship event. Though she would later admit to being on a number of illicit drugs, including "The Clear," "The Cream," and EPO, White was popped for having the prescription drug Modafinil in her system after winning the 100-meter and 200-meter gold medals at the 2003 World Track and Field Championships in Paris. The powerful stimulant is prescribed to people with narcolepsy, sleep apnea, or shift-work sleep disorders. Athletes without those conditions report feeling hyper-focused, energetic, and confident after taking the drug.

The Gouchers were none the wiser about what was happening with the doped athletes in the Trials. There was no way of knowing who was taking drugs and who wasn't, at least not until the drug tests came back. And it only seemed to damage your own psyche to consider everyone you lined up against a drugs cheat, so the couple simply chose not to think about it. They had come out of Wetmore's CU program, after all, a system where people seldom drank.

Although she had been dealing with a nondescript knee injury in the lead-up to the 2004 Trials, Kara was well trained, excited, and hopeful for the day. The race just so happened to fall on her twenty-sixth birthday, a fact that her supporters agreed would mean a little extra luck or, at least, an extra boost down the final stretch of the 5,000-meter event. Her family traveled from Minnesota to cheer her on, and Kara's mother, Patty, had even baked her a birthday cake, which she secured in her lap while the racers ran the oval 12.5 times in the preliminary round.

The women's A standard for the 2004 Athens Olympics was 15 minutes and 28 seconds, but as the laps ticked by it became apparent that none of the athletes would be putting in that effort; to Kara, even her slower pace felt more difficult than it should have. She fell farther behind the leaders with each lap. "I was the only person in my heat that didn't move on," she told me, "I got dead last."

Adam had won the previous Olympic Trials outright, and was now the man to beat. Since then, however, he had been dealing with myriad injuries, including low energy levels. Wetmore had referred him to Dr. Jeffrey Brown, an endocrinologist with a good reputation in Houston, Texas. Adam wrote a letter to Dr. Brown on April 26, 2004, in which he described both his pedigree and his maladies: "I am writing to you today in hopes that you may be able to help me with some problems that I have been dealing with for quite some time. Since competing in the 2000 Olympic Games, I have struggled with injury and fatigue. . . . [Wetmore] says that you are the best endocrinologist in the country . . . over the past four years I have repeatedly had blood tests come back with high CPK levels and low testosterone levels. I also have had low ferritin levels at times. Another concern for me is that I take Effexor for depression, and I worry that it dulls my bodies [sic] ability to compete. . . . With eleven weeks to go until the Olympic Trials, I am in need of some guidance."

A week later he was sitting on Brown's office couch in Houston. Adam later testified to USADA that in his first meeting with Brown, the doctor said "tell me everything you're on—whether legal or not" and asked if he'd ever been to the Bay Area, which Adam understood to mean the Bay Area Laboratory Co-Operative, or BALCO, which was believed to be distributing illegal performance-enhancing drugs to athletes.

Adam's medical records from 2001 to 2003 showed three blood draws that tracked his thyroid-stimulating hormone. They were all well below 2.00 (TSH levels below between 0.45 to 4.12 milli-international units per liter are considered normal). Regardless, Brown diagnosed Adam with hypothyroidism and within days had the young runner on thyroid replacement therapy.

WHEN ADAM ARRIVED AT HORNET STADIUM IN SACRAMENTO, CALIFORNIA, FOR THE Athens 2004 Olympic Trials, he was expected to once again earn a coveted spot on the American Olympic Team, or, at least, fight to the death for one. But Adam, after watching Kara finish last, lined up feeling deflated and was the nineteenth runner to cross the finish line, also failing to make it out of the preliminaries.

Wetmore, on the sideline, was visibly upset by the results, telling Adam, "Maybe that's as good as it gets."

"He was insinuating that I was done, that my career was over," said Adam. "Whether I misread it, or whatever, it really pissed me off and hit home." He was so psychologically distraught by the failure he couldn't even bring himself to watch the final 5,000-meter event, where Adidas runner Tim Broe won in a time of 13 minutes and 27.36 seconds.

While most of the athletes were out celebrating—either the relief that it was over, or, for fewer still, making the team—Kara sat crying in her hotel room with the blinds drawn. Before the Trials, Kara had already decided that she needed a coaching change, but after Adam's race and Wetmore's comment, they doubted even that pursuit. "We had to face the hard reality that maybe our best days were behind us," Adam admitted.

Most of the athletes tied to the BALCO investigation seemed to wither under the increased scrutiny of the federal investigation and either withdrew entirely or ran too slow to qualify for the team. The pressure seemed to be getting to Nike's superstar, Marion Jones, too. She finished fifth in the final of her best event, the 100-meter dash, missing the team spot. Her bodyguard escorted her out of the stadium after the race, allowing her to avoid answering uncomfortable questions. She returned to win the long jump event, however, before pulling out of the 200-meter semifinals.

All anyone could talk about at the Athens 2004 Olympic Games was, once again, doping. The day before the opening ceremony, the first president of the World Anti-Doping Agency, Dick Pound, brought up Jones specifically and told the press that if she was caught doping, "it's going to be a dark and deep hole into which she goes."

Nike's top sprinter, Justin Gatlin, won gold in the prestigious 100-meter distance in Athens. After the event, his Nike coach, Trevor Graham, admitted that he was the one who sent the syringe filled with the designer steroid "The Clear" to USADA, which set off the BALCO investigation. Around the time Graham sent the syringe, special agent Jeff Novitzky had found an unsent letter in Victor Conte's trash, addressed to USADA and the IAAF, in which he accused Graham of doping his athletes. It seemed the two brothers-in-arms had a falling out and had turned against each other, much to the delight of investigators.

"I was just a coach doing the right thing," Graham said about mailing the syringe. "No regrets."

Days later, he claimed he did it to "save the credibility of American sport," and that he got the syringe from Marion Jones's ex-husband, C. J. Hunter, the spurned lover who was attempting "to construct an athlete capable of beating Marion." (Hunter denied this.)

BACK IN BOULDER, THE GOUCHERS LISTLESSLY WANDERED THROUGH BORDERS BOOKstore and flipped through books on how to make a midlife career change. They were heartbroken over the Trials failure and wondered if they had what it took to win races at the highest level. After much introspection, they eventually came to believe that they needed to get back to work. If there was even an inkling of a chance that they could run at the Olympic level, then they needed to recommit to the sport that had brought them so much, including each other.

With Adam also looking for a new coach, they widened their search outside the immediate Boulder area—they could now go anywhere. Today, a top runner looking for a change could pick from any number of elite running programs across the country, but back in 2004 there were fewer choices.

They drove to see Kara's family in Minnesota and took a few days to visit the vaunted program in Madison that was led by the head coach of the University of Wisconsin's track-and-field team, Jerry Schumacher. Though he was happy to take on Adam, at the time, Schumacher didn't coach women. So Kara met with the university's women's track and cross-country coach, Peter Tegen.

The training they'd seen on the visit intrigued them. "It was exciting because their training was so different than Mark's [Wetmore]," said Kara. "They would sprint and do sit-ups, then sprint again. It seemed so interesting." And there was the added pull of being closer to Kara's family in Minnesota, something that she always hoped for. They left excited about the possibility and all but committed themselves mentally to a move to Madison, with Adam telling his grandfather that they'd most likely uproot and head east. In September 2004, however, the Gouchers visited Nike.

When the sun comes out in the Pacific Northwest it is truly one of the most beautiful, and green, places on Earth, and the Gouchers were treated to its full glory as they touched down in Portland. Salazar pulled up to the airport in a black BMW.

As he drove them to their hotel downtown, Kara and Adam recall, Salazar preemptively brought up old rumors. "You may have heard about me and this doping stuff," he said. "It's all bullshit. People are just jealous." He then explained that authorities no longer use the sort of drug test Mary Slaney had failed, and that being on birth control pills made the results unreliable.

The couple shared a glance but didn't think too much about it. Before that day, Adam had heard a few rumors about Salazar, but says, "It was not enough to make me be concerned about myself, because I know myself and I know where I stand and that would never happen."

Kara didn't know who Alberto Salazar was before this opportunity, but remembered hearing about Mary Slaney's failed drug test when it hit the news her senior year in high school, though she had no idea who her coach was. "You can call us naive, that's fine, but we bought it. In fact, we got to our hotel room and we were like 'What a great guy.'"

The next stop of the tour was the Nike house that had been oft-written about in the press throughout the early days of the program. When athletes joined the NOP and moved to Oregon, they were encouraged to live in the house, rent free. Odd PVC piping veined the green exterior, which looked unkempt in the plush Portland Hills that was dotted with older, meticulously cared-for homes. "It was a piece of crap," said Adam of the house. Inside there was the constant hum of the oxygen-scrubber and pump system that strained to deliver thin air to three bedrooms and the common area recreation room. There were constant admonitions to "shut the door!" because leaving one open for too long causes a loss of thousands of feet in simulated altitude. "You walked in, all the rooms were sealed, and it smelled so bad," added Kara.

There were bikes and elliptical machines inside, and the backyard had two underwater treadmills and two huge hot tubs. "We were like, no way we're going to live here," said Kara. "We were all in on the program, but we're not going to live here."

Over the next couple of days, they toured the Nike corporate campus, went on runs, and talked to the staff. They met Dr. Justin Whittaker, as well as coaches John Cook and Dan Pfaff. "I was really intrigued by Whittaker [a chiropractor and ART practitioner] and the idea of getting massages on a regular schedule," said Adam. "Dan Pfaff too; he really knew the body and was so obviously knowledgeable, he seemed like a genius." (No one knew it at the time, but Pfaff's days at the program were numbered, and by Christmas he would be gone.)

Kara offhandedly mentioned to Pfaff how intimidating and serious the program seemed, to which he responded, "You know what I think? I think you're scared that you're finally going to have everything at your fingertips and you're going to find out you aren't as good as you hoped you were."

"That's what made me feel like, we *have* to do this, because he's right," said Kara of Pfaff's psychological jujitsu. "If we go to Madison it's the same sort of life we were living, just in a new city. If we move here, we're actually professional athletes." Still, she had concerns about playing second fiddle to Adam's career. "I knew going in Alberto didn't want me. He wanted Adam. In Madison we would be of equal importance, fifty-fifty, and at Nike it would be seventy-five Adam, twenty-five me, but I still felt it was the better opportunity for us overall."

Salazar, who was coaching the boys' team at the Central Catholic High School in Portland during his spare time, told Kara that much of the programming he did for the boys could be applicable to her, and that he'd figure it out. He talked training specifics with Adam, while admitting to Kara that he didn't have much to contribute to coaching elite women, having only ever worked with one woman in the past, Mary Decker Slaney. But, as she recalls, Salazar didn't talk much about that early coaching relationship.

Long before coaching her, Salazar and Slaney had gotten to know each other through Nike's first attempt at an idyllic yet intense training environment for America's most-promising athletes. As it turned out, the Nike Oregon Project was something of a reboot of an old idea—and a team that Salazar and Decker Slaney were both members of in the early 1980s called Athletics West—that grew out of both a Nike failure and the American embarrassment at the Montreal 1976 Olympics Games.

IT WAS OBVIOUS, TO THOSE INSIDE THE SPORT, THAT THE 1976 OLYMPICS WERE OVER-run by doping-assisted regimes. The most glaring example might have been the female East German swimmers, who, with deep voices, broad shoulders, and five o'clock shadows, dominated the swimming events, winning eleven of a possible thirteen gold medals that year. The East German swim team hadn't had enough talent to win a single race at the previous Games in Munich, but were now, somehow, the best in the world. Most spectators seemed to deride this success with wry suspicion, but when the Americans voiced their criticism, they were considered sore losers.

US swimming star Shirley Babashoff, who was poised as a favorite for gold in five swimming events, suffered possibly the most egregious affront. The female Mark Spitz of her time, Babashoff would come in second place on four occasions during the Montreal Olympic Games, losing each time to an East German athlete. She later said that she never regarded her defeats in 1976 to be losses at all, saying she had been "beaten by men."

Thirty minutes before his second attempt at an Olympic marathon, on July 31, 1976, Frank Shorter took a deep breath, exhaled, and laced up his Nike running shoes. They had been made exclusively for him and his attempt to become the first American to win a second gold medal in the prestigious event. Usually, he would warm up in his trainers and then change into his racing flats moments before lining up, but for reasons that are still unclear to Shorter, even today, he decided to warm up in his race shoes.

As the heavy favorite and defending Olympic champion, Shorter was the marked man, the one to beat, the best in the world. His Nikes, however, didn't feel like world-beaters. As a matter of fact, they felt a little sloppy. He stopped to tighten them and then took off running again, but the shoes still felt loose. He stopped abruptly and turned the shoe on its side; the glue that held the upper part of the shoe to the midsole had failed and was delaminating.

Shorter had won the Olympic gold at the 1972 Summer Olympics in Munich wearing a mustache on his face and Adidas on his feet. Now in Montreal, he was clean-shaven and contractually obligated to wear Nikes, though he had a backup pair of Onitsuka Tigers—which would be renamed ASICS the following year—mailed out to him in Canada, just in case.

Institutionally confident since inception, Nike was eager to prove itself in Canada. This was their first time at the Olympics and they envisioned these Games as "Nike's grand debut, our Olympic coming-out party," as the company's enigmatic founder, Phil Knight, put it in his memoir: "We had athletes in several high-profile events wearing Nikes. But our highest hopes, and most of our money, were pinned on Shorter," wrote Knight. "You really weren't a legitimate, card-carrying running-shoe company until an Olympian ascended to the top medal stand in your gear."

When Shorter became the first American to win the 26.2-mile Olympic event since Johnny Hayes at the London 1908 Games, he became the father of the running boom. What began with the 1966 publication of *Jogging* by Phil Knight's Nike partner, Bill Bowerman, became a full-blown phenomenon after Shorter ran 2 hours and 12 minutes for gold in Germany.

Between the Munich and the Montreal Olympics, Shorter's performances inspired noncollegiate athletes—aka regular people—to buy trainers and start running as a competitive outlet, but also as a way to keep themselves fit. Sports science was beginning to emerge that made it clear: running improved your health. Shoe companies were, of course, taking advantage of the growing trend, which required little else than a reliable pair of rubber-soled trainers. As they became available in more and more sporting departments, independent running shoe stores sprouted up all around the country. Many of America's major city races began in this time, from Seattle to New York City. Atlanta's Peachtree Road Race started with just 110 participants in 1970 and now has 60,000 who line up each year.

But the citizen running scene was growing in tandem with the shadow of performance-enhancing drugs. "The modern Olympic movement had lost its last vestiges of innocence after Munich," Shorter told me. The Games had become a vehicle to demonstrate national strength and vigor, and the Eastern Bloc countries were upping the ante. The importance of winning the Games escalated into an "arms race" of performance enhancement. As the 1950s gave way to the 1960s the specter of drugs spreading throughout the Olympics became a full-blown crisis for western countries, and most of all the International Olympics Committee (IOC).

Drug testing seemed to be the essential step to returning the Games to what its founder, Pierre de Coubertin, had envisioned. A vision described well by Rob Beamish and Ian Ritchie, in their 2006 book, *Fastest, Highest, Strongest*, as a "Games to bring the youth of the world together in a chivalrous bond of brothers-in-arms so that in the cauldron of athletic competition they would develop the character traits needed to lead Europe out of its spiritual and moral decline."

A year before, Bill Rodgers defeated Shorter at the World Cross Country Championships, and a three-way rivalry began. Young upstart Alberto Salazar fought both men at the famed Falmouth Road Race in 1977, beating Shorter but losing to Rodgers. These three men would push the boundaries of the marathon for the next ten years, breaking the United States marathon record five times.

The Games of 1976 were the first time the country of Canada

had ever hosted an Olympics, but the cosmopolitan city of Montreal was well suited to impress with its boutiques and glass-and-neon malls full of imported European fashion.

All Olympic athletes of these Games were technically amateurs, and therefore unable to be paid by shoe companies or their countries as professionals. At least, that's what the rules dictated. But the entire enterprise had already changed from one where everyday working-class athletes asserted their autonomy and physical individuality on a somewhat level playing field to one where countries supported their ostensibly amateur athletes financially and pharmaceutically in a global geopolitical chess match.

In this era before professionalism, an athlete could not benefit financially in any way from their sport. If they did, they risked losing Olympic eligibility. They could not be paid appearance fees to show up at races. They could not be paid by shoe brands to wear and promote their products. In blatant disregard of the international Olympic rules, European nations were secretly providing support to their Olympians under the guise of jobs the athletes didn't have to show up for. "Employment" that enabled them to live comfortably, without other time-consuming responsibilities, so they could focus full-time on their training, which would inevitably allow them to perform better.

Athletes of this era knew that the European Olympic powerhouses they competed against were blatantly skirting the Olympic rules of amateurism. Many American Olympians, some of whom were living below the poverty line, grew so frustrated that they began finding their own creative ways around them too. In their 1992 book about the early years of Nike, J. B. Strasser and Laurie Becklund liken these Olympics to "a giant trading floor, an auction house where athletes were bought and sold on silent bids."

"We felt like pawns being moved around by the masters of war," recalls Shorter. But it wasn't just the athletes. If you knew what to look for, the 1976 opening ceremonies showcased the financial agreements being made all around. The seven thousand Olympic officials were dressed head-to-toe in Adidas, while the Canadian team was outfitted in Puma. Converse reportedly paid $170,000 in "donations" to the US Olympic team for the right to advertise that the team had "selected" Converse, even though the brand didn't make a track spike. The shoe companies knew that the world would be watching on television and that this was their chance to showcase their products in front of millions of potential customers.

These Olympic Games were supposed to have been Steve Prefontaine's Games, and feature his chance to vanquish his old nemesis, Finnish runner Lasse Virén, and finally win gold. Prefontaine was a record-setting American track star but had faltered at the 1972 Olympics in Munich, Germany. Virén won the 5,000-meter event. Prefontaine finished a disappointing fourth place, behind three runners who all wore Adidas' three stripes on their feet, and almost six seconds slower than his Olympic Trials qualifier. His untimely death in May 1975 meant he wouldn't have another chance. This spurred his good friend Shorter to want to work with Nike.

"Nike was Steve's thing," Shorter told me. "But once he passed away, they were looking for a prominent American runner, and that was me. It was more to carry on Steve Prefontaine's legacy." In the spring before the Olympics, Shorter had agreed to a clandestine contract with Nike that would pay him $15,000 for the year. All he had to do was run in their sneakers.

Now trapped in the warm-up area, about to run in front of more than 67,000 fans, his toe showed through his busted Nikes. There was no way to patch the shoes in time, and he couldn't risk footwear that had little chance of even making it out of the stadium in one piece, let alone the additional twenty-six out-and-back miles to the finish line.

He remained calm but knew full well that with every second that elapsed he was at risk of missing the start of the Olympic marathon—the one he'd thought about incessantly, and trained for religiously, for the past four years. (He would later tell *Sports Illustrated* that he averaged seventeen miles a day for the entire decade of the 1970s.) He had to get his backup shoes from his room, but now that he was checked into the Olympic corral, officials wouldn't let him out and then back in again.

He scanned the track for someone, anyone, who could help. He spotted Bruce MacDonald, a US race-walking coach. They were staying in the same suite in the Olympic Village, and Shorter knew MacDonald would have access to his room. Within moments the walking coach was off and running.

MacDonald easily located the shoe box in the room, since it had been graffitied with messages of encouragement by US customs, and raced back to the stadium. With just minutes to spare he threw Shorter's orange Onitsuka Tiger Obori marathon flats over the fence. As the clocked ticked down, Shorter quickly tied his trusted shoes.

Back in Oregon, Phil Knight had gotten up early to watch the race, and after his morning coffee sank into his recliner with a soda and a sandwich close at hand. As the cameras panned to Shorter and his orange shoes with the giant swooping A stripe along the side, Knight slid out of his recliner, onto the floor, and began crawling toward the television. "No, NO!" he cried.

For Olympic marathon events, the athletes don't line up in specific lanes, but rather crowd the start line. Shorter joined the pack with just seconds to spare and had to settle for starting the race on the outside, third row back. When the gun's trigger was pulled it ignited a flash of fire and a plume of smoke, which sent electrical impulses from the amygdalae of the world's best marathon runners, through their spinal cords to their leg muscles, animating them from a standstill to a sub-five-minute mile in seconds.

Sixty-seven competitors from thirty-six countries rushed down the track, with the odd mustache and painters' cap mixed into the crowd, as was the style of the day. "I watched in horror as the great hope of Nike took off in the shoes of our enemy," Knight wrote. He switched from soda to vodka.

Shorter's tall, proud stride was a thing of beauty, with the scarcely noticeable exception of his left arm, which barely moved.

(He attributes this to his days running to school with his books under his arm in the days before every kid had a backpack.) He was twenty-eight years old with a quintessential runner's physique, a tall, thin body that he had molded over years into the perfect expression of endurance and efficiency. He wore tight blue shorts and a long red USA tank top that left his rib cage and body fat percentage exposed. Always considering the tangent, he quickly made his way to the inside lane of the track. The group ran the oval once before exiting to the Canadian streets. It was an overcast Saturday with rain expected later in the race; Shorter hated running in the rain.

He knew most of his competitors from the international running scene and expected fellow American Bill Rodgers to compete for the win if he could overcome a recent hamstring injury. Then there was Virén, who, after winning both the 5,000- and 10,000-meter events in the Munich 1972 Olympics, had become an international superstar and, publicly at least, the exemplar of the Olympic ideals of honest fair play and hard work. But the athletes thought they knew better. Virén's uneven, inconsistent rise to the top of the podium suggested to his competitors that something was off.

"We sensed that Virén was cheating," Shorter told me all these years later. "Virtually all the distance runners on the Olympic-level circuit strongly suspected he was blood doping, but there was no way to prove it." (Virén has always denied these allegations.) There was no way to prove it because, in these early days of the fight for clean sports, blood doping hadn't yet been identified as an illegal procedure. In 1967, during its annual meeting before the Mexico City Olympics, the IOC defined "doping" for the first time as "the use of substances or techniques in any form or quantity alien or unnatural to the body with the exclusive aim of obtaining an artificial or unfair increase of performance in competition." An athlete's own blood is anything but "unnatural," though the process of extracting, storing, and reinfusing it is "an unnatural technique with the exclusive aim of obtaining an artificial increase in performance." But at the time the process was still not deemed illegal nor was there a test for it.

Viren was never caught cheating. But Shorter was unconvinced: "The athletes knew something wasn't right. In the eyes of many athletes, Virén was cheating morally."

Just the day before the Montreal marathon, Virén had performed brilliantly, and for a second Olympic Games, won gold in both the 5,000- and the 10,000-meter races. After the race he took his victory lap holding a Tiger shoe in each hand high above his head, soaking in adulation (and the following day's newspapers began the long tradition of media complaints about the "commercialization" of the Olympics).

If Virén could also win the marathon he would become only the second Olympian to take gold for the three distance events in the same Games, a feat that hadn't been accomplished before or since the legendary Czechoslovakian runner Emil Zátopek pulled it off during the 1952 Summer Olympics in Helsinki, Finland.

On the road in Montreal, Shorter went out hard, exploiting his 10K speed before settling into a more manageable pace. He and Virén clipped along at well under five-minute miles, covering the first ten kilometers in under thirty minutes—a world-record pace.

By mile fifteen it was now just Shorter and a virtually unknown East German steeplechaser named Waldemar Cierpinski, who Shorter had mistaken the entire race for an athlete from Portugal named Carlos Lopes. Cierpinski appeared to Shorter to be on an easy walk in the park, as if he'd "parachuted in or stepped off the sidewalk." There was no characteristic marathon-anguish in his face, even when, approaching the twenty-first mile, he decided to drop Shorter for good. "He just made it look so easy," Shorter recalled.

With just three miles to go, Shorter had reeled him back in, but as soon as Cierpinski realized it, he looked back at Shorter and took off. Sprinting a sub-five-minute mile toward the stadium, Shorter was so focused on finishing as strong as he possibly could he didn't even notice it had started to rain.

As he neared the stadium, Shorter heard the roar of the huge crowd, welcoming the first runner to enter the track, a sound he alone was uniquely used to unjustly hearing. During his gold medal run four years before in Munich, an imposter dressed as a runner ran through the tunnel moments before Shorter arrived, stealing his ovations. He would become the first US runner to win the marathon gold medal in sixty-four years, but the confused audience treated him like the runner-up.

Now on the Montreal track, Cierpinski, not sure where the race ended, mistakenly ran an extra four-hundred-meter loop. He finished with a time of 2 hours, 9 minutes, and 55 seconds. Shorter crossed the finish line 50 seconds later. The two men embraced in an awkward congratulatory hug, Shorter looking far worse for wear. It wasn't until Cierpinski asked Shorter, "*Sprechen sie Deutsch?*" did he realize it wasn't Lopes who beat him. Shorter, drifting around, contemplated what had just happened in a calorically deprived daze before being ushered to drug testing.

Elsewhere in the stadium was Kihachiro Onitsuka—the "Shogun of shoes," as Knight called him—the owner and founder of Onitsuka Corporation. It was the Shogun who had given Knight his first job in the shoe business, and the idea, the pie-in-the-sky dream really, that it could be possible to have everyone in the world wearing athletic shoes all the time. Not only had Knight betrayed him as a business partner, but now Knight was trying to make his vision of the sneaker-dominated future come true with Nikes, not Tigers. After the race, Onitsuka reportedly let out a deep sigh of relief. The most famous runner of the time had chosen to run in Tigers despite the fact that Nike was illegally paying him to wear their shoes.

06

IT WON'T BE PRETTY

NIKE'S PROMOTIONAL EFFORTS AT THE 1976 OLYMPICS HAD FAILED. THE $5,626.90 they had given away in product paled in comparison to the reputed budget of $7 million that Adidas had invested. There were no gold medal podiums or victory laps with Nike shoes held high in admiration, and to add insult to injury, not only did Frank Shorter not win but his Nikes had fallen apart before the race even started, leaving him with no other choice than to run in a rival's shoes.

Rankled by the Olympic failure, Nike's third employee, another one of Bowerman's Men of Oregon named Geoff Hollister, suggested the company create a running club to assist elite post-collegiate athletes. America's best, Steve Prefontaine and Frank Shorter among them, had lived in squalor and, at times, survived off food stamps to sustain themselves while chasing their Olympic dreams.

Knight, Bowerman, and Hollister conceived of a professional running team designed to provide equivalent support to what the Eastern Europeans were receiving—and one that, hopefully, would give America's world-class athletes a fighting chance. Receiving such a financial assist from a corporation, however, brazenly broke the amateur rules of the time, but Knight, working out the budget, reckoned it would only cost about "the price of an ad in *Sports Illustrated.*"

Nike's first large-scale attempt to nurture athletes by integration of commercial marketing, vast private sector resources, patriotism, cutting-edge medicine, and high-performance sport technology was called Athletics West. They chose not to include Nike in the team name because Knight didn't want to tarnish the patriotic image of the team with commercial interests. Instead, they chose "Athletics" to connote track and field, and "West" to distinguish the team from Eastern Bloc countries.

There was no greater honor for the country's most talented postgraduate runners than to be asked to join the Oregon-based program. Anyone offered a spot on the team was, for the first time, officially a professional runner, even if international rules didn't technically allow for such a distinction. But the future of professionalized running was unstoppable, and none of the associations that ran Olympic or national sport would ever pursue suspensions for any athlete involved in AW, as it came to be called.

The first attempt at a comprehensive training environment for their athletes was launched in the fall of 1977. Knight chose Nike employee Nelson Farris to manage the team. Bowerman's successor at South Eugene High, Harry Johnson, was picked as the coach. The charter members were encouraged to move to Eugene with the promise of a basic financial stipend, health insurance, and travel expenses. A year in, Nike began construction on the AW training center, as well as new corporate offices in West Eugene.

Coach Johnson's first acquisition was Craig Virgin, a talented runner who had first made a name for himself by breaking Prefontaine's National High School record at the two-mile distance. Virgin wore the first pair of Nike track spikes in international competition, handmade by Dennis Vixie and given to him by Geoff Hollister. A subsequent photo of Virgin running in the race was distributed to Nike stores without his knowledge or approval, making him, at sixteen years old, Nike's first poster boy. "They never asked permission," said Virgin. "That's the way they operated then, and now."

Virgin went on to become a collegiate standout at the University of Illinois, winning the 1975 NCAA Cross Country Championship and becoming a nine-time All-American. He finished just behind Frank Shorter at the USA 1976 Olympic Track and Field Trials in the 10,000 meters, but failed to make the final event at the Montreal Games.

After he graduated in 1977, Hollister flew him to Eugene to meet with Knight. He thanked him for wearing Nikes, and then took the young runner sailing, where he told Virgin about the idea to start a Nike-sponsored track team. One night they had dinner with Coach Johnson and his wife and discussed the concept further. They wanted to create a Nike-funded club that was based in Eugene and would be organized along the lines of an Eastern European subsidized track team. Virgin, who was excited to take his training efforts to a new level, moved to Oregon in mid-August 1977. Other gifted Americans followed, including Jim Crawford, an army veteran who ran a 3-minute-and-59.6-second mile, and mid-distance specialist George Malley.

"Hollister called me in Europe in August then a few times in early September," Malley wrote on the popular LetsRun.com message boards, of his time with the team. "He sounded shady and desperate but his persistence wore me down so he flew me to Eugene to hear them out." At the meeting with Johnson and Hollister, Malley was unimpressed with the two men, who, he wrote, "could not stop talking about themselves." He added that "neither of them cared about the sport or other people except when they could benefit from it/them personally. I could see that from day one Hollister and Johnson's sport wasn't anything like the sport that I signed up for." But the excitement of being in the vanguard of the professional running movement and the idea that his career didn't have to end at age twenty-two was ultimately too enticing. Malley signed on with the team in September, and by June 1978, it consisted of fourteen athletes. They received a guaranteed minimum income of $1,000 each month and a part-time job. "A thousand dollars a month was good money in 1977," Malley said, "especially in Eugene's awful economy. Up until then, no one was paid openly to run by the shoe companies, and those who were paid under the table were few and far between. It was a good idea, but a very poor implementation."

Malley and Virgin both described Johnson's coaching style as "dictatorial," and questioned his experience level, since he had only worked with high school athletes. "Johnson plagiarized from the ideas of famous coaches (Lydiard, Holmer, etc.) that he had read in running books, then put them all together into one 'program,' then threw in his own numerology metric on top of that, and wrote out complicated schedules. He didn't know a damn thing about training or human physiology," said Malley. "It was a lot like saying, *I like ice cream, I like tacos, I like sushi, and I like apple pie*, then throwing it all into a pot and serving it every day. It tasted like shit."

By the winter, internal tensions between coach and athlete were high, but the team raced well that first year, with most of the members improving, setting new personal records during the national championships and a European racing tour that followed. Virgin became the team's first national champion, when he won the 10,000 meters at the AAU Track and Field Championships, with AW teammate Jeff Wells taking the bronze. "Most all of us were in the top five of the year-end US seasonal list of mid- to long-distance times in 1978," said Virgin. "But, morale and attitude continued to deteriorate and the team atmosphere became tense and uncomfortable. They were ignoring this undercurrent of discontent."

Virgin and Malley weren't the only ones finding it increasingly frustrating to work with Nike in 1978. Knight had written what he called the "Nike Principles," the guiding ethos of the company, which he handed out to all new employees. Listed among the tenets were:

3. Perfect results count—not perfect process. Break the rules; fight the law.
4. This is as much about battle as about business.
9. It won't be pretty.

BILL BOWERMAN CHAFED AGAINST NIKE'S AGENDA AND WIN-AT-ALL-COSTS MENTALity, and his distaste for the way the corporation was run had grown impossible for him to hide. Prefontaine's former girlfriend, Mary Creel (née Marckx), who worked as Knight's secretary at the time, once spoke of Bowerman's disenchantment. "I think he was always uncomfortable coming to Nike. He was less and less happy with the corporation it became," she said. "Phil had a deep-seated fear of Bill, but at the same time he was essentially Bill's boss. It seemed to me that Phil's attitude was 'I have to tolerate you because I'd be nothing without you, but it's uncomfortable.'"

In the summer of 1978 Bowerman wrote a letter to a friend in which he described the widening ideological divide between himself and the company he founded. He wrote about a meeting in which the businessmen discussed Nike's objectives. "You looked at Buck and he said, 'Make money.' I concur that without it you can't swim, but I said, 'Mine continues to be to make the best possible product at a price that will achieve the objective but keep the customer coming back.' I hear lip service to that, but I observe us not only standing still, but distributing a lot of crap."

In a serendipitous turn of events for Nike, the Amateur Athletic Union (AAU), an organization founded in 1888 to assure common standards in amateur sports, who had, since 1894, overseen US Olympic athletes, eased its pursuit of those in violation of the amateur rules of sport. A chorus of complaints, often led by Bowerman and prominent athletes like Shorter and Prefontaine, had undercut the organization's credibility and authority. Critics claimed they were a power-hungry regime more likely to spend funds on fancy hotel rooms for their officials than on the athletes in their charge. Their regulatory framework was outdated, excluded women entirely, and offered no recourse or due process to athletes who were often capriciously suspended or made ineligible.

In 1978, the US Congress held a series of hearings on the issue of sports governing bodies and the lack of athlete representation in them. In November, President Jimmy Carter signed the bipartisan Amateur Sports Act, which achieved two important things: it established a voting stake for athletes alongside the presidents and athletic directors on all of sports legislative bodies, and it broke up the AAU's power as the national Olympic sports governing body by creating the United States Olympic Committee and reestablishing independent associations. With the AAU effectively neutered, Nike, seeing opportunity, began sponsoring more races and investing in more athletes directly.

After AW's 1978 European racing tour, both Virgin and Malley returned to Eugene fed up with what they considered to be the reckless training methodology employed by Johnson, and quit the team. Virgin says that Johnson was so upset by this that he launched a campaign of intimidation, in which he told Nike that if Virgin left and they continued to support him, that he'd quit as head coach, while threatening Virgin that if he moved back to Illinois he'd be cut off financially by Nike.

"I really had no choice. I could not survive post-collegiately on just a few pair of shoes and sweats," Virgin said. "They still believed in Johnson's coaching system at that point in time. So, I had become 'expendable.'" While Virgin was on his way out, he met a new Nike employee named Dick Brown. According to Virgin, Brown "was an exercise physiologist with a master's degree who had been hired to test and monitor all the AW athletes' blood and other physiological signs."

Malley, too, suffered a rough exit. "They were furious. They reneged on their bonus commitments and began a malicious whisper campaign," he said. Malley drove to Beaverton to tell Knight what was going on directly, which only proved to inflame the issue. "Johnson and Hollister, true to form, never spoke to me again."

In 1979, a twenty-year-old Mary Decker (not yet Slaney), inspired by Prefontaine, moved to Eugene and befriended many of the AW team members. Known for her pigtails and her speed, she quickly set herself apart on the racing scene and became the first female admitted to the elite Oregon squad. Many of the members' wives passionately protested, but she was voted in on her running ability nonetheless later that year.

In the early 1980s, Joan Benoit and Alberto Salazar joined the team. Benoit had proven herself in Cape Cod, winning the 1976 and 1978 Falmouth Road Race and earning her spot on the AW team as an athlete with seemingly unlimited potential. The team was now more dispersed, with athletes allowed to live wherever they liked rather than upend their lives for a move to Oregon. So Benoit was allowed to stay in Maine and receive her training remotely.

In their research for the book *Swoosh*, J. B. Strasser and Laurie Becklund found evidence that the program was testing performance-enhancing drugs and blood doping. Strasser, who had also been the company's first advertising director and wife of the late Rob Strasser, told me she reviewed meeting minutes from a 1979 Nike management meeting in which Dick Brown was mentioned for studying steroids as part of his program to help athletes recover more quickly from workouts.

Once Brown was on staff as the team's exercise physiologist, Nike had the ability to monitor athletes' hormone profiles. Armed with this data, AW offered its athletes medical tests to detect the serious side effects of anabolic steroids, monitoring liver function in particular.

Brown began submitting bills listing "injections" on AW invoices. In 1982, employee Tom Sturak called Brown to express his concern that those invoices could be proof that Nike was using banned drugs. From then on, the invoices listed "adjustments," instead of "injections."

Jeff Johnson, who was in charge of "all things running" at Nike through this time, told me he had no direct knowledge of steroid use, but said he could envision Brown feeling an obligation to catch up, at least in the knowledge about what the East Germans were using, especially if they felt they were being outgunned or they weren't operating on a level playing field.

Johnson's attitude at the time was that Nike would never become involved with performance-enhancing drugs, "but, it's fair to say that the athletes who compete on the world level take a slightly different view, for these are a different breed of human being," he told me. "I'd bet my life that Joanie [Benoit] wouldn't even consider that kind of thing, but I'd bet my life that Mary [Decker] would be crazy enough to consider that kind of thing."

Through this period with Nike, Johnson became disillusioned with the direction his beloved sport was taking; to him, paying athletes against AAU rules was a "ten" on the worry-scale, whereas Nike's steroid use was only an "eight."

For his part Virgin believes that everyone on the Athletics West team from 1977 to 1980 was probably PED free. "From '81 on, I have my doubts," he told me. "I heard from other AW team members that it wasn't mandated, but it was offered. They said, 'You are going to do it, we want it to be supervised, and here is a doctor that is going to help you.'"

Ron Tabb, an elite marathoner on the AW team, who was married to Mary Decker from 1981 to 1983, corroborated this. "Living in Eugene you heard rumors, who's doing this or who's doing that. Usually you'd hear about Winstrol and Anavar, those were the two biggies back in those days," he said.

He and Decker were offered drugs in the Nike team physiologist's office in either late 1981 or early 1982. "We had a conversation with Dick Brown about the Eastern Europeans and the performance-enhancing drugs they were using and how to combat that and compete on their level. His proposition was that if we wanted to do that he was willing to monitor and regulate it.

"Mary was intrigued with the prospect. I know she was open to the proposal, but I told Mary I wasn't interested in that. I wasn't interested in seeing her look like some of the European women, how they were buff and looked like they could kick your ass." Neither of them took Brown up on the offer at the time.

"Some of the guys in the club, distance guys at that time, I knew for sure, for a fact because I trained with them, were doing blood doping," Tabb told me. "I had another friend who did accounting for the athletes and these guys were writing their steroid usage on their taxes, as a tax expense. Alberto, I think, for the most part, was clean, but I wouldn't be surprised if he weren't clean."

After he left the team, Virgin signed with Adidas for $12,000 a year, and continued to compete against Nike's AW athletes. Based on that experience, he now believes doping on his former team became endemic around 1981. "Alberto's jump from '81 to '82 was so . . . ah . . . impressive," he told me. "It was such a dramatic improvement. If there was a time that Alberto was on, I'd say that that was a time, late '81 to '82 and early '83. It would be consistent with the Alberto I know for him to experiment with just about everything under the sun to try to improve his performance."

(The now deceased Dick Brown always denied the claims.)

Tragedy struck in June 1986, when one of the AW athletes collapsed after a run and died in the Athletics West Track Club offices. Time has obscured many of the details, but we know Jeff Drenth, a twenty-four-year-old distance runner from Charlevoix, Michigan, was joking with friends while he waited for a massage. He stepped into the bathroom in the intervening time, and never came back out alive.

Drenth, who had attended Central Michigan University and was a member of the previous three United States cross-country teams at the world championships, reportedly had a history of heart arrhythmias. Just over a month earlier he'd seemed the picture of health and athleticism when he and AW teammate Joan Benoit won a ten-mile race, called Trevira Twosome, on April 26 in New York City's Central Park.

The Lane County medical examiner listed the cause of death as a heart rhythm disturbance, but people have openly doubted this explanation, assuming instead that the performance-enhancing drugs the team was allegedly experimenting with must have contributed to his death. Malley, for one, has publicly called it "valid speculation."

People who knew Drenth don't believe drugs were involved, however. His brother, Walter Drenth, who is now the director of the men's and women's track-and-field and cross-country programs at Michigan State, told me he had talked to his brother about performance-enhancing drugs, specifically.

"Would you ever stoop to that?" Walter once asked him.

"No, I'm never doing that," Jeff replied.

"It was the loss of a wonderful person," said the AW team psychologist, Scott Pengelly, who told me he'd become so close to Jeff through this period. "I wanted to be very caring and vigilant about what happened after his death because we wanted to find out why and how this happened, because there were going to be rumors about the various reasons that led to his death." At the time, rumors were rampant about what elite athletes were doing to improve performance. He had fluids and tissue sent to two places—UCLA Medical School and the National Institute of Health—to be tested. Initially the results came back without any determination, Pengelly said, but further testing showed the likelihood that Jeff's exceedingly low levels of iron stores, and not PEDs, contributed to the under functioning of his autonomic nervous system and his sudden death.

Two months after the tragedy, Nike shuttered the Athletics West program entirely in favor of regional teams.

07

NOTHING TO LOSE

EXITING COLORADO FOR PORTLAND IN OCTOBER 2004 TO WORK WITH THE NIKE OR-egon Project was tough for the Gouchers. Just talking about leaving behind their mentor, friend, and longtime coach, Mark Wetmore, still brings solemn expressions to their faces. In their post-Trials frustration, the couple had mentioned a possible change to reporters, so Wetmore read about their decision to leave in the newspaper before they had a chance to speak with him about it. They agonized over the decision, and when they broke the news to him personally, he told them, "If you do this, there is no coming back."

There was no risk to going to Nike. Salazar told them, "Just give me six months, no commitment, no hard feelings." They didn't sell their house in Colorado and they kept the line of communication open with Tegen and Schumacher. "So, it just felt like we had nothing to lose," Kara said. "If we hated it we'd just move back, it wouldn't cost us anything."

They packed up about 90 percent of their belongings, then put the rest in storage. But by the time they drove their two cars, with one cat each, from Boulder to Portland, they had decided they were "all in" on the Oregon Project.

When they arrived, in late 2004, a few of the first-generation NOP athletes were still around. The couple was friends with Marc

Davis, whom they relied on to help find them a home in Portland, not too far from Nike headquarters. The town house they moved into looked better online than in person, however. It had paper-thin walls and no working heat, and without furniture the couple was forced to sleep in sleeping bags on the floor.

They weren't wholly prepared for the instant chaos they had just signed up for, however. Each day was consumed with the rigors of elite endurance athletics. A typical twenty-four-hour period began with a group training session, physical treatment (like massages), run drills, a quick lunch, a second run workout, more treatment, then a third run on an underwater or space-aged antigravity treadmill. They'd stuff burritos in their faces at nine, at the end of a taxing day, and wonder if all the effort was worth it.

Their first workout of the day was usually with the entire NOP team on the Michael Johnson Track at Nike headquarters. Salazar would show up to lead the morning session in his customary Nike sweat suit and yellow Livestrong bracelet, which, at the time, was still years from symbolizing a global scandal.

The summer before the Gouchers arrived, Nike had been looking for a way to commemorate Lance Armstrong's fifth consecutive Tour de France victory. Scott MacEachern, Armstrong's personal Nike representative, thought it should involve something people could wear as a public show of support for Armstrong's philanthropic nonprofit, the Lance Armstrong Foundation (which he launched in 1997 to provide support for people affected by cancer). Together with Nike's advertising agency, Wieden+Kennedy, they conceived of a silicone wristband with the word "Livestrong" on it, after seeing similar bracelets become popular among NBA players. They would sell them in Niketown stores, for a dollar each, to people who wished to show their commitment to curing cancer, much like the red ribbons for AIDS and the pink ribbons for breast cancer that were also popular at the time.

The Armstrong team hoped the May 2004 launch of the Nike bracelet and the "Wear Yellow" campaign would divert the public gaze away from a new exposé, first published in France and titled

L.A. Confidentiel: Les secrets de Lance Armstrong, that was a collection of the most damning accusations yet leveled against, arguably, Nike's most famous athlete (Michael Jordan had retired for the last time in 2003). The book's authors were Irish journalist David Walsh and French sportswriter Pierre Ballester. Walsh, in particular, had proven to be a longtime fly in the ointment for Armstrong, and a man that was hard to discredit owing to his journalistic reputation and his trophy case, which was quickly filling with Sportswriter of the Year awards.

Nonetheless, Armstrong derided Walsh as a "fucking little troll" and "the worst journalist I know," adding that he would lie, steal, and threaten people to make the story more sensational.

The scheme worked. The bracelets sold out within weeks and the American public didn't seem to notice the new book. The wristbands became a successful fund-raising initiative—eventually selling an estimated eighty million units—for Armstrong's foundation. They also became a wildly popular fashion accessory and a cheap way for people (including the author of this book) to virtue signal. Bill Clinton, Gwyneth Paltrow, Ben Affleck, and Serena Williams wore the yellow bracelets, as did Armstrong's cycling rival Ivan Basso. (Many of Armstrong's 2004 Tour de France competitors wore them during the race.) Oprah Winfrey wore a yellow band and even helped sell them on her website. Of course, Salazar was not alone in his fashion choice on the Nike campus, where it seemed like almost everyone wore one.

As the year progressed, Nike had a more immediate emergency to deal with, as their star runner, Marion Jones, was once again caught up in doping news. That December, Victor Conte, the founder of BALCO—which had been raided by federal agents—emailed reporters, "Everyone in the entire world is finally going to learn that there is no Santa Claus, Easter Bunny or Tooth Fairy in the world of sport . . . The Olympic Games and, in fact, all sports at the elite level are a fraud and the world is going to find that out in great detail." He then did interviews with both *ESPN the Magazine* and ABC's *20/20* where he outed Nike's highest-profile runner as a drugs cheat. In response to Conte's allegations, Jones filed a $25 million lawsuit, denied ever taking drugs, and reminded everyone that she had never tested positive.

Unbeknownst to the Gouchers, many of the BALCO cheaters had done workouts on the picturesque Michael Johnson campus track that they now called home.

Salazar, using Nike's munificence, was able to lure some of the best coaches in the country to join him on the Oregon Project. John Cook, who had been the head track-and-field coach at George Mason University for nineteen years, was one of the first on that list. He sat down for a meeting with Salazar, Phil Knight, and John Capriotti to discuss the Nike Oregon Project. During this meeting Knight made it patently clear that everything the program did would be "by the book," warning the men that "'we've already had one situation,' referring to an incident that goes way back. He made it very certain that if anything like that were to occur again, we'd all be gone," Cook told *Runner's World*. He also said that taking the job was a "huge mistake," but "the money was really good. It was a budget like I'd never had in college." Cook got to bring along a couple of compatriots that he was eager to work with—Vern Gambetta and Dan Pfaff.

As the NOP was shifting focus from marathon athletes to those who would compete in middle-distance races and speedier track events, the three men were tasked by Salazar with making their runners stronger, more agile, more durable, and, of course, faster.

Pfaff had worked his way up from coaching high school to assistant positions in the college ranks before taking over as head track-and-field coach at the University of Texas at El Paso. Moving to Louisiana State University in 1985, he coached the team to multiple national championships. Two years in a row, in 1989 and 1990, his men's and women's teams both won the national championship.

Gambetta had played football at California State University, Fresno. In 1968, his senior year of college, he first noticed the school's runners wearing "these weird-looking shoes." He asked the team's assistant coach, Gene "Red" Estes, where he could get a pair.

Red, who was high school teammates with Phil Knight and ran in college for Bill Bowerman, told Gambetta, "My teammate from college brings these in from Japan."

That same year, Red had his old coach come in to speak at their first-ever track-and-field clinic. "Bowerman was a captivating individual," Gambetta told me. "He was crusty and he loved the sport, and that came through. When I walked out of there, I said I want to be like him. I want to be a track coach."

Gambetta worked his way up through the ranks, taking his first teaching and coaching job at La Cumbre Junior High, in Santa Barbara, California. He coached next at the high school level, before being hired by the University of California, Berkeley, to head their women's team. Next, he became a consultant for the Chicago White Sox and the Chicago Bulls professional teams and, after a couple of years, he was asked to be the director of conditioning for the White Sox. In 1984, he reengaged with track and field. "I started coaching high school again and was one of the founders of the track and field coaching education program," he said. "I consulted on injury prevention, rehabilitation, development of strength training programs, and speed development."

Cook and Gambetta would meet for coffee to talk shop. Cook had been a paid consultant for Nike in the past. "He and Capriotti were real tight and he got to know Salazar," Gambetta told me. "He just told me, 'Oh, Salazar's the greatest, he's such a hard worker and all this stuff.'"

Pfaff joined the team in September 2004; Gambetta in January 2005. By this point in the program, Salazar, though still invested in Adam's success, had begun to look beyond collegiate national champions for potential talent. He now believed that beginning to work with athletes after college, even certified stars like Adam Goucher, was simply too late. "We feel if we can get this group to do well, it can set the framework for a larger junior project," said Salazar. "I believe we've got to go farther back down that pipeline, identify promising kids, and train them with world-class methods. The rest of the world is doing that. The Kenyans, Ethiopians, and Moroccans are all doing that, and we know how talented and successful their runners are. How can we think we can bypass that source?"

Improbably, there was a young soccer player down the road named Galen Rupp who had just the running pedigree Salazar was looking for. His mother, Jamie, had been an accomplished athlete, winning two Oregon State high school cross-country championships and posting a sub-five-minute mile as a freshman.

By eighth grade Galen was already running the 1,500 meters in 4 minutes and 32 seconds. Salazar, as the coach of the Central Catholic High School in Portland, Oregon, first met Galen in 2000. During a fall meet-and-greet barbecue, where new students and their families meet the school's coaches, he was approached by a woman and her shy fourteen-year-old. Jamie Rupp's son, she explained, was a very good soccer player—Galen was on the Olympic Development Team as a freshman, which meant he was one of the top seventeen players in the state—but she thought he might have some talent as a runner.

Salazar personally knew Galen's Central club soccer coach, Jim Rilatt, because he had also coached Salazar's son Alex. When they spoke about Rupp, the soccer coach told Salazar that during their sprints "the other guys can barely see him; he's so far ahead" and that he was covering their 200-meter sprints in under thirty seconds.

"There's no way. You're measuring wrong," Salazar told him.

"No, I'm measuring it right," Rilatt replied.

Soccer was important to Rupp and he dreamed about playing in college and maybe even in post-collegiate pro leagues. "But then, lucky for me, I crossed paths with Alberto," Rupp told *Runner's World* in 2012.

The high school track team Salazar inherited had been last for several years in a row. Rilatt told him he could borrow some of the kids he didn't need full-time to help fill out the team, and since Galen was a freshman he got to come over, run some meets, and miss out on a soccer practice here and there. Salazar convinced Rupp that if he put soccer aside for one month and dedicated himself to run-training, he might be able to win the Junior Olympic Nationals.

Rupp agreed, and working with Salazar for a short training block before the Nationals, showed great promise on the track. "He ran the national junior cross-country meet and won it against kids who've been running for years," Salazar told *Competitor Radio* in 2011. "The rest is history and he just kept improving and improving."

Around the same time that Kara and Adam arrived in 2004, a nineteen-year-old from California and a certified high school phenom named Caitlin Chock moved to Portland to join the team. Like Rupp, Chock had set the US high school record in the 5,000 meters the previous summer, during her senior year, breaking the twenty-five-year-old time set by Mary Shea, of Raleigh, North Carolina (Chock finished the race a full one minute and twenty-six seconds ahead of second place).

She briefly attended the University of Richmond, in Virginia, but left after one month because it wasn't a good fit. Unsure of what to do next and with her running career hanging in the balance, she found herself drifting, until Salazar called.

Caitlin was only the second female athlete to be admitted to the Oregon Project, and she became fast friends with Kara. But more apparent was her instant connection with Salazar, even moving into his home to live with his family while she trained. To assure that she kept her collegiate eligibility, she paid the Salazars $500 a month in rent.

The NCAA "promotes amateurism to create a level playing field for all student athletes." At eighteen and nineteen years old respectively, Rupp and Chock were essentially coming out of high school and going right into the pros. But Salazar saw to it that they weren't ostensibly breaking any of the NCAA eligibility rules. Individuals who plan to attend college, or at least leave the option open to run in the NCAA, aren't allowed to be paid for, or profit from, their sport in any way. Rupp and Chock were risking their future collegiate eligibility by joining the Nike Oregon Project, but both of them felt like the opportunity was too good to pass up.

Salazar and the Nike employees involved in the Oregon Project, therefore, had to support Rupp and Chock without being seen as enriching them. They couldn't receive salaries or sponsorship deals. They couldn't have lunch bought for them. Free coaching, on the other hand, is something that every high school and college athlete receives, and therefore didn't ruffle the feathers of the NCAA.

EVERYONE SEEMED TO REVERE ALBERTO SALAZAR ON THE NIKE CAMPUS, AND PHIL Knight was no exception. He had a building named after Salazar, after all. Coach Salazar quickly became a father figure to Rupp, who was just eighteen years old when the Gouchers began training in Beaverton. As a mentor, their coach urged team members to be closer to God, and to attend church, which they often did together.

Salazar had been born in Havana, Cuba, on August 7, 1958, into the virtual civil war of the Cuban Revolution. His father, José, was an acolyte of Fidel Castro and fought for him beside Che Guevara in the liberation army. José worked as a civil engineer, and for a time designed projects for the revolutionaries in their rebuilding efforts, creating plans for national parks, public beaches, and hotels.

He lived his life by the strict doctrine of Cuba's prevailing religion, Roman Catholicism, and proposed a plan to build a new church. The men in power, however, had already determined that religion ran contrary to the Marxist philosophy to which they ascribed. Guevara himself rejected the idea, returning the plans to José with a pithy note that read, "There is no room for God in the revolution." After appealing to Castro himself and receiving the same sentiment, José, feeling betrayed, decided he could no longer remain in a godless Cuba, and fled with his family for the United States.

"My father's deepest allegiance was to God, not the revolution," Salazar wrote in his biography, recounting how, on his father's side, "eleven generations preceding me have distinguished themselves through acts of faith. These acts have tended to be dramatic rather than contemplative, bound up with death, honor, and blood loyalty and, often, with violence."

Defectors from the turbulent island nation to America were becoming commonplace in the revolutionary era. The Salazar family, with a two-year-old-Alberto, first arrived in Florida in 1960. José took work at a construction company in Manchester, Connecticut, before eventually settling in the Boston suburb of Wayland, Massachusetts.

Salazar's family life was tumultuous. His biography begins with a kindergarten-aged Alberto beating up another boy who had struck him with a vine. "It seems fitting that my earliest childhood memory entails strife, conflict, and intense emotion," he wrote. "Our house was characterized by yelling and screaming," of an intensity "unlike that of other American families." Steeped in the religion of his parents, a young Alberto began to believe that God, and "forces beyond my understanding and perhaps beyond my control," were working through him. He would become the most devout of the four Salazar children, and as his biographer put it "the one most in need of grace."

Salazar idolized his older brother Ricardo, who was also a runner and would often organize the neighborhood kids to compete in races around the block or around the house. "He bought a stopwatch and we'd do like a fifty-yard dash out in front of the house, so basically we just started playing at racing," Salazar told *Competitor Radio* in 2011.

The Boston area was thick with runners of all stripes through the later part of the 1970s and remained a hot spot for the sport even after the initial booming had subsided. Salazar, owing to the fact that he "wasn't good at anything else," as he put it, devoted himself to running. Though his running form lacked the concentric bounce of many champions, and has been compared to "a man running on hot coals," his times didn't lie—Salazar was certifiably fast.

Ricardo would regale his younger brother with stories of Prefontaine's courageous battles on the track. In an effort to mimic the Oregon athletes coached by Bill Dellinger, Salazar began running twice a day and eating peaches and toast as his pre-race meal, just like his idols on the U of O track team, like Prefontaine.

As a high school junior, he befriended Greater Boston Track Club member Kirk Pfrangle, who started bringing him to training sessions to work out with older, faster athletes like Bill Rodgers, just months before Rodgers won his first Boston Marathon. By the time Salazar was a senior, he ran as fast as anyone else in the club for 5- and 10K, and the group nicknamed him "the Rookie."

That same year, Salazar tied Craig Virgin's world's-fastest 5,000-meter time by a sixteen-year-old, running 14 minutes and 14.6 seconds at the National AAU Championship in Knoxville, Tennessee. Though he didn't win the race, he was the youngest runner in the field, and the scrawny youth proved he could hold his own against stronger, more mature athletes. The running world began to take note.

One evening that spring, Steve Prefontaine lost his life in a car accident. After successfully orchestrating a track meet at Hayward Field, he had attended a post-race party with the other athletes at the home of Geoff Hollister, Nike's third employee. Later, he drove his friend Frank Shorter back to Kenny Moore's, where he was staying while in town for the race. Frank and Steve talked for a while about athletes' rights and how the unsanctioned track event earlier in the day would propel the movement toward professional running—then Prefontaine drove off.

On a precarious corner on Skyline Boulevard near Eugene's Hendricks Park—which Prefontaine had likely driven and run hundreds of times—his car crossed the yellow line, jumped the curb, struck a large rock outcropping, and flipped over. The gold MGB he had purchased with his Nike money landed on top of him and squeezed the air out of his lungs before anyone could help him.

There were no witnesses to the accident, and many, including the police, believe that Prefontaine died because he was driving drunk. Others believe that he must have swerved to miss an animal or another car. Though Shorter admits they had been drinking that night, he insisted that they knew their limits and that running was always the priority. "We drank, but we weren't drunk," Shorter told me. "We had a rule and it was very simple. We never drank so much that we couldn't get up and run as well as we needed to the next morning. That's why we left the party early, because we were running ten miles the next day." No alcohol was found on the scene. One report claimed that among the broken glass and twisted metal was a John Denver tape, *Back Home Again*, lying next to Prefontaine's body, which has led some to speculate maybe he took his eyes off the road to change the music.

The first officer to the scene, Sgt. Richard Loveall, told the Eugene *Register-Guard* on the tenth anniversary of Prefontaine's death, "He died because he consumed too much alcohol and drove. That was the truth of the matter." The official police report lists his blood-alcohol level at 0.16 percent, which is twice today's legal limit (though in 1975 the limit was 0.10). Loveall would later tell the *Los Angeles Times* that what stuck with him all these years after the accident was "the pungent smell of alcohol" at the scene.

No matter the actual cause, at twenty-four years old the brightest light in running had been extinguished. He died holding seven American distance records from two miles to six miles. "Of course, what he really held, what he captured and kept and would never let go of, was our imaginations," Knight would say years later.

Salazar was crushed by the news. "The legendary Pre, my hero and role model," he said, "who brought passion and meaning to running, a style that would later infuse and inspire Nike and guide so much of my life—was gone."

SALAZAR WON THE STATE CROSS-COUNTRY CHAMPIONSHIP HIS SENIOR YEAR OF HIGH school and placed fifth in the World Junior Cross Country Championships in Chepstow, Wales. Though he was recruited by universities all over the country, there was no question where he'd go to college. The East Coast kid would become an Oregon Duck, like Prefontaine had before him, running on both the cross-country and track teams.

As a Duck, Salazar helped the team win the 1977 NCAA Cross Country Championship team title, and the following year he won the 1978 Individual Cross Country Championship. The next challenge for the twenty-year-old would be the prestigious Falmouth Road Race on Cape Cod, made popular by Olympians Frank Shorter and Bill Rodgers. In the 1977 race, Salazar had finished second to Rodgers, the Greater Boston Track Club star who had been passing victories back and forth with Shorter since the second running of the event in 1974. Shorter would have to sit the 1978 event out, recovering from surgery after having a bone spur removed from his ankle. But the race had grown exponentially in the last five years and didn't lack for talent.

The record at the time for the 7.1-mile course was 34 minutes and 16 seconds, an average mile pace of 4 minutes and 53 seconds. En masse the runners lurched off the start line on Main Street in Woods Hole promptly at 10:00 a.m. and ran east along the tricep of Cape Cod, toward the finish in Falmouth Heights. It was at most 78 degrees Fahrenheit, but the humidity lingered at 70 percent.

Rodgers wore the number one bib and orange ASICS shoes; Salazar, who for his finish the previous year wore the number two bib, sported a white tank top with the race's title sponsor "Perrier" emblazoned across his chest, and Converse shoes with red-striped, shin-high socks. Rodgers surged on the downhills and at the five-mile mark found himself overheating badly while running next to Salazar. "I was ready to say, 'It's your race,'" said Rodgers, "but when I looked around, he was ten yards back. I couldn't believe it."

As the temperature soared, Salazar ran himself near to death trying to catch up, and completely fell apart. Rodgers went on to win, as did a young woman from Cape Elizabeth, Maine, named Joan Benoit. "With a mile and a half to go everything went out of me," Salazar said later. "The world looked strange. It was fuzzy and had dim patches. People passed me, four or five in the last half mile. I can't remember anything after the finish. I woke up in a bathtub full of ice." The young upstart finished in tenth place but collapsed with heat stroke at the finish line, with his temperature reportedly hitting 108 degrees Fahrenheit at one point. He was rushed to medical treatment and given intravenous saline. But it appeared to be too late, however, and his last rites were read. The race doctor put him in a plastic kiddie pool filled with ice water, which revived him, and likely saved his life.

He believed the entire incident to be evidence of his exalted position in the eyes of the creator. "God spared my life," he said after the race. Surviving Falmouth was the first and most meaningful entry in his personal toughness ledger that seemed to prove his self-worth. "I knew intellectually that the world didn't revolve around me, and that on a larger scale running around a track amounted to very little," said Salazar, "but I felt that I had been chosen for a special destiny, and I had little patience for the obstacles standing in my way."

Newly motivated to become the best runner in the world, Salazar experimented with various tools to enhance performance, including a gangly scuba mask that used crystals to scrub oxygen out of the air on intake, which presumably made each breath more like breathing at altitude. He spent time in Kenya's Rift Valley to train for the 1980 Olympics, only to return early when he heard the United States would boycott the Games. During a layover in New York's LaGuardia Airport on his way back to Oregon, Salazar ran an eleven-mile training run around Queens, the largest of the five city boroughs.

As a college senior with a single track season of eligibility left at Oregon, he decided to enter the New York City Marathon as his first attempt at the distance. During the media scrum around the event, Salazar boldly told reporters that he planned to set a record and beat the reigning champion, Bill Rodgers, in the process. Rodgers was one of the world's best marathoners at the time and at thirty-two had won the Boston and New York marathons four times each. Although Salazar had run the second-fastest time of the year for the 10,000 meters, Rodgers had the record time on the New York City course, set in 1979: 2 hours, 9 minutes, and 27 seconds.

"I prepared only five weeks for my first New York Marathon," Salazar said. "I didn't read any training manuals. I had run Falmouth to the point of death, and I hadn't been afraid . . . I was twenty-two. What did I have to fear now? I knew I was the toughest runner in the race."

Despite the fact that he had never run longer than twenty miles in training, Salazar had a new superpower bestowed by his near-death experience at the Massachusetts race. He no longer feared death, and he was confident that he would be the fastest man to line up in New York. During a pre-race exchange, he disclosed that he felt capable of running 2 hours and 10 minutes. Upon hearing his remarks, Rodgers responded, "There are twenty miles out there after Salazar is used to stopping. He's going to learn something in them and I'm firm on one point: no rookie is going to beat me." This back and forth amounted to the most trash-talking that had ever taken place in the world of elite marathoning.

It was a blustery October day when the men ran through the cold wind and swirling leaves of New York City. Fourteen miles in, Dick Beardsley, who was in the midst of a record-setting streak, stepped in a pothole and went down, taking Rodgers with him. By the time they regained their sub-five-minute pace, Beardsley reconnected with the pack, but Rodgers had lost more ground, and was now a distant tenth.

Salazar was six feet tall, 144 pounds, and wore a yellow-and-green University of Oregon singlet with short green shorts. As he passed mile twenty, only Rodolfo Gómez of Mexico and John Graham of England remained in contention. "The only guy in the field I knew I'd have a hard time beating if it came down to the last half or quarter mile was Rodolfo," Salazar said. In an effort to drop him, he ran the twenty-first uphill mile in 4 minutes and 57 seconds, which scraped Graham off the pace but not Gómez.

Hunched into the wind the athletes entered Central Park and Salazar accelerated. This time, Gómez could not respond in kind. Salazar crossed the finish line alone in 2 hours, 9 minutes, and 41 seconds—a blistering 4 minutes and 56 seconds per mile—a New York City Marathon record, and the eighth fastest marathon ever run.

The cocky rookie collapsed into the arms of marathon staff as they ushered him away from the finish line. "I really hurt at the end. But when I felt bad, I thought everyone else must feel worse," he said. Salazar took 28 seconds off Rodgers's time and joined him as the only other American ever to go under 2 hours and 10 minutes. Gómez came through 33 seconds too late in second. The next day, the front page of the *New York Times* declared, "The Streets of New York Get New Marathon King." The subsequent issue of *Sports Illustrated* read "The King of New York" and featured a tight shot of Salazar's profile adorned with the victor's wreath, his large smile and chipped tooth augmenting his pride.

Now a record holder in the vaunted marathon, Salazar was the hottest commodity in running and a household name, though he struggled to be content. "I remained hungry and unsatisfied. All of my success, I felt, wasn't true success," he wrote in his biography. "I still hadn't unequivocally proven that I was a man of honor like my father."

Salazar hired an agent from the International Management Group (IMG) named Drew Mearns to help him negotiate with the prominent brands in running, Adidas and Nike. Even though he had attended the University of Oregon, the birthplace of Nike, Salazar claims in his book that signing with the brand wasn't a given. Although Bowerman had retired, the University of Oregon was still a top-ranked team with deep ties to Nike, though the brand wasn't yet the market-dominating force it is today. Eventually, both companies offered him a base salary of $50,000, with large cash incentives written into the contract for big wins and national or world records. To sweeten the pot, Nike also promised to create a line of performance-running clothes with Salazar's name on them. But it was Geoff Hollister who pushed the deal in Nike's favor when he mentioned the possibility of employment with the Oregon brand once his professional running career was over. Salazar, who had graduated with a business and marketing degree, could envision himself working for the brand for the rest of his life.

In January Salazar set a new US road record, running 22 minutes and 4 seconds for five miles. He then won a two-mile indoor race in Portland, before returning to New York City the next week for the 1981 Millrose Games, where he broke the American indoor 5,000-meter record.

In the days leading up to the 1981 New York City Marathon, Salazar, the once shy young athlete who had only kissed one girl in high school, now saw himself as "a distance-running version of Joe Namath or Muhammad Ali." He wore a black leather jacket to interviews and told the *New York Times*, in his thick Massachusetts accent, that he averaged 106 miles a week of running for the previous twelve-week block of training. It "had gone smoother than he imagined," he said, and he was "running longer, faster, more aggressively, and with less effort than last year," and offhandedly mentioned the use of an experimental mask that restricted his oxygen intake. He then boasted unequivocally, "I make no bones about it, I make my living off running."

This was the twelfth year of the event, which had grown to a record number of participants with 14,496 starters lining up for the five-borough race (they could not accommodate, and were forced to turn away, more than twenty thousand athletes who registered). It was the first time the event would be broadcast live, as it happened, on network television.

Salazar brashly told the media that he wasn't just after the win, but he also aimed to set a new world record for the distance. There was considerable controversy over who held that time, however. Australian runner Derek Clayton had run 2 hours, 8 minutes, and 34 seconds in Antwerp, Belgium, on May 30, 1969, but many experts suspected the course was short, and subsequent city construction made it impossible to reevaluate. The next fastest time was Holland's Gerard Nijboer's 2 hours, 9 minutes, and 1 second set in Amsterdam in 1980. Salazar said he meant to break both times, settling once and for all who the best marathoner on earth was.

For a world record to be possible, he would have to average a pace of 4 minutes and 54 seconds per mile, 12.26 miles per hour, an incomprehensible speed for such a long distance considering the world record for a single mile at the time had been set just two months prior by British athlete Sebastian Coe, who ran 3 minutes and 47.33 seconds. Any interruption in his forward pressure could take him off the aggressive pace and put the world-best time out of reach.

A major city marathon is a massive undertaking of planning and organization. Roads are closed and barriers are erected in early morning darkness. ABC's cameras offered the bold among the 2.5 million spectators a tantalizing new opportunity—to dart out into the road and be seen on television by friends and family. It was a cloudy, overcast day, with temperatures near 55 degrees Fahrenheit, ideal for a fast marathon. Police on foot and on motorcycles tried to keep order but risked causing their own catastrophe. One of the elite female runners, US Olympian Julie Brown, was tripped by a marathon streaker at mile eight, but remarkably managed to stay upright and press on. Rodgers was a no-show at the start line, apparently unable to come to an agreement on appearance fees and compensation by sponsors.

In an effort that the *New York Times* opined, "rivaled any of running's legendary feats," Salazar separated himself from José Gómez by throwing in a 4-minute-and-43-second mile and pulled far enough away at Eighty-Sixth Street to solidify the win. With fans screaming "world record, world record!" at him along the way Salazar broke the tape in 2 hours, 8 minutes, and 13 seconds, running the distance faster than anyone ever had. In only his second marathon, and at just twenty-three years old, he was now the fastest marathoner the world had ever seen. The second-place finisher, Jukka Toivola of Finland, crossed the line 2 minutes and 39 seconds later.

Financial terms for athletes were still a clandestine affair in 1981, but the *New York Times* estimated that Salazar was likely to make a considerable purse in under-the-table money on top of the known $14,000 reportedly paid by the sponsors. Nike then upped his salary to an astonishing $250,000—likely more than any runner had ever made from a single sponsor. In December, Salazar married Molly Morton, whom he met at the University of Oregon. The two Ducks made a formidable pair, as Molly was an impressive athlete in her own right, and held running records in the 3,000, 5,000, and 10,000 meters. They were married in an Episcopal Chapel in Portland, and to appease Salazar's father, held a dual Episcopal-Catholic ceremony. After a honeymoon in the Bahamas, the couple returned to Oregon and their black American pit bull named Toby.

Molly managed the family finances while Salazar dedicated his every waking moment to being as fast as humanly possible. And he now felt a loyalty to Nike and Knight that he likened to his father's dedication to Castro. "Beside being patriotic, my father was driven by a sort of chivalrous impulse, the hunger to subsume himself both to a noble cause and to a leader he trusted and admired. I understand this impulse well," Salazar wrote in his biography. "I would pledge loyalty to a leader whose vision and talents were, in their way, as far reaching as Castro's: Phil Knight, the genius behind Nike, a company that has achieved its own sort of global revolution."

He joined the Athletics West team in the early 1980s, but with their checkered past, he now distances himself entirely. In his biography he wrote, "Other than wearing the red and white AW singlet, getting an occasional massage, and being tested by the exercise physiologist Jack Daniels, however, I didn't have much to do with the program."

Nineteen-eighty-two proved the high-water mark for Salazar's career. In March he won the silver medal at the IAAF World Cross Country Championship. Then in April, at the Boston Marathon, Salazar famously battled talented journeyman Dick Beardsley stride for stride for more than two hours (Salazar in a pair of Nike American Eagle racers and Beardsley in custom New Balance 250 racing flats). Salazar only drank two cups of water the entire race for fear that the added weight would slow him down. Crystallized salt streaked his face when he out-sprinted Beardsley for the win by just two seconds; the victory earned him a $5,000 bonus from Nike.

He once again collapsed after crossing the finish line. He had lost ten pounds and his core temperature now read 88 degrees Fahrenheit. Fearing it would take too long for him to drink the requisite liquid electrolyte concoction his body required, paramedics jammed an IV in his arm and fed him six quarts of intravenous fluids.

"I viewed every marathon as a test of my manhood," Salazar told *Runner's World* while reflecting on the event. "It wasn't enough for me to win the race; I wanted to bury the other guys."

It seemed at the time like these two young men would provide America with many more endurance duels—and a rivalry for the ages. Instead, neither man would ever run a marathon that fast again, and both of them would decline into prescription drug use, the full accounting of which we may never know.

After a farming accident crushed Beardsley's leg, he became addicted to opioids and struggled with substance abuse for years. Salazar's decline was less abrupt. He was now ranked the number one marathoner in the world, but secretly worried he had done irreparable damage to his body. After the 1982 "Duel in the Sun," as it came to be known, he didn't feel the same energy for exercise, and whenever he returned to his normal levels of training he inevitably got sick. Even for a well-paid, professional athlete, happiness can be surprisingly elusive. Salazar's doctors diagnosed him, at different times, with depression, anemia, and asthma, among other ailments. Years later, a doctor would determine that his chronic health problems were largely due to his overheating at the 1982 Boston Marathon (though it's not the only theory his doctors have favored over the years).

While not feeling his best, in October Salazar somehow made it a three-peat win, and once again defeated Rodolfo Gómez at the New York City Marathon. But he spent much of 1983 feeling lethargic and fending off repeated bouts of chronic respiratory infections. With his powers diminishing, he ran one final spectacular race at the national level in the 1983 American championship in the 10,000 meters. Salazar in his Nike AW kit and Craig Virgin in a coordinated yellow singlet, shorts, and shoes bearing the Adidas three-stripe motif battled until the final straightaway, when Salazar pulled ahead and won the event. However, in the subsequent World Track and Field Championship 10,000-meter race, he finished in last place.

In April 1983, he finished fifth at the Rotterdam Marathon, his first loss in the distance he had come to dominate. Then, at the Fukuoka Marathon in December, another fifth place. He performed well enough at the Trials to make the marathon team but ran a relatively slow 2 hours and 14 minutes at the 1984 Los Angeles Olympics, in the California sun, and finished far behind and out of medal contention in fifteenth place.

Later that summer Salazar told the *New York Times* that he knew his best running years were ahead of him. "God places things in our lives and expects us to handle it," he said. "I can handle it."

As was his bent in trying times, Salazar beseeched heaven and prayed for guidance, "Lord, please let me know if I should continue. Are you telling me to stop, or if I stopped now would I be turned away from you? Lord, show me your will, let me see it clearly, one way or the other."

But the laurel wreath withered. Then his world record was taken from him.

His 1981 New York City Marathon time had been deemed "pending" for years as a team of experts kibitzed over what, exactly, to do. They suspected the distance was short of the 26 miles and 385 yards of the official marathon. The New York City course Salazar covered in 2 hours, 8 minutes, and 13 seconds was reevaluated by the Athletics Congress records committee and was found nearly 170 yards short, more than 70 yards over the allowed margin of error at the time. Salazar's asterisked record was stripped from the list of world-record marathons. The *Los Angeles Times* reported that Salazar "reacted with a mixture of anger and amusement." On his personal IAAF page there is an asterisk by the time that reads "* Not legal." (An IAAF representative told me, "The time was initially thought to be a world's best, but the course was found to be only 42.047 km, a shortfall worth about twenty-eight seconds.")

In the prime of his athletic career, Salazar now seemed to be cursed with an unshakable lethargy. He manically cast about for other ways to get back his old vigor and return to the podiums that fulfilled him. He bought an ultrasound machine hoping it would tame his tendinitis. He employed a Finnish massage therapist, who sometimes lived with him above his garage. He spent time in a hyperbaric chamber that super-saturated its air with oxygen, hoping that it would increase his body's healing capabilities. Nothing seemed to work.

In 1985, Salazar's coach, Bill Dellinger, invited Victor Conte—the not-yet-notorious performance-enhancing drug dealer—up to Oregon to work with his athletes. Conte was just starting out in the supplement business and was ostensibly working on creating personalized vitamin and mineral concoctions for athletes. He would run their blood and urine through a battery of tests in a $250,000 machine called an inductively coupled plasma (ICP) spectrometer. The high-tech, industrial device was the size of a pickup truck and was used to detect toxic materials in soil, as well as analyze welds on atomic bomb casings to identify defects. By analyzing an athlete's blood, urine, and hair, Conte claimed that the ICP could detect the mineral deficiencies that were causing athletes to compete below their potential. Then he'd sell them the supplements that would get them into what he considered the appropriate range.

All of this was considered pseudoscience nonsense by the medical and sports nutrition communities, however. Conte was not a chemist, doctor, or nutritionist and had never worked as a medical professional in any capacity. He was, however, a tireless self-promoter who had a steadily growing reputation as an astute self-made scientist who could make athletes better. He pushed a magnesium and zinc supplement product that he created, called ZMA.

According to Conte, even minute deficiencies of zinc or magnesium were cause for concern and required supplementation if an athlete was to perform at their genetic best. He claimed that his tiny magnesium and zinc pill would "not only make you stronger, but would enable you to run faster, jump higher, hit harder, swim better, sleep sounder, and even have better sex." Conte now asserts that in these early years of his business they were still nothing more than a legitimate supplement dispenser, and that it wasn't until years later that this business provided a necessary cover story for the illegal PED distribution enterprise.

The system Conte would develop has been compared to that which the East Germans deployed to win all those Olympic medals: ply the athletes with performance-enhancing drugs on a strict schedule so they could avoid being tested when they were "glowing," a term meant to indicate a time period in which they had used a banned substance recently enough to fail a drug test. Central to the scheme was pre-testing the athletes, just to be sure there were no surprises. If the test showed they were glowing, then the athlete would withdraw from the event to avoid being tested at all, citing an injury, a sickness, or an asthma flare-up. Conte also employed doctors without scruples as "consultants" who would prescribe drugs, as needed, that were either useful as performance enhancers themselves, or would help cover up the use of more illicit drugs, like the female fertility drug Clomid, which helped mask steroid use.

"There was never any discussion or involvement with any type of PEDs between Alberto and myself," Conte wrote in an email to me. "I was simply helping coach Bill Dellinger to provide some nutritional support for some of his elite runners." His team took Salazar's sample and flew it to Los Angeles, where they tested it at the BALCO Laboratories. "I put together an individualized supplementation program for him based upon his test results," Conte said. "I recall that he needed zinc and magnesium and iron and possibly a few other minerals and trace elements."

SALAZAR'S MALAISE CONTINUED, AND HE WOULD FAIL TO MAKE THE 1988 OLYMPIC team. Now, with an uglier shuffle than normal and unable to run more than four or five miles at a time, Salazar abandoned the sport that had brought him fame and fortune. Marathoners often improved into their midthirties, but Salazar had shined brightest at twenty-four, and was washed up by twenty-six.

For the next decade he tried to busy himself with other things. Outwardly he seemed to adjust to a post-running life as a husband with three children and a successful business. He bought into a local Eugene restaurant and managed its daily operations, but he privately struggled and was often cruel to the staff. He considered suicide. Eventually, he grew disenchanted with selling alcohol to young people. "I wanted to earn a living with something more satisfying," he said.

Searching, he made a religious pilgrimage to Medjugorje, a Balkan village in Bosnia and Herzegovina in 1987. For years his father had told him about a miracle that happened to him there, in which the chain of his rosary beads turned from silver to gold. When Salazar visited the village the same miracle happened to him. Overnight, his chain of rosary beads also turned from silver to gold. Salazar says he was transformed. He returned to Oregon, sold the restaurant, and moved to Portland. He began attending church every day, observed the sacraments, fasted twice a week, and denounced his arrogant ways. He then called in the Nike favor promised to him all those years before by Hollister.

The offer stalled at Knight's desk. "The problem is that he's so fond of you, Salazar," Hollister told him. "He regards you as part of the family. Phil is worried about what might happen if your job here doesn't work out. He doesn't want to jeopardize that relationship." But Salazar was able to allay Knight's fears, and he was hired by Nike in May 1992, working in the sports marketing department.

In the meantime, to help solve his health issues, Salazar began consulting with a former national-class distance runner and sports physician in Portland named Paul Raether and Eugene endocrinologist Jan Smulovitz. They surmised that the years of intense training, punctuated with episodes of heat exhaustion, had "suppressed [his] body's endocrine system." They suggested he try the prescription drug fluoxetine, which goes by the brand name Prozac. The selective serotonin reuptake inhibitor (SSRI) antidepressant affects chemicals in the brain and is used to treat major depressive disorders, the eating disorder bulimia nervosa, obsessive-compulsive disorder, and panic disorder. It's a legal substance in world athletics with the exception of shooting, for which it is banned. Finally, the veil of sadness he had been cloaked in was lifted. And although his doctor could not properly explain the mechanism, Prozac also seemed to resolve the worst symptoms of the exercise-induced asthma that he had been struggling with.

Although news articles at the time credited the Prozac, to much disgust in the media, Salazar was also taking the corticosteroid prednisone, which he hoped would revive his deficient adrenal system. The drug became popular among cheaters in the Lance Armstrong era for its potent ability to revive a tired athlete, allowing them to sleep less and train more. It's a powerful systemic anti-inflammatory drug, usually prescribed for arthritis, blood disorders, skin diseases, and cancer. Taking it during a hard training block or a Tour de France, for example, makes an athlete feel invincible because it decreases the immune system response to various insults, and effectively tamps down the inevitable inflammation of hard training that makes one feel sore and achy. But more than anything, the stimulating effect of the drug allows one to train hard, sleep less, and still feel the drive and energy to complete the hours of intense training, day after day, without getting tired.

The use of prescription drugs to improve and even prolong an athlete's career was not illegal per se, as long as the athlete could prove they had the medical condition that warranted the use of the chemical. Properly prescribed, the athlete then had a plausible excuse for the ancillary benefits. This is another one of the unfortunate ethical gray areas that persist today, because it leaves enough space for the unscrupulous to convince their doctors to prescribe them both legal and illegal substances that will ultimately improve their athletic performance, even if they don't truly need the drug for medical reasons.

Another drug Salazar took around this time was testosterone, though he was not openly talking about it to the media. Testosterone isn't a gray-area drug; it's an unequivocally illegal substance for any competitive athlete, under any circumstance. Outside of sport, the hormone is only approved for use in men with clinically "low testosterone levels in conjunction with an associated medical condition," such as genetic irregularities or side effects from chemotherapy.

He's now admitted to taking testosterone as early as 1991, while he was still a competitive athlete and three years prior to his final victory at his first ultra-marathon. (This was also the year that taking testosterone without a valid prescription became a federal crime.) As the 1992 Olympic Trials approached, Salazar applied for a therapeutic use exemption (TUE) for the steroid but was denied.

It's likely that some combination of Prozac, prednisone, and testosterone allowed Salazar to feel close enough to his old self again to begin training in earnest at thirty-four years old. He spent much of his training through the cold winter months that preceded the race in the basement, running 6-minute miles on a treadmill while chanting the rosary to himself. A thicker, more muscular Salazar won the prestigious 89-kilometer (55-mile) Comrades (ultra) Marathon in South Africa in 1994. The race alternates directions each year, and Salazar won the tougher, uphill direction, with a time of 5 hours, 38 minutes, and 39 seconds.

"I can unequivocally declare that my Comrades victory was a miracle," he said. "After Comrades, I arrived back in Portland feeling the divine hand. It seemed clear that God intended me to be an ultramarathoner, to win races all around the world and thereby spread the truth of the spirit."

Runner's World editor-in-chief Amby Burfoot thought the news about Salazar's victory, all these years later, must be a joke. He simply couldn't believe someone could come back from the depths Salazar had reached to win another competitive running event. He wrote in the magazine that after more than thirty years in the running media business, "May 31, 1994, stands out as easily the most amazing" performance he'd ever seen. This only proved to intensify the rumors that had always swirled around Salazar throughout his running career.

Frank Shorter, too, remembers him trying almost anything that would give him an advantage. "Salazar was always willing to try the newest thing," he said, recalling him running with a backpack during training and the rumors that he was one of the first to try DMSO, dimethyl sulfoxide, a lotion that horse trainers used to reduce inflammation in thoroughbred racehorses. "It's basically an industrial solvent that they started to use for rehab, because they thought it helped you heal faster." When I asked Shorter how other athletes knew Salazar was taking it, he replied "Oh, everyone knew, plus you could always tell when someone was on it because they would smell like sulfur, like rotten eggs."

08
THE CLEANEST

NO ONE KNEW WHAT IT WAS LIKE TO HAVE A CHEATER SNATCH YOUR DREAMS OUT from under you better than Frank Shorter did. But he wasn't the only one who suffered the consequences of performance-enhancing drug proliferation. The Soviet Union, a country of more than two hundred million people, collected the most gold medals at the 1976 Olympic Games, with forty-nine, but East Germany shockingly came in a close second, with forty. The country of just seventeen million people managed to win forty gold medals, proving their superiority to the world's democratic states, most of all their capitalist neighbor, West Germany, who only won ten golds. State-sponsored, scientifically assisted sport systems were integral to world-class, high-performance success in the postwar period.

American athletes, however, were not innocent bystanders through this time. Though they lacked the systematic guidance of the Eastern Bloc, they too were experimenting with performance-enhancing drugs.

"We knew in '72. They [the IOC] knew in '68, because the weight men were starting to do it," Shorter told me one sunny day in the fall of 2018. "I think the two places where it started were in the Soviet Union and the US."

He told me a story he's never shared publicly. It was the summer of 1973, in Bakersfield, California, and Shorter, Prefontaine, and a University of Houston runner named Leonard Hilton were jogging back to their hotel from the stadium that hosted the National Amateur Athletic Union Track and Field Championships. Hilton had just run the fastest mile ever by a Texan, and the fastest mile of his life, at 3 minutes and 55 seconds, and in doing so won the national championship in the mile.

"We're jogging and Leonard says, 'Shit I don't know if I could have done it if it wasn't for those steroids,' and Steve and I are going, whaaaat?" Shorter recalled. "He'd been injured in February, couldn't run at all, had to recover, went on steroids, then first weekend in June, wins the mile in the national championship."

Steroids weren't yet illegal at this point, so I ask Shorter if he or Steve Prefontaine had ever used them. He tells me no, that it felt morally wrong, then continues, "And to be honest, personally, I was too scared. I knew too much. I knew enough about long-term health and I saw what it could do to people. So, again, it's not a holier-than-thou thing. I'm just trying to explain why I never did. I mean, Leonard Hilton died at fifty-two."

When I ask about Pre, he cuts me off, "No, no, no. He was a naturally big kid, like Jim Ryun [the first American high school runner to break 4 minutes for the mile], he was a strength runner, and his improvement did not indicate it. How fast he got was a natural progression, it was not exponential. For me that was always the first indication, exponential improvement and huge jumps."

Shorter kept his suspicions about the East Germans to himself for twenty-two years. "For decades, I had been haunted by the memory of Cierpinski looking back at me at the twenty-one-mile mark in Montreal, then turning and rocketing away," said Shorter.

In November 1991, the Berlin Wall came down and thousands of previously confidential documents containing sensitive material were uncovered by a German molecular biologist and professor, Dr. Werner Franke, and his wife, a former Olympic discus thrower and shot-putter, Brigitte Berendonk. Their subsequent best-selling book, *Doping: From Research to Deceit*, showed that from 1966 on, hundreds of physicians and scientists, including top-ranking professors, performed doping tests and administered prescription drugs and unapproved experimental chemical compounds to as many as ten thousand German Democratic Republic (GDR) athletes. This is an arrangement that would be analogous to the United States' federal government using the Central Intelligence Agency to oversee their Olympians' medical treatments, supplementation, and training. (Berendonk referred to State Plan 14.25 as the "Manhattan Project of Sport.")

The government project, which was originally called "Research Program 08" and later "State Plan Research Theme 14.25," was against German law and antithetical to any ideal of medical morality, where "First, do no harm" is the overriding ethic. East Germany, according to Franke, created "monsters" in the name of ideology, adding, "These were not real people, just engineered experiments." Thousands of athletes were treated every year, including minors. If they asked questions, administrators claimed the substances were "vitamins." Though the athletes were commodities to be protected, anyone who continued to object was summarily thrown out of their sport.

Athletic preparation was an expensive and closely guarded state secret, and there was a cloak-and-dagger surreptitiousness applied to all facets of the program. The GDR wasn't just breaking new ground in drug use and administration, but also in training methodologies.

The country's most prominent scientists and physiologists worked tirelessly on methods of drug administration that would evade detection by international doping controls. And many of the damaging side effects were dutifully recorded, some of which required surgical or medical intervention at the time, and have caused ongoing health concerns for many athletes.

In early 1998, the United States Olympic Committee invited Franke to Colorado Springs. With his revelations in mind, they were trying to figure out how, exactly, they could recoup some of the medals lost to athletes who had clearly broken the rules.

Shorter, after hearing that Franke would be traveling to Colorado, sent him a message and asked for a meeting. In January 1998 the two men met in the lobby of The Broadmoor hotel in Colorado Springs. The respected West German doctor told the runner that he found evidence that the man who had beaten Shorter by just under a minute on that muggy day in July, all those years ago, was a drugs cheat.

The hard proof arrived at Shorter's house in north Boulder not long after their talk. On a sheet of athletes' names who were administered drugs, "(Wald Cierpinski) Marathon" was listed in the right-hand column, alongside his secret athlete code, number 62.

A note from Franke read, "This indicates that several middle and long distance runners already in the '70s were part of the doping program of the GDR and that Cierpinski was already on androgenic steroids in 1976."

Exposing the systematic doping of as many as ten thousand GDR athletes came with attendant risk. Franke and Berendonk were accused of being traitors by their countrymen and threatened with reprisals. But the couple persisted, and over time more and more disaffected former East German athletes began coming forward (many of them were suffering serious health problems due to years of haphazard performance-enhancing drug administration).

A decade after the dramatic 1988 disqualification of Ben Johnson from the Seoul Olympics—for having the banned steroid stanozolol in his blood—it was clear that elite, top-level sports had a growing problem. As the decade approached its end, three consecutive incidents made the issue impossible to ignore, throwing performance enhancement in world-class, high-performance sport back in the spotlight.

First, on January 8, 1998, the Chinese national swim team was detained by customs officials on their way to the world aquatic championships in Perth, Australia. One of the top swimmers, Yuan Yuan, was found to be transporting thirteen vials of the illegal synthetic human growth hormone (HGH). Two months later, the TVM cycling team was caught by customs officials in Reims, France, with

104 doses of erythropoietin in the team van. Just four months later, a Festina Team Tour de France car was stopped at the Franco-Belgian border with more than four hundred ampules of EPO and other performance-enhancing substances. The incident now referred to as the "Festina affair" put a glaring spotlight on the inefficacy of the IOC's testing—and just how inept at catching anyone they were.

Two important conferences followed that served as catalysts for real doping reform. The first was the International Olympic Committee's "World Conference on Doping in Sport," in Lausanne, Switzerland. On the agenda that chilly February day, "The Protection of Athletes: Rights, Health, and Responsibilities."

The second took place on the sprawling nine-thousand-acre Durham, North Carolina, Duke University campus. In May 1999, the University's law school assembled a panel of experts to speak about the growing specter of drugs in sport.

At this conference, the now-retired Nike athlete Alberto Salazar gave a presentation called "Locating the Line Between Acceptable Performance Enhancement and Cheating."

After laying out his list of "legitimate practices to improve performance," Salazar described the three types of athletes, "those with strict morals who won't do anything, those in the middle, and then the outright cheaters."

He discussed the many gray areas that an athlete must navigate, with diet, nutrition, and most of all supplements. "At what point do these supplements cross the line and too closely resemble or mimic a banned substance?" he said. "This is a very gray area that needs much clarification."

Explaining the athlete's rationale, they "feel they cannot compete any more with athletes that are doping, these supplements may present an alternative to doping. Their reasoning for using them may simply be that if it is 'natural' and not on the banned list, 'I'm going to use it because I know a lot of my competitors are using much more powerful banned substances.'"

He listed "legitimate" practices to improve performance, which included endurance training, weight training, massages, anti-inflammatory medications, asthma medications (and other medicines that may or may not treat sports-related ailments), surgeries for musculoskeletal problems, altitude training, skill-development training, and flexibility training.

"I believe that it is currently difficult to be among the top five in the world in any of the distance events without using EPO or human growth hormone," he said. "I can definitely understand how a good moral person might feel compelled to do so. That person might not even consider it cheating if they believe all their top competitors are doing it.

"This is similar to athletes taking appearance fees and prize monies prior to it being allowed by the governing bodies. It was against the rules, but everyone did it because they knew the rules were a sham and that everyone else was doing it."

The IOC, after first stating it would not rethink the medals given to East Germans, was now publicly willing to consider each medal, case by case. But rewriting the record books would prove to be impossible for both the American swimmers and Shorter.

A frustrated Shorter wrote a letter to President Bill Clinton's director of the Office of National Drug Control Policy, General Barry McCaffrey, who was tasked with the misguided and impossible chore of leading the nation's fight against drugs. McCaffrey asked the Olympian to create an outline of how a drug-control agency for sports would work. Shorter, who had become a lawyer after his running career, wrote a memorandum on how this agency should be organized and what its mandate should be.

In Washington, on October, 20, 1999, Shorter spoke before the Senate on "Effects of Performance-Enhancing Drugs on the Health of Athletes and Athletic Competition." He described how the level playing field of elite athletics had been chemically skewed and how taking illegal drugs is now the price of entry into the competition. "This is a bona fide plea on behalf of the world's athletes for help from the American government," he said. "Without major changes in the system that detects these drugs and imposes penalties, they will continue to be the price of advancement for every young athlete who aspires to emulate a sports hero and pursue his or her career to the highest possible level."

After a four-month deliberation the USOC Select Task Force on Drug Externalization recommended the creation of an independent agency. Congress recognized the United States Anti-Doping Agency as the official anti-doping organization for all Olympic, Paralympic, Pan American and Parapan American sports in the United States in October 2000. As Shorter had envisioned, it was created as an independent, nonprofit organization, governed by a board of directors, and not the USOC. After announcing their board, the organization's next press release declared that none other than Shorter himself would lead them as their first chairman. And it concluded with a line that now seems naïvely optimistic about their mandate, "We look forward to proving beyond any doubt that American athletes are not only the best in the world, but also the cleanest."

09
LOYALTY OVER COMPETENCY

ATHLETES WHO TRAIN AT A HIGH LEVEL FROM HIGH SCHOOL THROUGH COLLEGE AND on to the professional ranks run the risk of growing stagnant. Over time, an athlete's response to repeated training stimuli diminishes, and inevitably, improvements don't come as fast as they used to. Some of them will quit their sport; some will double down only to crash and burn physically, if not emotionally; others will change something fundamental to what they have been doing in an effort to break through and reach that elusive next level.

Nike saw in Salazar a new way of training America's best runners. A change in stimulus, focus, and training regimen that would spark new growth in their athletes. A new coach can facilitate all of those things, provided they're a good fit and the athlete and mentor get along well, and are basically aligned toward the same goal. To this end, full buy-in by the athlete is essential.

The Nike Oregon Project was, for all intents and purposes, a failure up to the point in 2004 when the Gouchers arrived in Oregon. "Right away Adam ran faster than he had in like two years," said Kara. "He ran thirteen minutes and twenty-something at an indoor race and I knew we weren't leaving. It's not like he was

PRing or anything, but he was back to being one of the fastest people in the country in just a couple months." Kara, however, had a lot further to go.

In August, she too went to see Salazar's new favorite doctor in Houston. As with Adam, Dr. Brown asked if she'd visited BALCO, before imploring her to "tell me everything you're on," illegal or not. He then gave Kara the TSH test, and diagnosed her with hypothyroidism, before prescribing thyroid medication. It wasn't long after that Salazar began seeing Brown himself, and making referrals to the athletes in his care to see the doctor. In short order, Rupp too was diagnosed with a thyroid condition. Eventually, Salazar employed Brown as a paid consultant to the Nike Oregon Project.

Salazar saw something in Kara, and he began telling her, "You're going to do amazing things in the five-K, but someday you're going to be a marathoner." She thought Salazar was crazy, no way was she going to run a marathon. Growing up in Minnesota, watching the Grandma's Marathon, Kara had seen what happens to the "psychos" who put themselves through that grueling test of endurance—nothing but pain and puke. Even as she became a local short-distance running star, the idea didn't appeal to her.

"I never wanted to do it," she recalls. "Never."

Adam and Kara both grew to appreciate the rhythm of their regimented workday. After a group workout on campus, run drills, a massage, lunch, a second run, then a session with the chiropractor Justin Whittaker, team members were then expected to arrive at their allotted time at Salazar's house to get a third run on a new NASA-inspired device called an antigravity treadmill.

In the 1990s, NASA scientist Robert Whalen was tasked with working on ways to allow astronauts to exercise in space. Gravity provides the resistance when a person runs on Earth, so spending extensive time outside of our planet's gravitational pull means no gravity to work against for long periods of time. As a result, astronauts return from space having lost muscle and bone mass.

Whalen proposed using differential air pressure in a bubble

around a treadmill on a spacecraft to mimic the Earth's gravity. His son, Sean, working on a graduate school engineering project, reversed the concept, flipped the air pump, and created a treadmill that removes weight from runners' legs. The antigravity treadmills allowed athletes to mimic running and all its benefits but lessen the impact on their hips, knees, and ankles by effectively removing up to 80 percent of their body weight. Someone who weighs 150 pounds, for instance, could run with the impact of just 30 pounds.

But in 2005, the device was still a novel concept. When Salazar heard about the AlterG he emailed Sean and had one shipped out. The Nike coach was the first person to use the fledgling company's prototype antigravity treadmill, which he kept in his garage, and the Nike Oregon Project athletes were the first to test its efficacy. Salazar always stressed how important running volume, or total time running, was to his athletes. And he estimated that with the AlterG he could add at least 15 to 25 percent more training volume to an athlete's overall program, while keeping them just as healthy, if not healthier.

Salazar would eventually buy six of the space-age treadmills for the Oregon Project, which were priced at $75,000 each. To increase cardiovascular stress on the athletes, Salazar put the AlterG in his garage in an altitude tent, so they could add oxygen deprivation to their weightless runs. (Three years later, the US Food and Drug Administration approved antigravity treadmills for use as medical devices appropriate for rehabilitation purposes and they are now widely used in both collegiate and professional sports programs.)

A plastic bubble enclosed the treadmill from the waist down. Runners had to wear a pair of tight neoprene shorts that zipped into the enclosure, around the waist, to create an airtight seal. The shorts, which were pulled up over running shorts, would quickly become sweaty, and there was only one pair.

"Well, who do you think got to go first?" Kara asked me rhetorically. "Galen got to go first, then Adam, then me and on down the priority list." Adam may have been Mr. Right Now, but it became clear early on that Salazar saw Rupp as the future. The daily AlterG schedule was just one small example of this.

THE GOUCHERS WEREN'T THE ONLY ONES TRYING TO GET SETTLED IN BEAVERTON IN early 2005. In rebuilding the program Salazar hired an all-star coaching staff to help him make America, and Nike, great again. But employment with the Oregon Project was tenuous no matter who you were, and Salazar was capricious. Sprint guru Dan Pfaff had joined Salazar's coaching staff in September and was gone by January.

"Pfaff was one of the main reasons we went there," said Kara. "Then, a few months after we get there, Alberto told us he had depression issues, that it's too rainy here, he missed his family, and that he might have some alcohol problems. I don't think any of that was real, I think he had a falling out."

This wasn't uncommon. Salazar would become enamored with someone, sometimes flying them in from across the country to work with the athletes, and then, all of a sudden, they were gone. It happened so often some of the staff and the athletes began calling it "getting disappeared" by Salazar. "It was very dramatic," said Kara. "As a group, we couldn't live without this person and then we never spoke to them again."

Five days after he resigned from his job with the New York Mets, assistant coach Vern Gambetta was on the Nike campus talking about joining the Oregon Project. "Even then, I thought something was fishy and I wish I would have backtracked right then," he told me. "But I love track and I wanted to go back." For a coach who cut his teeth on Bill Bowerman talks, the position seemed a little bit like destiny. Plus, John Cook had spoken so highly of Salazar, and, well, the salary was really good.

Gambetta arrived in January and lived in a hotel before Nike got him an apartment close to campus. They gave him a cubicle in the Mia Hamm building, upstairs, next to Salazar and just down from the hall from Capriotti.

He describes the program as haphazard and disorganized. "It was pure madness," he told me. "I asked to see the training plan. I do things in a systematic manner so I can decide when we're going to do running technique, etc. When I got there, there were no training plans."

Three weeks into the job, the team decamped from Oregon to train in Santa Fe, New Mexico, where the altitude—7,200 feet above sea level—makes running harder, but also stresses the body enough to create more red blood cells in the weeks that follow. It was wet and cold and Salazar didn't have a good idea of where to run, or what they would be doing. And even though the hotel is at 8,000 feet above sea level, he had Rupp sleep in an altitude tent at 14,000 feet.

While there, Salazar, who didn't have a computer, would frequently come to Gambetta's room to use his Sony VAIO laptop. "I'd look at it afterward and see what the search was for, and he was always looking for stuff, crap, you know, supplements on the internet," said Gambetta. "That was just three weeks into the job."

Galen Rupp matriculated as a freshman at the University of Oregon in 2004 and was performing well. There was only one problem—Salazar didn't have any faith that the head track-and-field coach was the right collegiate mentor for his young protégé. So Salazar and Cook helped orchestrate the firing of coach Martin Smith, a quirky leader who many of the Nike loyalists didn't think was the right fit for Rupp.

In this effort they came to loggerheads with Bill Moos, the university's athletic director. Knight and Nike had had a long and mutually prosperous twelve-year run with Moos in which the school's athletic budget grew from $18.5 million to $41 million. But he didn't want to fire his head coach, who was objectively good at his job. Knight threatened to withhold funding for the construction of the school's new basketball arena until both coach and director were gone.

Less than a week after he led the team to a sixth-place finish at the NCAA indoor championships, Smith was replaced by former

Stanford coach Vin Lananna, a devout "Nike guy." Moos would retire a year later, saying, "I created the monster that ate me." Knight then made a donation of $100 million—the largest donation in Oregon history—to the university.

"This extraordinary gift will set Oregon athletics on a course toward certain self-sufficiency and create the flexibility and financial capacity for the university to move forward with the new athletic arena," said Pat Kilkenny, the new athletic director. "Now we can roll up our sleeves and get to work on making the arena a reality."

AS ADAM'S FITNESS IMPROVED, KARA SUFFERED THROUGH VARIOUS INJURIES IN 2005, the worst of which was a sore right calf that became so bad that she could barely put weight on it. Before they knew what it was, Salazar came up with a treatment that he said they used when he was a member of Athletics West. He crushed up aspirin, applied it to Kara's leg, covered it with dimethyl sulfoxide (DMSO, the horse anti-inflammatory that makes your breath smell bad), and then wrapped it with Saran wrap.

When Kara started to complain that the itching had turned to burning, Salazar told her to leave it alone for at least a day, otherwise it wouldn't properly absorb into her system. By the time he cut the Saran wrap off, she had a second-degree burn that turned into blistered skin and lasted for weeks.

"We trusted he knew what he was doing and he didn't," said Kara. "It was something we laughed about back then. We'd joke with him, 'Alberto, you're not a doctor!' He doesn't know what he's doing. He thinks he knows what he's doing, but he doesn't. I mean, he literally burned me."

Back then, no one saw anything sinister in Salazar's often misguided and elaborate attempts to improve performance of his athletes; he was just a "mad scientist." Like the time he made a custom orthotic for Kara, using jelly and duct tape.

First, he took the insoles out of Kara's shoes. He covered her feet in jelly (yes, peanut-butter-and-jelly jelly). He then had her stand on them. When she stepped off, wherever jelly remained indicated a space between her foot and the insole. Salazar thought this space must be filled in for Kara to run properly. So he stacked other insoles on top of one another, cut away the parts that weren't needed, and added pieces that would fill in the gaps, before wrapping the entire mess in duct tape and sticking it back in Kara's shoe for a test drive.

First she had to clean her foot off. "So he'd call his dog over to lick the jelly off my foot," said Kara. After each iteration, and dog clean-up job, he would send her back out on the tennis court and up the hill behind his house for a test run. "I ran world cross with this homemade orthotic that Alberto made with jelly and his dog."

After running in the orthotic for months, Kara happened to show it to Justin Whittaker. He was dumbfounded. "This is so heavy," he said. "I can't believe you were running in this. Can I have it?"

"This sort of thing happened all the time," Kara said. "People think he's so high tech, but that just shows, he doesn't actually know what he's doing."

A couple months after the skin burn, Kara got a diagnosis that the pain in her leg was caused by another bout of compartment syndrome. Salazar told her to get surgery.

Gambetta didn't think she needed it, and he told her he thought Salazar was running her too hard and that she needed a more systematic approach. But he wasn't the boss, and he knew if he pushed Salazar too hard, he'd be fired. So he focused on what he could affect and continued to work on Kara's running mechanics. He had her in the pool and in the weight room, taking her through circuits that would strengthen the ancillary muscles that allow an athlete to maintain proper form when fatigue sets in. On Easter Sunday Salazar left Gambetta a voice message: "I don't know what you've done with her, but her running mechanics are better than ever."

Regardless, Kara consulted with one of Salazar's doctors and got surgery for compartment syndrome in March. The procedure relieves the painful pressure by cutting open the sheath that wraps the muscle, called fascia. The surgery was successful, but would set Kara back for months.

The night before the Oregon Twilight Meet in Eugene, the team went out to dinner. Rupp was aiming for a new junior record in the 10,000 meters for athletes nineteen years old and under, a record held by Salazar's close friend Rudy Chapa. The two had been teammates on the track and cross-country teams at the University of Oregon in the late 1970s. Chapa's celebrated record of 28 minutes and 32.7 seconds was from the 1976 Drake Relays. It had weathered decades of attacks from America's best juniors.

Salazar dictated nearly every move that Rupp made throughout his day at this point. When he got out of bed, when he ran, how far he ran, what he ate, if he napped and for how long, and, of course, what supplements or drugs he took.

Before ordering his meal, Rupp called Salazar. "What should I order to eat?" he asked his coach. No one present can remember what it was, but it must have worked, as Rupp performed spectacularly the next day at the event, getting stronger as the race progressed. He set a new junior standard by nearly 17 seconds of 28 minutes and 25.52 seconds, a record that still stands as of the writing of this book.

Rupp was on prednisone at the time. Some people on the team knew; they had talked about it and even joked about it openly. The Monday after the record, USADA's hotline received a new tip: "The American record was broken on Saturday by Galen Rupp. This athlete was on oral prednisone."

Two months later the nineteen-year-old raced his country's fastest adults at the NCAA Men's Outdoor Track and Field Championship. He placed second in the 10,000 meters, a huge achievement for an athlete of his age, of any age.

In July, Rupp broke another junior record—this one slightly less revered—when he set a new US junior record in the 3,000 meters in Lignano, Italy. He bested Gerry Lindgren's time by running 7 minutes and 49.16 seconds (as of the writing of this book, this is still the North American and Pan American junior record).

Adam was also flying. In February, he placed second in the 4K at the US Cross Country Championships, punching his ticket to the world championship event in France. In July, he ran the fastest American 5,000-meter time of the year in Heusden, Belgium—a new personal best of 13 minutes and 10.19 seconds.

Just months earlier, Adam had to skip the USA 8K in New York so that he could allow some persistent issues to abate. He was having trouble with fibular head dysfunction, which would lock up his foot and cause intense nerve pain in his calf. The event trainers taped him up and it seemed to help.

"The gun went off and when I pushed off at full strength I felt the most horrendous, painful, ripping through my leg," said Adam. "I damaged the nerve. It was so excruciating that I did one loop at 2K and I dropped out. That was the first time in my life I ever dropped out of anything."

Treatment seemed to help at times, but Adam would never fully recover from this spate of cascading injuries. Instead he learned to deal with them. Blood work from the Nike lab also showed that he was anemic, a death knell for an endurance athlete since it indicates a lack of iron in the blood and a decreased ability to transport oxygen during exercise.

Things didn't go to plan at the June 2005 US Championships in Los Angeles. Tim Broe caught Adam off guard by taking the race out faster than he had anticipated, leaving him in eighth place. After the race, Adam was interviewed by a reporter he knew from the *Denver Post.*

"I thought I was going to make the team," he said.

"Oh, really?" the reporter replied, as if shocked by the statement.

"I'll run faster than all those guys by the end of the year," he replied.

Racing overseas, Adam won a 3,000-meter race in Cuxhaven, Germany, and then another in Lignano, Italy, all while risking being "shot" by his ankle dysfunction. But he was fit, still running extremely well, and happy with his performances. Then, in Heusden, Belgium, Adam ran 13 minutes and 10 seconds for the 5K, a PR for him, and one of the fastest American times in recent history.

By the end of their first racing season with Salazar, both Gouchers were running their careers' fastest times. And they had also comfortably moved into a new house in Portland.

"It was a huge step in the right direction, especially from the year before," said Kara. "I mean, we didn't even make the final in 2004, and now, I think I ran the fourth or fifth fastest American time and Adam ran the fastest, besides [Bernard] Lagat, by the end of the year."

Around this time, a Nike-sponsored professional runner named Lauren Fleshman had an experience with Salazar that was telling. Though she was never on the Oregon Project team and wasn't technically coached by Salazar, when she turned to him for advice he was gracious and helpful.

In 2005 Fleshman told Salazar about her worsening respiratory issues. She had recently been tested for asthma, but due to environmental conditions, her doctor couldn't determine whether she had it or not. So Salazar set up an appointment, during Portland's allergy season, with one of his Oregon Project doctors.

To ensure that she failed the lung capacity test, Salazar told Fleshman his protocol, which he had successfully used with other athletes, including Adam. On the day of the appointment he drove Fleshman to a local track near the doctor's office and had her run laps to agitate her lungs into something that would trigger an asthma attack. He then had her run to the building and charge up the twelve flights of stairs to the office, where they were waiting to test her. Fleshman failed the asthma tests and was prescribed the corticosteroid Advair and the rescue inhaler albuterol.

Salazar was excited for her. This was great, he explained, because now she could legally use the drugs to enhance her performance. He told her that the glucocorticoid in Advair may get systemically into her body and bestow a legal performance advantage.

The moment she left the doctor's office, Fleshman had misgivings about using illness as a performance enhancement. The drugs would eventually give her oral thrush, a white rash inside her mouth that is caused by a yeast fungus, but Salazar wasn't concerned. He told her to stay the course. The doctor had instructed Fleshman to take the drugs during pollen season, then to titrate her dose way down or abstain from taking it altogether the rest of the year. Salazar advised her to take the highest dose possible all year round.

Then, Fleshman received a USADA pamphlet in the mail, which described clean sport. "There was just something about it that made me feel very clearly that that approach to my inhaler was wrong," she said, "that the spirit of the sport did not support that. Turning illness into an advantage was not right . . . And it felt dangerous."

GAMBETTA WAS THE NEXT COACH TO LEAVE THE OREGON PROJECT. HE HADN'T EVEN lasted six months. Salazar called a team meeting to explain what had happened. "We've only been there for six or seven months at this point," said Kara, "and Alberto tells us Vern is having a mental breakdown. That he's gone crazy." Gambetta had confronted Salazar about Rupp's use of the drug prednisone during the 10,000-meter junior record, Salazar told the team. He then showed them Rupp's doctor's note and his prescription for the drug.

Cook was now the last coach standing, though he would leave later in the same year, after eighteen months with the program. He says he's not a disgruntled former employee; he left on good terms and went on to coach other Nike athletes for the next decade. There were just some things that he didn't like about living in Oregon and the way the Project was run. "I don't want to comment too much on that, and I don't want to accuse anyone," he later told *Runner's World*. "But I think it's pretty obvious that drug testing can be circumvented in pretty much every corner.

"What I would say is, there's no stone left unturned. If there's a way to get better, it's done. Is it the prednisone? Is it the inhalers? Is it the cryotherapy? The idea of the program is to avail the athletes of every opportunity to stay healthy and recover to get to the pinnacle.

"I don't think everybody has a thyroid problem. I don't think everybody is asthmatic. One person, maybe, and he may need prednisone and this and that. That may all be legit, but like I was telling you earlier, I can get you to fail that test in a heartbeat. And that's just a small part of the equation."

Cook doesn't believe Knight would want the team to circumvent the rules, however, saying, "He's loyal to certain people, and he backs his people pretty much to the hilt."

"The other thing that bothers me: I knew Steve Prefontaine pretty well, and I knew Bill Bowerman a little bit, and respected him a great deal. I don't think they'd be happy with what's going on in the sport right now. I'm not saying you have to go back to the old days. I'm not that stupid. But that part of the sport is gone, and that's the part of the sport I like."

The coaches Salazar hired and fired at this time, Pfaff and Gambetta, are now barred from speaking specifics to anyone about their time with the Oregon Project by strict nondisclosure agreements.

"Some of the coaches I brought in, like Vern Gambetta and Dan Pfaff, got fired for various reasons because they disagreed with some things," Cook said. "Those guys were my amigos, and I just didn't feel comfortable."

When I reached Dan Pfaff over email, he responded, "I have a [sic] NDA so can't assist. Sorry." When I pressed, he wrote, "Sorry. Legal counsel and previous experience tells me to stay away," before he stopped responding to calls and emails entirely.

Gambetta, like Cook, feels a sense of responsibility to the sport but fears potential retribution. "Nike and the Oregon Project are ruining the sport that I love," Gambetta told me. "They've completely corrupted the values of the sport. They bought USA Track and Field and they have a history of supporting drug coaches and drug athletes. And it's not just Salazar. They have no ethics. I saw it up close and personal."

10

YOU HAVE NO IDEA

DESPITE ALL THE CORPORATE ENERGY AND ATTENTION NIKE HAD PUT INTO THE OR-egon Project, by 2006 it was more often Deena Kastor, Ryan Hall, or Meb Keflezighi who were grabbing the headlines. American running had a significant breakthrough in April, at the Boston Marathon, when Keflezighi, Brian Sell, and Alan Culpepper finished third, fourth, and fifth, respectively. In addition to Sell's fourth place, his team, the Hansons-Brooks Distance Project, also had athletes finish in tenth, eleventh, fifteenth, eighteenth, nineteenth, and twenty-second place. It wasn't exactly international domination, but it was legitimate progress and the biggest contribution to the best showing American athletes had had in a major marathon in decades. Although Hansons-Brooks were using a wholly different strategy than Nike, they shared the exact same goal—make American athletes internationally competitive again.

They too looked to East Africans for guidance. "We don't buy the genetic argument," Kevin Hanson told *Runner's World* in 2007. "We've spent a lot of time looking at the Kenyans and trying to figure out what they are doing right. And then we looked at the Japanese, and they are also very, very good at distance running. And one thing the Japanese share with East African runners is group training."

In the summer of 1999, two years before the genesis of the

Nike Oregon Project, the brothers purchased a small house for the runners to live in free of charge and signed their first three "athletes nobody had ever heard of"—Clint Verran, Kyle Baker, and Jim Jurcevich. (Since its inception, the brothers have invested between $200,000 and $250,000 a year of their own money into the program.) Within a year, Verran had improved from an American marathon ranking of fifty-ninth going into the 2000 Olympic Trials, to finishing in eleventh place.

The Hanson group would rely on the Made-in-Michigan belief that hard work alone—not technology, black magic, or pharmaceuticals—was the key to returning America to the top of the podium. "It's actually not a big secret," said Keith Hanson. "It's just about hard work and group training. That's it."

IN FEBRUARY OF 2006, WHILE WORKING THROUGH VARIOUS INJURIES, ADAM BEGAN to turn things around. He defeated Ryan Hall at the US Cross Country Championship in the 4K and punched his ticket to the April World Championship race in Fukuoka, Japan.

After a few days of recovery, Salazar had Adam back on the track and running 1,000-meter laps as he looked on, stopwatch in hand. Adam started experiencing shin pain during this workout, so Salazar had him remove one of his track spikes and put one regular running shoe on instead. That would fix his shin issues, the coach thought, and Adam finished the workout with a different shoe on each foot.

At the IAAF World Cross Country Championship in April 2006, Adam finished in sixth place. He considered this palpable progress marked by the fact that he was just eight seconds behind possibly the best runner of all time, Kenenisa Bekele. The only other American in the top twenty was Hall, who placed nineteenth.

That night at dinner, everyone was in a jovial mood, especially Salazar, who would usually have some alcohol with his meals.

Counterintuitively, Salazar told his athletes that it was fine to drink while they were training. On this night, they enjoyed a glass of wine with dinner. It was fun. It was relaxing. And though they didn't drink while training under Wetmore, they opened up to it. Salazar held court. He now had a critical data point: eight seconds behind the best of all time. That seemed like a span that his program of marginal gains could easily whittle down. "This is the most fulfilled I've ever felt as a coach," Salazar told the group, before catching himself, "well, except for when Galen broke the high school 5K record, but other than that, this is the most fulfilled I've felt as a coach."

That summer, the group traveled to London so that the Gouchers could chase the world championship's qualifying standard for the 5,000 meters. Before the event, organizers switched the women's 5,000-meter race to a 3,000-meter race. Kara had already run under the 3K women's standard, so Salazar began looking at other events. He couldn't find a 5,000-meter race but called around and found a 10,000-meter event in Helsinki.

"Ohhh no," Kara told Salazar, her mind flooding with images of her first and last 10,000-meter race in 2003 at the Mt. SAC 10K where she was forced to walk. "No way. You have no idea."

But Salazar was persistent and convinced Kara to try some laps, at 10K pace, on a nearby track. "This is what you need," said Salazar.

Eventually, Kara agreed. They traveled together to Finland, and for the first time, Salazar's attention was squarely focused on Kara. Normally, in her presence, his concerns were with Adam and his training.

Salazar told her that the first mile would feel slow. But when it came time for the actual race, the front pack went out at a blistering 4-minute-and-58-second-mile pace. Kara gritted her teeth and held on for dear life. At the exact point in the event where she should have been falling apart, just past her farthest race distance of 5,000 meters, Kara began to feel great. She was flooded with unexpected confidence until she heard Salazar yelling at the other competitors as they ran past, "They're gonna die!"

God, shut up, Kara thought.

Many of the runners began to fade. Now, as Kara was shoulder to shoulder with fellow American Jennifer Rhines, Salazar screamed to her, "If you can take it, take it!"

Kara pushed past Rhines and finished the race in third.

Afterward, Rhines said, "Do you realize you just became the second fastest woman in American history?"

"In what?" Kara responded. New to the distance, she had no ready memory of what a really fast 10,000-meter time even was.

"At the ten-K!" said Rhines.

Deena Kastor's American record time was 30 minutes and 52.32 seconds, and Kara, in her first time running a 10K, had just run 31 minutes and 17.12 seconds—now, the second fastest American time at the distance. Salazar excitedly hugged her after the race.

"That's when we started to get close," Kara told me.

Adam, however, was proving to be a nuisance for Salazar. Their personalities clashed often. "I challenged him on stuff," Adam told me. "I wanted to know why we did what we did in training. I was always curious. And he didn't like to explain it, he'd say, 'because,' and that was basically it. He didn't like being challenged on things."

One of those things was Salazar's preferential treatment of Galen Rupp, who by now was running at the University of Oregon and was subject to the rules that govern amateur athletes. Watching Salazar pay for Rupp's groceries at training camps, in particular, infuriated Adam. He thought Rupp was being treated like a professional athlete, which was unfair to all the other college athletes he'd line up against—and against the NCAA rules.

"Honestly, we were never going to be," Adam told me of his relationship with Salazar. "Those first couple of years I was getting healthy and running well, so I put up with him. He put up with me because he needed me there to legitimize the program."

BY 2006, ALL OF THE "BIG THREE" HIGH SCHOOL PROSPECTS FROM THE CLASS OF 2001 had become successful professional runners. When Alan Webb decided to forgo collegiate running entirely to try his hand in the professional ranks, Nike aggressively pursued him to become one of their athletes. Dathan Ritzenhein signed with the brand after a stellar running career at the University of Colorado where he won the individual cross-country title in 2003 and was named the NCAA Division I National Cross-Country Athlete of the Year.

Ryan Hall, however, was a harder sell. When I asked him about being courted by Nike he emailed me, "I'm going to have to defer to my dad's words of wisdom to me as a kid, 'If you don't have anything nice to say don't say anything at all.'"

While Webb focused on shorter distances, Ritzenhein and Hall were poised to take up the mantle left by Frank Shorter and Alberto Salazar as America's next great marathon champions. Ritzenhein lived and trained in Boulder, the endurance mecca of the Americas, and was coached by Brad Hudson. In 2006, he was the hottest prospect in running. Ritzenhein's marathon debut at the 2006 New York City Marathon was heavily anticipated. Reports claimed that the twenty-three-year-old was being paid appearance fees in the $200,000 range, just to show up at races (a figure his agent has disputed).

The hype must have been overwhelming, but Ritzenhein seemed to be taking it all in stride as he cruised the first twenty miles through the five boroughs in the shadow of the leaders. But the wheels began to wobble in the Bronx and by Central Park, Ritzenhein was in an internal battle to just make it to the finish line. He was the second American that day, finishing behind six Kenyan athletes, in eleventh place overall, 4 minutes and 3 seconds behind the winner—an eternity in the marathon.

As big as the expectations were for Ritzenhein, it was Lance Armstrong who garnered most of the media attention on the thirty-sixth running of the famed New York City race. This was his first athletic event since he retired from professional cycling the year prior, after his seventh and final consecutive Tour de France victory in 2005. He was admittedly less prepared than he would have hoped, citing his nonprofit responsibilities and jet-setting with Hollywood actors as distractions.

As a young man, Armstrong had been a competitive triathlete before his record-setting cycling career launched him into infamy, so he wasn't a total novice on foot, and few humans knew better what it took to achieve a fitness goal. Most experts and marathon training guides encourage runners to extend their long runs to at least the twenty-mile mark in preparation for the grueling 26.2-mile distance on race day. Armstrong said he was only able to run about forty-five minutes a day, with his longest run being sixteen miles.

For race day, a dream team of Nike athletes paced him through the event.

Salazar accompanied Armstrong for the first ten miles, both men wearing Nike Livestrong shirts with the numbers 10/2 on them. In 2005, in addition to marketing the yellow Livestrong bracelets, Nike began selling a Lance Armstrong line of cycling gear and clothing that commemorated the date that Armstrong was diagnosed with cancer, October 2, 1996. Nike rebranded it "his carpe diem day, a day to overcome adversity and reaffirm life." (By now parody bracelets had already begun popping up that read "CHEAT TO WIN," instead of "Livestrong.")

A television camera, dubbed the "Lance Cam," was dedicated to following Armstrong and his entourage throughout the event. Salazar brought him water, kept him abreast of his split times, and encouraged him along.

"It was very hard to hold him back," said Salazar after. "For him, cardiovascularly, it was very easy, he could talk much better than I could talk during the race."

Then it was Olympic gold medal champion Joan Benoit Samuelson's turn to pace. Armstrong had said he thought he could run a 2-hour-and-30-minute marathon if he trained properly, but that he was shooting to at least come in under the 3-hour mark today, a goal that pundits reveled in pontificating about before the event.

As the race turned toward East Harlem, around mile eighteen, Armstrong's legs became leaden. Then his shin splints began to throb. "That's when I started to feel helpless," he told reporters after. "I thought, *Uh-oh, maybe I should have trained a little harder for this. I think I'm in trouble.*"

Ever the good pacer, Samuelson gave Armstrong landmarks and goals to reach, like people or buildings up ahead. "I have no doubt that he would have finished if we weren't there to help him," she said, "but I think breaking the three-hour barrier would have been questionable. I think the marathon is a mental game, and he's got that down pretty well."

Another Olympic gold medalist and Nike athlete Hicham El Guerrouj joined the Armstrong armada just past the twenty-mile mark as the group's focus intensified on breaking three hours. Armstrong stayed focused and managed to cross the finish line twenty-four seconds ahead of the three-hour mark, which allowed him to raise more than $600,000 for the Lance Armstrong Foundation. With a time of 2 hours, 59 minutes, and 36 seconds, he placed 868th overall (a 6-minute-and-51-second per mile pace).

"I can tell you, twenty years of pro sports, endurance sports, from triathlons to cycling, all of the Tours, even the worst days on the Tours, nothing was as hard as that," said a flushed-faced Armstrong at the post-race press conference. "And nothing left me feeling the way I feel now, in terms of just sheer fatigue and soreness."

WATCHING A RACE ON TELEVISION IN 2006, IT OCCURRED TO SALAZAR THAT A RUNner's form probably matters more than he'd previously considered. As Ethiopian Kenenisa Bekele, the world record holder in the 5,000 and 10,000 meters, streaked across his screen, Salazar noticed some peculiarities of his gait that he was not seeing in the Oregon Project athletes in his stable. For one, Bekele retracted his trailing leg extremely quickly, pulling his foot to his glutes, rather than allowing it to float up behind him.

"I thought, *Is that just coincidence?*" he said. "Or could that perhaps be part of why he's so good?"

He called former Nike athlete and four-time Olympic gold-medal winner Michael Johnson. The sprinter, who retired in 2001, had parlayed his clean image and world-class pedigree into consulting positions for speed athletes including NFL prospects, and was running his own training facility in Dallas. Johnson laughed and told the coach, "Alberto, that's Sprint 101 biomechanics!"

Experts believe that retracting the rear foot quickly—in other words, bringing the heel to the butt faster during the gait cycle—creates power and shortens the distance the foot has to travel. This, in turn, means more strides per minute and faster running speeds.

Salazar reconsidered his career in light of his own less-than-perfect biomechanics. "The way I ran, it wasn't sustainable. The attitude at the time was: if you were gifted with perfect form, great," he said. "If you weren't, you were just kind of stuck."

He worked with the Michael Johnson Performance Center's director of performance, Lance Walker, to measure the exact angles of Bekele's arms and legs through the stride cycle so he could apply them to the Oregon Project runners. The more video the two men broke down, the more Walker came to see the similarities in form between Bekele, the world's best endurance runner, and the sprinters he worked with. Good runners in both events have high thigh drive, where the femur appears to rise almost parallel to the ground, and extremely brief ground contact, which Walker said, is like "a pogo stick with a stiff spring."

While form and technique certainly matter a great deal, by the end of July 2006, sports fans were beginning to learn about another factor elevating the best in the world: testosterone.

Professional cyclist Floyd Landis had stepped into the void left by Armstrong to win the Tour de France in stunning fashion that summer. After a collapse on Stage 16 left him with a seemingly insurmountable time gap, the pundits had written him off. But the following day, Stage 17, he did the unbelievable, first telling the peloton that he was going to go for the win from the gun, and then actually doing so. Landis rode away from the best cyclists on earth and continued on by himself through the French mountains to an astonishing stage victory. He regained the overall lead two days later and would ride to the finish line on the Champs-Élysées as the 2006 Tour de France champion (and just the third American to ever accomplish the feat).

Landis had roughly four days to savor being considered the best cyclist in the world before his test results from that fateful day, Stage 17, came back positive for testosterone. His urine sample showed his ratio of testosterone to epitestosterone was eleven to one, nearly three times higher than the allowable limit of four to one. Landis became the first man in the race's 103-year history to be stripped of his title due to a failed drug test.

Days later Nike athlete and fastest man in the world Justin Gatlin was exposed for failing a drug test for exogenous testosterone. This was not Gatlin's first doping violation, however. The runner had had a false start to his career when he failed a PED test for amphetamines in 2001. Gatlin was able to get the two-year ban reduced to one after convincing an arbitration panel that he failed the test because he was taking Adderall to treat a well-documented condition of attention deficit disorder.

After his suspension, Nike signed him to a lucrative contract. Gatlin began working with embattled Nike coach Trevor Graham, and his career began to take off. He won gold in the 100 meters at the 2004 Summer Games in Athens, a race he then followed up with the sprint double by winning both the 100- and 200-meter events the following year at the IAAF World Indoor Championships. In May 2006, Gatlin, wearing a yellow Livestrong bracelet, tied the 100-meter world record time of 9.77 seconds set by Jamaica's Asafa Powell, then told reporters, "I am the best of the best because I am the Olympic champion, the world champion, and the world record holder now."

Outwardly, Gatlin was the new, clean face of the track world. He was well-spoken and seemed genuinely believable when talking to reporters, especially about doping. When asked if he would turn someone in who he knew was cheating, he replied, "Yes, I would."

It seemed that the media and the fans were even giving him the benefit of the doubt on the 2001 failed PED test. *ESPN* magazine wrote that he "might finally change track for the better."

On April 22, twenty-four-year-old Gatlin was part of the winning 4 x 100 meter relay team that competed at the Kansas Relays in Lawrence. After the event, anti-doping agents analyzed his sample with a new, labor intensive, and costly testing method that had been recently developed. Officials had learned through the revelations in the ongoing BALCO doping scandal that there were now substances, namely Victor Conte's The Cream, that were custom-made to evade the testosterone-to-epitestosterone ratio tests. The new method measured the carbon isotope ratio (CIR) to detect if any were derived synthetically (i.e., through exogenous testosterone). Gatlin's urine passed the ratio test, but was found to have exogenous "testosterone or its precursors" in it.

"I cannot account for these results, because I have never knowingly used any banned substance or authorized anyone else to administer such a substance to me," Gatlin said in a statement.

On its website, US Track and Field's executive director, Craig Masback, posted that the organization "is gravely concerned that Justin Gatlin has tested positive for banned substances. Justin has been one of the most visible spokespersons for winning with integrity in the sport of track and field, and throughout his career he has made clear his willingness to take responsibility for his actions."

Gatlin's coach, Trevor Graham, told the *Washington Post* and the Italian sports paper *La Gazzetta dello Sport* that the massage therapist was to blame, claiming that he had rubbed Gatlin with testosterone cream and that's why he failed the drug test. The theory seemed fairly ridiculous on its face, but its propagation would kick off an unbelievable drama that continued to play out for years to come.

It was Graham who fanned the flames of the BALCO investigation when he mailed a syringe containing "The Clear" to the US Anti-Doping Agency in the summer of 2003. Now, that trip to the post office was threatening to take him down with it.

The coach's proclamations about being clean were becoming increasingly harder to believe as more of his athletes tested positive for banned drugs, including six world champions and Tim Montgomery, a former 100-meter world record holder. Although he never tested positive, evidence in the BALCO case proved Montgomery had used PEDs. He fought the USADA ruling through arbitration in December 2004, lost, received a two-year ban, and retired from the sport.

The massage therapist Gatlin's team pinned the failed drug test on was forty-two-year-old Nike stalwart and well-respected masseur Chris Whetstine. Based in nearby Eugene, he had worked with elite athletes for more than a decade. Whetstine traveled with Marion Jones from 1998 to 2001 before beginning to work closely with Gatlin in 2003. And though he worked with other athletes and celebrities—most notably Sean "Puff Daddy" Combs in his preparation for the New York Marathon in 2003—Whetstine was known in sports circles as the "Nike guy," and then "Gatlin's guy."

Graham's conspiracy theory was predicated on the idea that Whetstine had some sort of grudge against his star client. Graham provided no evidence but seemed to describe events out of a Hollywood thriller: an angry and vindictive massage therapist sneaks a tube with a crooked "S" on it out of his pocket, then proceeds to rub the mysterious cream on the fastest-man-in-the-world's inner thigh.

Behind the scenes, Gatlin and his lawyers cringed at the conspiracy that Graham put forth (though Gatlin would eventually use this defense in his arbitration case against USADA). The running community was incredulous. Meanwhile, Whetstine's first public statement thanked Alberto Salazar and track meet promoter Tom Jordan for their behind-the-scenes support: "While I am choosing to allow my attorney to speak publicly regarding the specific issues, I am comforted by the continued support given me by members of the running community, specifically Tom Jordan and Alberto Salazar. It is at times like these that we find out who our friends are."

Salazar then did an interview with ESPN, telling a reporter that Whetstine continued to work on Gatlin during his lead-up to the June US Track and Field Championships, after they had found out about the failed test, which wouldn't make any sense if they knew he'd rubbed an illegal substance on Gatlin. "Trevor is a world-renowned, legendary coach," said Salazar. "Right then and there, if he sees somebody acting strange and putting something in his pocket, he's going to put a stop to it and figure out what the heck is going on. It wouldn't go any further than that. And Chris would never get close to Justin again. You can't help but laugh. It's just preposterous on so many fronts. I guess anything's possible, but it just doesn't make sense. There's no way in the world."

During the 2006 US Track and Field Championships, on the night of June 22, Whetstine says he was accosted by drunk Nike employee Llewellyn Starks outside of the host hotel. Starks, who had been Marion Jones's and Tim Montgomery's agent before being hired by Nike in 2004, worked in sports marketing for John Capriotti. The police report claims the two men argued before Starks began striking Whetstine. He says he absorbed the attack without retaliation and that he suffered a broken nose, a dislocated thumb, a sprained ankle, and a concussion in the altercation. Whetstine sued Nike and Starks for $3.9 million, arguing that the injuries were debilitating and caused a loss of income due to not being able to perform massages with a broken thumb. He also says that he's dealt with lingering brain damage due to Starks's blows to his head. There was an undisclosed settlement to the lawsuit in 2009. Sources familiar with the case say that Nike lawyers drew the proceedings out for so long that Whetstine was unable to work and almost lost his home to foreclosure.

When I reached Starks by phone, he was happy to talk about work, but as soon as I asked him about Nike's vice president of global athletics marketing, John Capriotti, he became agitated. And when I asked him about Chris Whetstine, he hung up the phone.

In August 2006, after USADA announced Gatlin's eight-year ban (which would be dropped to four upon appeal), Nike finally suspended Graham's contract. Earlier in the same week, another of

Graham's athletes, LaTasha Jenkins, had an A-sample test positive for the steroid nandrolone at a track meet in Europe. Nike finally ended their contract with Graham but remained vague in their public explanation. "I can't divulge the details," Dean Stoyer, a Nike spokesperson, told reporters. "All I can say is we're terminating the contract."

11
EVEN DYING WON'T KEEP HIM

ON A SATURDAY NIGHT IN 2007, SALAZAR MADE A CALL TO A SPORTS PSYCHOLOGIST he'd heard good things about named Darren Treasure. Through his private practice, the British-born Treasure had worked with NBA stars, NFL players, and athletes on the US women's soccer team.

Salazar asked Treasure if he would come to Oregon and sit with his protégé, Galen Rupp, who had been struggling through injury setbacks.

He also envisioned Treasure helping Kara, whose mind had always been at war with her legs. She studied psychology in college and had even worked with a few sports psychologists in the past. Salazar told her this might be a chance to wrangle her psyche into working for her instead of against her. "Look, I think you should just sit down and meet with him," he said.

In March, she indeed sat down with him, and like a doctor asking for a patient's medical history, Treasure had Kara start at the beginning. He asked her why she started running. She detailed her extensive high school and college successes, and the injuries and psyche-sabotage that accompanied them. She told him about her first 10,000-meter disaster at the Mt. SAC Relays, then her Helsinki 10,000-meter success. Salazar had been encouraging Kara to leave the shorter distances behind and focus on 10,000-meter races and longer, but she worried that she wouldn't be able to live up to the new expectations at longer distances.

She frequently doubted that she was as good as her competitors were, often midrace, a pattern of thinking that she knew could become a self-fulfilling prophecy. The longer the race distance the more time one's mind has to undermine the day's goals. She questioned whether her body, or her mind, could withstand the new demands of longer distances, and was scared to disappoint everyone.

Through tears she told Treasure, "I know I'm really good, I just haven't been able to show it. Every time I go to the starting line, I doubt myself. I sabotage myself." Surprisingly, the therapy proved a great relief for Kara, who noticed an easing of her psychological distress after the session. *This might actually work*, she thought.

"First, we had to establish belief," Treasure told *Runner's World* in 2010. "Belief that she could actually run a competitive ten-K."

Salazar and Treasure began working closely together, the sports psychologist reinforcing what Salazar had been telling Kara—that Helsinki wasn't a fluke, she was that fast. Not only was Kara destined to do amazing things in the 10K, but she would eventually be a marathoner, they said. As Salazar gradually increased Kara's training load with the goal of performing well at the USA Track and Field Championships in three months, and ultimately a medal performance, in August, at the IAAF World Championships in Osaka, Japan, she met with Treasure a couple times a week.

"We realized very early that Kara was capable of handling an awful lot of volume and intensity from Alberto," said Treasure. "She has an incredible ability to handle the pain and discomfort that come with those longer distances. Alberto and I came to the conclusion that she could go to some places that very few athletes are capable of going to."

The men gave her phrases and words of affirmation to repeat, which some psychology research shows will improve self-esteem. For this season, Kara's word would be *fighter*, and she'd use that as a prompt in the race to fight all the way to the finish.

"I am a world-class runner," she'd say out loud, "and I deserve to be here."

IN MAY 2007, FIVE MONTHS BEFORE THE PREMIERE OF *KEEPING UP WITH THE KAR-* *dashians*, which would keep much of America transfixed, Nike launched a web-based show called *Keeping up with the Gouchers*. The three-part series was published online and followed the professional running couple through a typical day on the Nike campus.

"Meet Adam and Kara, your friendly neighborhood kick-ass distance runners," began the voiceover of each episode. "Injuries almost forced them to hang up their spikes, but now they're hungrier than ever."

The show began in the Gouchers' bedroom as Kara opened the curtains, and Adam lumbered out of bed to brush his teeth. "The one thing I think people would be surprised is how aggressive my goals in life are for running," Kara said while the couple drank coffee in their Portland home. "It is my passion to go and run for the US. I want to go and do something that's memorable."

The camera crew followed them to work, on the Nike campus, where Salazar assessed where the two athletes were currently positioned in the distance hierarchy. "Adam is ranked third, but could slide back, while Kara's more comfortably in the top two in the country," he said. Their coach looked healthy and happy, dressed head to toe in black, from his Nike hat to his Nike shoes, belying the fact that his heart was in distress and would soon attack.

The 2007 USA Outdoor Track and Field Championships began June 20, in Indianapolis, Indiana. As Kara lined up with running star Deena Kastor and Katie McGregor, who had won the 10,000 at the 2005 USA Outdoor Track and Field Championships, she called upon her training to block them out. She knew Deena was going for the Olympic A standard in the 10,000-meter race from the gun and would likely take it out fast (this was Deena's last season on the track, as she would devote herself entirely to the marathon the following year). Kara was to stay focused on her goal—to qualify for the world championship team in Osaka—not to beat Deena or even win the race. When Deena steamed away from the pack after 1,200 meters, Kara had to let her go.

"That was the hardest thing we've ever asked Kara to do," Treasure said after the race. "She wants to compete. When she toes the line, she's there to win." The strategy worked. Kara finished second place and qualified for a trip to the world championship.

Back in Beaverton, on June 30, 2007, Salazar walked with athletes Galen Rupp and brothers Jared and Josh Rohatinsky toward the Lance Armstrong Fitness Center to start the morning's weightlifting drills. Just two months prior, the coach had undergone a complete physical. His family lineage included heart disease and high blood pressure, both of which he was acutely aware of and was taking medication to treat. His EKG looked good, but based on his family history, Salazar's doctor suggested an echocardiogram, which he never did get around to scheduling.

Still, he was the picture of perfect health. Even at forty-eight years old, he managed to run anywhere from twenty-five to thirty miles a week at a "a very relaxed, seven to seven-thirty pace." He was famous for reliably dropping down and doing intense sets of sit-ups and push-ups while in the presence of journalists, and he claimed to have just 4.9 percent body fat.

Before Salazar reached the Lance Armstrong building, he began to feel faint. He told Rupp, "Hey, I'm starting to get dizzy. I'd better get down on one knee." That was the last thing he remembers before he collapsed on the grass.

Rupp stayed with him and called 911 on his cell phone, while the brothers ran in different directions for help. Jared ran toward a football training camp that was assembled at the far end of the Nike fields. Josh ran into the Armstrong building to find a defibrillator. The group returned to Salazar in a panic. He was blue and didn't appear to be breathing.

Finally, paramedics arrived and took over. All told, it took four charges from the resuscitation paddles to electrically induce Salazar's heart to start beating again.

"This was about thirteen or fourteen minutes after I first went down," said Salazar in the weeks after the incident. "And that's pretty amazing, because I've learned since that you don't have a very good chance of surviving in good health if you go more than five minutes without a pulse."

As he came closer to consciousness in the hospital, Salazar said he remembered hearing voices floating around him and the rhythmic whine of what he assumed was an oxygen machine. As the fog of confusion lifted, he noticed a rosary in his hand and a crucifix on the table next to the bed. He was in the hospital.

Phil Knight, ignoring the NO ADMITTANCE sign posted in the cardiac unit, walked around the halls until he found Salazar's hospital room. Lifting his head off the pillow the Nike coach managed "a pained smile," as Knight put it. Before Knight left the bedside, Salazar reached out for his hand and said, "If something happens to me, promise me you'll take care of Galen."

"Of course," Knight said. "Of course. Galen. Consider it done."

In Salazar's first major interview after the heart attack, writer Amby Burfoot was shocked by how the usually bellicose Salazar seemed different. "He kept telling me to ask all the questions I wanted. Nothing was off the record," Burfoot wrote in *Runner's World*. "I've known Salazar for thirty years, but I haven't always enjoyed interviewing him. He was once the fiercest, most combative runner I'd ever met. He bristled at prying questions."

Salazar's heart attack confirmed his belief that God had put him on Earth to do something special. He told writer John Brant, his eventual biographer, that the fact that the infarction didn't cause more damage to his brain or heart was a miracle (yet another after his surviving the heart attack), one that allowed him, after having a stent and a defibrillator surgically implanted in his chest, to return to work and even run shortly after being discharged. He was so grateful he had been spared, he told Brant.

Salazar's heart attack shook many members on the team, who were taking stock of what, exactly, they were doing with their lives. "In the first weeks after Alberto's heart attack I felt really demoralized," Kara said. "'What's the point of spending my life running around in circles?' I asked myself. But as I thought more about it, I realized that all of this had happened for a reason. Running was my gift, and I wanted to be worthy of it. I started training with a renewed focus and passion."

It also shocked the running world. Not since Jim Fixx, author of *The Complete Book of Running*, was felled by a heart attack and died in July 1984, had runners been forced to reckon with the fact that jogging wouldn't inoculate them from heart disease (the leading cause of death in the United States). Within a year after Fixx's death, the number of marathon races in the US dropped from the 1980 high of 208 to 130. What the public wouldn't find out until late 2019 was that Salazar's ostensible vigor had been secretly propped up for years by the drug testosterone, which he'd first used in 1991. His physician, Dr. Jan Smulovitz, diagnosed him with hypogonadism and began prescribing him the drug in 1994. Salazar has now admitted that he took it continuously from at least 1995 to early 2006.

He was just one of many aging athletes who tried to regain some of their former vigor by taking the drug. In 2002, prompted by the alarming increase of men using testosterone in the apparent "absence of adequate scientific information about its risks and benefits," the National Cancer Institute and the National Institute on Aging asked the Institute of Medicine to conduct a review of the knowledge related to the drug. Finding a dearth of historical research to draw from, they commissioned researchers to take the steroid through its paces with the gold-standard, double-blind, randomized, placebo-controlled clinical trials. The most disturbing of the findings was the cardiovascular risk: in certain men, testosterone accelerated coronary atherosclerosis and possibly increased their chances of a heart attack.

Nine days after his hospital stay, during which time all of his athletes came to visit, Salazar was back on campus in his black Nike jumpsuit and yellow Livestrong bracelet. In his first week back, he managed to work four hours a day, building in time for rest and naps. In his second week, he increased his work output to six hours, then back to a normal eight-hour day. Aside from the pacemaker—that protruded from below his clavicle, over his left pectoral, and not far from the Nike swoosh tattoo on his shoulder—he looked no worse for wear.

TO ASSURE A CLEANER WORLD CHAMPIONSHIP, THE INTERNATIONAL ASSOCIATION OF Athletics Federations (IAAF) announced that it was stepping up efforts during the 2007 event in Osaka, Japan. They had conducted 885 tests in 2005 at the previous world championship meet in Helsinki; for Osaka, they would test more than one thousand samples, though they remained intentionally vague about which tests they would use, and when, exactly, they would employ them, so as not to tip their hand to the chemical athletes attempting to use illegal substances to improve their performance.

"Anyone considering cheating should be aware we will use every available method to catch them," said IAAF president Lamine Diack, "and that should we choose to do so, we can store their samples for testing at a later date."

By the first day of championship racing, August 25, 2007, on the heels of an Under Armour campaign targeted at young girls, Nike launched a new marketing effort called "ATHLETE" featuring many of their female professionals, including Mia Hamm, Gabrielle "Gabby" Reece, and Picabo Street.

Nike's market research showed that young female athletes across America still felt "unequal" when it came to respect in sports, so the company gathered eighteen female athletes and one man (high school coach Bill Ressler) in a Los Angeles high school gym and gave them topics on which to ad-lib into a fifteen-foot-long megaphone with a three-foot-wide mouthpiece. They demanded that female athletes everywhere be respected for their athletic abilities.

"Are boys bigger, stronger, faster? Yes," said Gabby Reece, a

Nike-sponsored beach volleyball player, in the commercial. "Is that all that has to do with being an athlete? No."

Olympic snowboarder Gretchen Bleiler then stepped to the megaphone and said, "The half-pipe doesn't care that I'm a girl," before streetballer Alvina Carroll closed the thirty-second ad with, "It's not a girl thing. It's not a boy thing. It's a skills thing."

Nike, who owned 19 percent of the women's United States and European footwear and apparel market, first aired the group spot, then released commercials with individual athletes Bleiler, Carroll, Street, and Hamm on ESPN and MTV. For their celebration of female athletes, Nike was applauded in the *New York Times*, with a piece titled "Nike Puts Women Back on the Pedestal."

"It's time to nudge the conversation," Reece said in the *Times*. "Women's professional sports have plateaued."

IN OSAKA FOR THE WORLD CHAMPIONSHIP, RACE DAY WAS OPPRESSIVELY HOT, WITH morning temperatures starting in the midseventies and the humidity around 80 percent. In their preparation for these championships, Salazar had the Oregon Project athletes run in thermal training tops the team called "sauna jackets."

"It was a glorified garbage bag, basically," Kara told me. "I'd wear it to get hot, sweaty, and uncomfortable because we knew it would be that way in Osaka."

On the bus ride to the venue, Salazar told Kara that she had improved so much that he believed she was ready to place top five in the world.

When it came time for the 10,000-meter race, she spent most of it in the middle of the field but says it was harsh inside the pack, with athletes using any means necessary—elbows and shoves included—to secure positions or create space for themselves to run.

With one lap to go, in fourth place, she had thoughts of resignation and couldn't envision herself on the podium. Fourth place in the world would be pretty good, after all. But then she saw Britain's Jo Pavey just ahead.

"I knew I was on the edge of dying. I was going to go by her and hope the move breaks her will," said Kara. "Some folks were telling me afterward that I should've waited until the final straight, but I knew that if I didn't go, that I might break." Moving into third place as the laps ticked off into single digits, she looked up at the scoreboard and saw that the two athletes in front were telescoping away.

"And Kara Goucher is in third place for the United States!" screamed the television announcer. "This would be a shocking result if she can claim the medal!"

Kara ran a 65.2-second final four hundred meters to finish in third place becoming the first American woman to medal at a major global championship in a distance race on the track since Lynn Jennings took bronze in the 10,000 at the Barcelona 1992 Games. In doing so, she silenced the critics who didn't think an American could earn a medal of any kind on the world stage.

"I thought I was going to pass out I was screaming so hard," said a proud Adam after the event. "It was unbelievable."

On the television broadcast they asked her after the race if this had been the greatest moment of her life. "Ah, running, yes," Kara said. "Meeting my husband is the greatest moment in my life, but nothing compares to this in running."

During the media scrum Kara credited Salazar and the fact that he employed Darren Treasure to convince her to become a 10K runner. "Alberto's the most comprehensive coach I know of," she said, "even dying won't keep him from being here." The ascendant bronze-medal performance in Japan also solidified her as one of the world's best runners.

Two days later, on the morning of the men's 10,000-meter final, Salazar, who had turned forty-nine earlier in the month, began having heart attack symptoms again. He felt tired. He was dizzy. And it seemed to be getting worse. He told Team USA's physician that he thought he was having another heart attack, and they rushed to the hospital. While there, doctors deduced that Salazar's blood pressure medication, and not a cardiac event, was the cause of his symptoms.

"I left the hospital and got to the stadium just as Galen was warming up for his race," Salazar said. "I don't think my condition affected him too much, but it couldn't have helped."

Rupp was on a roll, having improved his 5,000-meter personal best to 13 minutes and 30 seconds, coming from behind to just barely beat one of Jerry Schumacher's stars from the University of Wisconsin-Madison, Chris Solinsky. Nine days later he vanquished a stacked field at the Payton Jordan Cardinal Invitational race in California, winning the men's 10,000-meter race, and setting the American-born NCAA Collegiate Record at 27 minutes and 33.48 seconds. At the Pac-10 conference meet, just two weeks later, Rupp won both the 10,000- and 5,000-meter events, and his team, the Oregon Ducks, took home the team championship title. And before school let out, he would place second at the NCAA Track and Field Division 1 Championships in 10,000 meters.

Osaka was Rupp's international racing debut. Although hopes were extremely high, he only managed an eleventh place in the 10,000 meters, two spots behind American Dathan Ritzenhein. Ethiopian Kenenisa Bekele won the race in 27 minutes and 5 seconds.

On the last day of events, Adam lined up next to some of the fastest men in the world at 5,000 meters, including Ethiopian Tariku Bekele, Kenyan Eliud Kipchoge, British Mo Farah, Kenyan-American Bernard Lagat, and Australian Craig Mottram. As the two groups merged, Adam settled into the back of the fifteen-man group where he remained until 11 minutes and 40 seconds in, when he began to move up with less than a thousand meters to go. Farah took the lead and pressed the pace. The pack splintered, leaving Adam in no-man's-land, just off the front group. On the final lap, first Kipchoge, then the rest of the pack streamed past a fading Farah. Lagat overtook Kipchoge for the win, with American Matt Tegenkamp a surprising fourth place. Adam was eight seconds off the gold-winning time.

After the race, his mounting frustration spilled over in a heated discussion with Salazar. Though he didn't know it at the time, Adam's career was over.

"After your race in Osaka, he was done with you," said Kara. "He never was excited about you ever again after that."

THE BRONZE MEDAL IN OSAKA HAD EFFECTIVELY RAISED KARA'S PROFILE. ADIDAS reached out with a contract offer that was considerably more than what Nike was paying her, but her loyalty ran deep.

"Why would I ever leave Alberto and Nike?" she said. "Nike had been with me through thick and thin. They were there for me when I was injured all the time. We were one hundred percent loyal to them. I mean I thought I'd be at Nike forever, and that I'd eventually work there."

The increased attention meant she was also being drug tested more often (seventeen times in 2007) and offered considerable money to show up at races, an agreement called "appearance fees." Race directors offer monetary incentives to high-profile athletes as professional bait to get them to show up and race at their events. Bigger athletes bring more attention, more attention brings more race entries, more swag sold, and more money.

One offer in particular caught her eye from the Great North Run, the largest half marathon in the world and a popular British event that starts in Newcastle and ends in South Shields, England. It was a startling amount of money to just show up and start running. But that voice was still there, despite her continued work with Treasure, Kara's self-sabotaging internal dialogue kept creeping back in, reminding her that the shorter races were hard enough and surely the half marathon would be unbearable.

"You have nothing to lose," Salazar told her. "You're in great shape. You can run five-tens to five-fifteens. Worst case, that's going to be top five."

If Kara went, she'd have to contend with the Great North Run's course record holder, the venerable British runner Paula Radcliffe. The thirty-three-year-old was just coming back to racing after a nearly two-year hiatus due to injury and childbirth the previous January.

The longest distance Kara had ever run before today was 10 kilometers, or 6.2 miles. The Great North Run would push her into unexplored territory. Salazar told her to "run five-minute-and-ten-second miles and to stay with the leaders unless they take the pace under 5 minutes and 5 seconds, then let them go." But that wouldn't be a concern.

Just a month off her bronze medal performance at the World Championships in Osaka, Kara began separating herself with a couple of sub-five-minute miles around the halfway point of the race. Seven miles in, Radcliffe was fading behind her. Kara stayed focused and never looked over her shoulder to see where the competition was.

Radcliffe continued to fade as Kara set one US record after another, though she had no idea. Running on feel, she blazed through the ten-mile (50 minutes and 59 seconds) and twenty-kilometer marks (63 minutes and 33 seconds) faster than any female US athlete before her ever had (though the course had too much downhill to certify them as official records). Kara crossed the finish line more than a minute ahead of Radcliffe. She had conquered the distance and won the race, but she also ran the fastest female time of the year for a half marathon, finishing in 1 hour, 6 minutes, and 57 seconds.

After the race Radcliffe said, "I knew I was in shape to run under seventy but I wasn't expecting to get dropped as well. I had seen Kara's results, but I was surprised she was running that fast. It wasn't that I was running slowly, just that she was going really well. My pride's taken a little bit of a bashing. I came out here wanting to win the race, but it is good to be back."

Kara then asked Radcliffe if she could get a photo with her.

Five weeks later, Kara found herself riding along in the pace car out in front of the New York City Marathon leaders. There, pressing the pace in front, was Radcliffe, who would go on to win New York for a second time, in 2 hours, 23 minutes, and 9 seconds.

As athletes are wont to do, Kara wondered how she'd fare in a race of this distance against women of this pedigree. She'd just beaten the winner a few weeks ago, after all. As the crowd went wild with applause for Radcliffe, Kara thought, *I want that to be me. I'm coming back next year. And I'm racing.*

IN OREGON, ON OCTOBER 10, SALAZAR WAS ONCE AGAIN RUSHED TO THE HOSPITAL with heart symptoms. And again, doctors determined that he wasn't having another heart attack. But during his examination they realized that one of his arteries was 90 percent blocked. He went into surgery and received a stent to permanently hold it open.

Salazar told the *New York Times* that because of what seemed like ongoing heart issues, he would not take on new athletes. Behind the scenes, he reached out to the University of Wisconsin track-and-field coach Jerry Schumacher to possibly become his successor at the Oregon Project. But Schumacher would prove to be a hard sell; he loved Wisconsin, had a family to consider, and was having success working at his alma mater.

A month later, in November, Adam had surgery on his ankle. Doctors removed a bone fragment, cleaned up some cartilage, and then shaved down a bone spur. A permanent screw was placed in his navicular bone to correct the dysfunctional joint.

Injuries and disagreements with Salazar were peaking, but Adam tried to keep his thoughts to himself at this time. Kara was running too well under Salazar to break from their mercurial coach.

"Kara was running phenomenal so there was no way we were going to go anywhere else. I had no other options in Oregon," said Adam. "I did my part, but I was basically running by myself from the '07 World Champs on. I wasted my career from that point on. But we weren't going anywhere, and it didn't matter that I had a lot of potential left because Alberto was done with me and I was done with him. So I was trapped."

IN 2007, DANNY MACKEY WAS RIDING HIGH. HE HAD QUALIFIED FOR THE US OLYMPIC Trials and he had just taken a dream job at Nike, the most influential brand in the sport he loved.

Mackey grew up on the South Side of Chicago, obsessed with Michael Jordan and the local big-three sports teams, the Bulls, the Bears, and the Cubs. He attended Andrew High School and was mentored by hall-of-fame coach Joe Mortimer. He ran PRs of 14 minutes and 46 seconds in the three-mile cross-country event and 1 minute and 58 seconds for the 800 meters. He then ran for Eastern Illinois University where he says he was again blessed with great coaches, John McInerney and Tom Akers. Mackey studied business with a minor in biology and managed to run solid collegiate PRs of 4 minutes and 16 seconds for the mile and a 15 minutes and 18 seconds personal best in the 5,000 meters, but says he was injured or sick more often than he was healthy in his six years of collegiate eligibility. This got him interested in the "why" behind injuries, performance, and training philosophy.

He attended Colorado State University for graduate school in exercise physiology, where he was also a volunteer coach for both the cross-country and the track-and-field teams. He published biomechanics research in the National Strength and Conditioning Association journal that examined how intrinsic and extrinsic factors influence muscle contractions.

In 2006, Mackey, who admits he broke every scientific rule out there when it came to training for a marathon, qualified for the US Marathon Trials by running a time of 2 hours, 21 minutes, and 38 seconds. "Not a rocking time, but considering I'd never raced over a ten-K, and wasn't able to train enough because of school and work," he said, "I was ecstatic." It earned him an invite to join the Hansons-Brooks Distance Project, the professional running team based in Michigan that was consistently placing American contenders back toward the front.

Mackey wasn't there long, however, and now jokes that he holds the record for the shortest time in the house owing to his lack of training base. "That was rough because what the Hansons have going on is great," Mackey told me. "I just would have wasted their money."

Armed with a master's degree, Mackey moved back to Illinois to teach anatomy and physiology classes at Parkland College and ran professionally for the Saucony shoe company, while trying to find his dream position as a professional running coach.

He applied to more than 213 coaching jobs before he hit send on two applications at Nike headquarters: one was for a biomechanist and the other as a perception researcher.

The Nike job was a long shot, Mackey admits, likely harder to get than coaching jobs he'd already been denied. Plus, he didn't know anyone who worked for the corporation and he'd never even been to Oregon when he applied online.

Five months later, Nike called, and Mackey began working on the Beaverton campus in the fall of 2007 as a perception researcher, which operated on the psychological side of shoe innovation. Though the pay was shockingly low, at just $45,000 a year, it was his dream job, working for the brand that had sponsored his childhood heroes, the Bulls guard Michael Jordan, and the Cubs second baseman Ryne Sandberg. Campus was lush and beautiful, professional athletes came and went, and in his role, Mackey worked closely with many of them. "If I saw LeBron James I wouldn't freak out, though I did once get in trouble with Michael Jordan," he said. "I kind of lost it a little bit, but that is a childhood idol of mine."

Mackey had qualified to compete in the Olympic Trials in the marathon and was still sponsored by Nike rival Saucony. The morning before his second day on the job he went for a run on the wood-chipped berm trail that surrounds the campus, over the white bridge with the orange swoosh on it. When he arrived at work that day he was pulled into the office of his boss's boss, the Nike Sport Research Lab director, Mario Lafortune.

"Hey, were you running around campus this morning?" Lafortune asked.

"Yeah," replied Mackey.

"What shoes were you wearing?"

"Saucony."

"You can't wear that."

Before taking the job, Mackey was just barely scraping by as a part-time teacher. "I was broke, so I told him, 'I don't have any money and I'm running like a hundred and thirty miles a week.'"

"I don't care. Figure it out," Lafortune said.

Mackey let Saucony know he couldn't run for them anymore, and thought, as an elite athlete, maybe his new employer would be excited to help him with his dream of making an Olympic marathon team. He reached out to Josh Rowe, who worked for John Capriotti in the sports marketing department, hoping to score some Nike clothing and shoes to run in. Rowe never responded.

"My friend from Naperville Running Company, Kris Hartner, ended up giving me some Nike gear from his running store," said Mackey. "I was kind of shell-shocked because I was twenty-six years old and broke, but I thought, *That's cool, Nike really takes their brand seriously.* But, obviously there can be a dark side to that too."

While at Nike, Mackey helped Chris Cook, a coworker and senior developer, resurrect the Bowerman Athletic Club, named for the now iconic track coach who passed away in Oregon in December 1999. Nike was everything Mackey dreamed it could be. The lab was fully funded with every physiology and biomechanical measuring device and tool money could buy. "I was totally drinking the Kool-Aid," Mackey told me. "I really liked working there."

He would work his way up into a lead position for the athlete insight group within the Nike Sport Research Lab. Part of his job description included servicing athletes who had issues or injuries. Mackey would adjust their shoes or their equipment in order to get them back on the field or court or track. "For instance, if someone had an Achilles injury, or say, Tiger Woods had something he needed adjusted, he'd come in and we'd do an assessment and we'd help them out."

Over time, however, the sheen wore off a bit and Mackey found Nike to be an extremely political and cutthroat work environment. He realized, for instance, that if he came up with an innovative idea he should keep it to himself. Nike is known for patenting everything that is conceivable in the world of footwear ideation, and Mackey worried that his patent ideas might be stolen.

Eventually, Salazar got wind of just how fast Mackey was and offered to help the young marathoner and fellow employee. Pulling him into the Oregon Project fold, Salazar gave Mackey an altitude tent to take home and sleep in, the stress from which would cause his body to increase its production of the endurance-boosting red blood cells. But the tent proved too much for Mackey, who was also running 130 miles a week and working a stressful full-time job. He started to feel lethargic.

In a meeting, Nike physiologist Loren Myhre asked Mackey how his training was coming along. Mackey told him how he had begun to crater. At twenty-six, he was otherwise perfectly healthy, but he was exhausted and his blood work showed an underactive thyroid and low testosterone.

Myhre suggested he see one of the team doctors, either Dr. Jeffrey Brown in Houston, or, locally, Dr. Kristina Harp, and get on thyroid and testosterone medication. Mackey, taken aback, asked, "Isn't that cheating?" Myhre brushed it off, saying that Alberto does this with all his athletes, and that they would use just enough to "get you into the normal range."

Mackey's busy work schedule kept him from traveling to Texas, so he visited Harp's office south of the city. She prescribed him both thyroid medication and testosterone.

Leaving the office, Mackey couldn't shake the feeling that this was cheating, so he made an appointment with an endocrinologist. This doctor told Mackey, "Yeah, you can't take this. You're just overtrained."

Desperate for answers, but fearful of losing his job, Mackey confided in another doctor, Cory Hart, who had experience working with professional cyclists. Dr. Hart told him that what he'd been asked to do was called "micro-dosing," referring to the process of taking a smaller quantity of a drug in an effort to gain a performance boost, but not one so egregious that it would cause a failed drug test. (A process that Victor Conte has said was "like taking candy from a baby. That's how easy it is for smart chemists and advisers to circumvent WADA testing [with micro-dosing].")

"If that is what the physiologist is saying Alberto does with all his athletes, then something is going on there," said Hart.

Mackey started to pay more attention to what was going on around him. "Keep in mind," Mackey told me, "I was working at Nike. I thought everything they did was awesome. I was a huge fan of Alberto Salazar. And I thought I was going to work there for the rest of my life."

Then he saw the blood panels.

Before the Trials, Dr. Myhre was showing Mackey how to review athletes' blood work. Any value out of range appeared in bold text on the paperwork so they could be easily identifiable. Mackey remembers two of the athletes' testosterone values being flagged as too high. Knowing this was likely additional evidence of anti-doping violations, he called USADA and left a message on their hotline. "It's crazy listening to it now," said Mackey. "My voice is shaking, and I say, 'I've seen this stuff and I don't know what to do with it.'"

LATER IN THE YEAR, MARION JONES'S INVOLVEMENT IN A CHECK-FRAUD SCHEME would trigger her downfall in the BALCO drug case. On Friday, October 5, 2007, Jones pleaded guilty to making false statements to IRS Special Agent Jeff Novitzky when she repeatedly claimed that she had never taken PEDs. Her ex-husband C. J. Hunter testified against her, leading to Jones's conviction on two counts of perjury. She was sentenced to six months in jail.

By December the IOC stripped all of her Olympic medals and vacated her performances. Lamine Diack, the president of the IAAF, said that Jones would be "remembered as one of the biggest frauds in sporting history." Conte pleaded guilty and eventually served four months in prison and four under house arrest for charges of distributing steroids and money laundering.

With the 2008 Olympic Games approaching, analysts were breathlessly predicting a resurgence of American men on international podiums. The US Men's Olympic Marathon Trials were held on November 3, 2007, in New York City. The results of the event determined qualification for the American Olympic team at the 2008 Summer Olympics, held in Beijing, China.

Not only was a chance at Olympic glory on the line, but the race offered the athletes a total of $250,000 in prize money in addition to covering travel and lodging costs for those who achieved the Olympic Trials A qualifying standard. For an American athlete to get the chance to line up in Beijing in August, they not only had to finish the race in one of the top three spots, but they also had to run the Olympic A standard of 2 hours and 20 minutes, or faster, for the 26.2-mile distance.

The New York City Marathon Trials course started at Rockefeller Center and took the men through five loops of Central Park. Nike had a full squad of their sponsored athletes lining up, including Dathan Ritzenhein, Meb Keflezighi, Alan Culpepper, Josh Rohatinsky, Dan Browne, Abdi Abdirahman, Matthew Gonzales, Jason Hartmann, and Mbarak Hussein. Nike employee Danny Mackey also lined up, swoosh-adorned from head to toe.

"I felt ready to run two hours and seventeen minutes to two hours and eighteen on that course," said Mackey, "but my stomach had a different agenda."

After all the spilled ink and hand-wringing about the resurgence of American distance running, two of the Big Three—Ritzenhein and Ryan Hall—seemed poised to make a mark. Although they would be competing only against their fellow Americans, this had to be the first step toward an Olympic medal in the event and their times would be a meaningful barometer of whether an American would be a medal contender in China or not.

On a course that many thought was slow and difficult, it was Hall, the twenty-five-year-old ASICS-sponsored athlete, who would dominate the day. He led the race in its entirety with Ritzenhein gradually disappearing from view as the miles ticked by. Ritzenhein would run a personal best time of 2 hours, 11 minutes, and 7 seconds to finish in second place, behind Hall's 2 hours, 9 minutes, and 2 seconds. Hansons-Brooks athlete Brian Sell rounded out the

American 2008 Olympic Team, finishing third, 33 seconds behind Ritzenhein. Danny Mackey's stomach contributed to his eighty-fifth place and a time of 2 hours, 28 minutes, and 45 seconds. (Later, ASICS athlete Deena Kastor would win the women's Marathon Trials, in a race in which not a single Nike professional runner made it to the finish line.)

12

AM I WORKING FOR THE NIKE MAFIA?

KARA'S POTENTIAL SEEMED LIMITLESS IN EARLY 2008, AND THE GOUCHERS WERE steadily becoming the "first couple of distance running." They were sponsored by the biggest athletic brand on earth, paid handsomely to run, and were palpably in love. With the US Olympic Trials beginning at the end of June, *Runner's World* put the Gouchers on the cover of that month's issue. The couple, legs shaved and scantily clad in their small Nike kits, looked vibrant and happy.

Kara had started the year off with the fastest mile of her life, clocking 4 minutes and 36.3 seconds to beat Sara Hall in a thrilling race at Madison Square Garden in February. In June, with just over a month to go until the Trials, the Oregon Project headed back to Park City for what was becoming their yearly pilgrimage to the Beehive State. The training camp served as a way for them to get away from the minutiae of daily life in Oregon and focus exclusively on training. The small ski town, up-canyon from Salt Lake City, was ideal because of its altitude, eight thousand feet above sea level. At this time, the team included Adam and Kara, Josh Rohatinsky, Amy Yoder Begley, and the wunderkind, Galen Rupp.

Before they left Portland, Adam had invited a friend named Pete

Julian out to help with the incidentals of the training camp and to "hold a stopwatch for the team." Julian had become friends with the Gouchers through the running community around the University of Colorado, where his wife, Colleen, ran in college. Julian, who was thirty-seven years old, had grown up in Ashland, Oregon, where he was part of a high school state championship team coached by his father. In college he was a four-time all-American (three in track, one in cross-country) at the University of Portland. Julian then became a professional runner, racing for Adidas for the next decade. He made the US World Cross-Country team in 1997 and 1998, and took bronze at the 1999 Pan Am Games in the 10,000 meters.

By 2008, Julian had transitioned into coaching and was running the upstart program at Metropolitan State College in Denver. As the head track and cross-country coach, he had summers off, which meant that when Adam called, he jumped at the chance to help out with the vaunted Nike team. Plus, after the year he'd had—one in which he raced-directed the US Cross-Country championship, was diagnosed with stomach cancer, and adopted a child from Nepal—he welcomed the change in pace and scenery.

He was still recovering from his third cancer surgery when he drove by himself from Colorado to Utah. The procedure had only been partially successful, as doctors removed some, but not all, of the gastrointestinal stromal tumors. "I was really having a hard time," Julian told me. "And for me it was an escape. It was fun because Adam and Kara were close friends." There also seemed to be a kinship between him and Salazar around the health scares that had plagued them.

The team stayed at the exclusive Black Bear Lodge in the Silver Lake Village, at 8,100 feet and just south of downtown Park City, where rates average $200 in the summer months and $2,000 during peak ski season. Julian shared a condo with another team assistant, Julius Achon, a retired middle-distance runner with a harrowing backstory. Growing up in northern Uganda, he was abducted by the Lord's Resistance Army and put on the frontline of his country's bloody civil war at just twelve years old. He eventually escaped and a year on entered and won his first official race, barefoot. Achon would go on to win the 1,500-meter Junior World Championship in 1994. He then attended George Mason University where he ran for John Cook, who later hired him as an assistant coach for the college team. It was Cook who helped get him a job with the Oregon Project, pacing the athletes through their workouts.

The group fell into a rhythm while in the village: eat, sleep, train, repeat. Salazar had an AlterG shipped out from Oregon and set it up in one of the rented condos where the athletes could do their second or third runs without the extra pounding, before getting massages and having dinner. In their free time, they'd cook and watch TV; Rupp would play a Nintendo Wii or would fly a miniature helicopter he'd received for his twenty-second birthday around the condo.

A couple times a week, Salazar would take Rupp down to Salt Lake City, though the reason for these trips wasn't disclosed to the rest of the group. Most of the team assumed it was to get body fat measurements, blood work, or possibly allergy tests done—though the frequency didn't seem to add up. Either way, there was a wall of separation between coach and protégé and the rest of the group. Supplements: Rupp got them first. Blood work: Rupp got it more frequently than the rest of the team. In some instances, Salazar himself gave Rupp his massages, a fact that baffled the other team members. The Oregon Project employed some of the best professional massage therapists in the world, why would his coach need to, or want to, do it himself? This pairing lent itself to endless snickering from the rest of the team, who repeatedly used the word "weird" in describing the relationship.

"Tomorrow, I'll go and be paced by Galen Rupp," Achon would joke. "*No one ran in front of Rupp.*" Achon had quickly learned not to challenge Rupp in the workouts, a transgression for which other pacers had been fired.

A distinct pattern seemed to be emerging as the eighth year of the program approached: a certain dysfunctional team structure where Salazar answered to no one. The only metric applied to his efforts were whether or not his athletes were winning medals. Similar to Nike's corporate culture, anyone in the Oregon Project who raised concerns or had objections was labeled a cynic and a nay-sayer. If they persisted with their complaints, they were usually marginalized, outright fired, or simply weren't ever called back. There was a constant pruning, of athletes, of doctors, of therapists, of journalist—though sycophants remained.

For those who endured, despite the turmoil, times were good. Even with the limited amount of success, Nike was still pouring north of a million dollars into the program each year (not including the athletes' salaries).

Kara had grown to love Salazar like a father, while Adam was finding it more difficult by the day to contain his distaste for the man. Part of the problem for Adam, whose sense of fairness categorizes all black or white, good or evil, was what he saw as flagrant and repeated NCAA rules violations by Salazar and Rupp. Salazar paid for Rupp's lodging, flights, meals, and incidentals, said Adam. And in turn, Rupp was essentially an amateur athlete living like a professional. While getting groceries in Park City, for example, Rupp put his food down on the belt next to Salazar's Corona beer, and the coach paid for it.

"Okay, I'm sure that's not a violation," Adam said in the store loud enough for everyone to hear.

"Alberto hated Adam by '08," Kara told me.

"I despised Alberto by '08," Adam added.

In early June, Kara and Amy Begley were in the Salt Lake City airport waiting to board their flight to Portland for the 34th Prefontaine Classic at Hayward Field, when they met a couple, John and Sarah Stiner. The Stiners, who were massage therapists and running enthusiasts, were also headed to the event to spectate. By that time, the team had gone through about five sports massage therapists already that they didn't like, so Kara wisely took the Stiners' business card.

Once they were back in Utah, Begley texted John Stiner to expect a call. The massage therapist could not have been more excited. John was a running super-fan who knew well Salazar's athletic accolades. In 1984 he skipped his own college graduation to watch the US Olympic Trials in Buffalo, New York. He climbed a tree at the race's seventeen-mile mark and took a photo of the lead pack as they streamed past, a sweaty Cuban-American leading the effort. At Salazar's induction into the distance running hall of fame in 2000, John had a friend take his photo with him to get it signed by the famed runner. As far as he was concerned, a chance to work with Salazar's group and the Nike athletes was a one-in-a-million opportunity that he could not pass up.

Sitting on the couch in Park City's Massage Now clinic, where he worked, he got the call. "If Obama had called me, I wouldn't have been more thrilled," John told me.

Salazar asked the couple to come up to Park City and work on the athletes. It would be a test run of sorts, with no commitment, but if they were exceptional at their craft there was the possibility of becoming part of the elite team.

"I'm just assuming you might be better than your wife," Salazar told John, "so you'll work on Kara Goucher."

The couple packed up their tables and headed to the Black Bear Lodge. Salazar met them inside the parking area under the building and helped them bring their equipment up to the condos. Rupp was there relaxing, curled up on the couch watching baseball in front of big sliding glass windows that opened up to the surrounding mountain panorama. John put sheets on his massage table and immediately got to work on Kara, while Sarah treated Begley, before working their way through most of the athletes that first day. After leaving later that night, John, in his excitement, took a photo of the check that Salazar used to pay them. The couple worked with the Oregon Project every other day for the next four weeks.

"You could see through Adam's skin, that's how fit the guy was. He was the leanest guy you'd ever meet. He was really hard, super stiff, hard to work on," said John. "Rupp was like working on a deer, he was so soft, like working on a woman."

After some time with the athletes, John told Salazar that in his professional opinion the runners weren't getting enough rest. He could tell by their skin quality. Stiner told the coach that they needed a better hydration strategy and to turn the altitude tents down. In the summertime, when air temperatures rise, the small enclosure of an altitude tent can become unbearably stuffy and hot, which would have an adverse effect on sleep—the foundation of recovery, according to most specialists. Without adequate sleep, for instance, sickness, hormone dysregulation, and overtraining are all more likely. Plus, the body recognizes oxygen deprivation as a stress, and when running 100-plus-mile weeks, this additional insult can push an athlete over the edge and into injury or overtraining. For these reasons, the Gouchers began unzipping their altitude tent at night without mentioning it to Salazar.

The day after he'd told Salazar that he was worried about the athletes' recovery, John arrived at the condos with a case of bottled water on his shoulder. As he slammed it down on the counter, Salazar looked askance. "He couldn't handle the fact that I had a strong personality, but, truthfully, I didn't care," said John. "I was about the quality of the care for the athletes themselves."

It was obvious to John that Salazar was obsessed with the athletes' weight. "He would pull out the scale and weigh them," said John. "I remember him busting Rupp's balls about not being light enough."

Still, the two men were friendly, and Salazar would often offer John one of his Coronas promptly at five in the afternoon, when he twisted the cap off his first. And although the massage therapist is a self-professed "big beer drinker," he always declined. He intended to keep this relationship professional.

Salazar could be capricious, but saved much of his ire for Begley, who could never be skinny enough in her coach's eyes. The previous March, after her dog died, he told her that her sadness was bringing the rest of the team down. A month after that, he warned her not to laugh at practice because it was annoying to him and others on the team. He even asked her to sign a contract that said she would not attempt to become friends with the other Oregon Project runners. They were to be seen as "business acquaintances" only.

At this training camp in Park City, he made Begley get on the scale in front of both Adam and Kara. At five-foot-four, Begley usually weighed between 106 and 116 pounds, and she'd performed well at both ends of the spectrum. When Salazar saw that she was 116, he berated her, saying that she'd blown her chance at making the Olympic team, as though this one metric was a tell-all for performance. She left the room, and when Salazar didn't have Begley's ears to voice his complaints, he made sure the Gouchers heard him loud and clear. (Begley later told the *New York Times* that Salazar's opinion could change in a matter of days. "If I had a bad workout on a Tuesday, he would tell me I looked flabby and send me to get weighed. Then, three days later, I would have a great workout and he would say how lean I looked and tell me my husband was a lucky guy. I mean, really? My body changed in three days?")

"Amy was treated so terribly. I was relieved it wasn't me, but I look back and I'm disappointed at who I was," said Kara. "It was everything being on that team—every aspect of your life controlled—and I was just relieved that I was the chosen one, that I was the favorite and she wasn't."

As camp wound down, and everyone headed to their respective homes, Kara and Begley told the Stiners the news. "Did you hear?" they said. "Alberto said, 'Clear your schedules.' We're taking you guys to Eugene with us for the Olympic Trials." Salazar then called John Stiner to let him in on the excitement. "We're going to be bringing you guys in," he told him. "We'll make the arrangements. You're gonna be part of the crew, but first, I need you to clean up a few things at the lodge."

He told John they had left some items in Park City that they'd need to ship back to him in Portland. The next day, the Stiners drove to the condos and started boxing everything up. The place was littered with Rupp and Salazar's belongings, which made the room look like the inhabitants had been abducted in the middle of the night.

Salazar called to walk him through the condo to make sure he got everything. "Listen, I don't want you to get the wrong idea,"

Salazar said, "but in the bedroom is a tube of AndroGel. It's for my heart; it's all fucked up" (Salazar denies saying this but does not deny using the topical testosterone cream).

Under a two-tone Nike T-shirt John found the tube of AndroGel, a steroid lotion and a (primarily male) hormone that is applied to and absorbed through the skin for people with serious medical issues. The word "steroid" has a connotation of powerlifters and bodybuilders adding mass of muscle, but for endurance athletes, the drug is illegally used to recover faster and feel better after hard training.

Salazar then told him that there were vials in the refrigerator that he needed to keep cold through the shipping process. No problem, John said, but his mind thought otherwise.

In the refrigerator John found a small Cordura cloth bag with two detents used for holding vials securely. Inside were two small glass vials with what appeared to be homemade labels on them. Before getting off the phone, Salazar reminded John two or three more times to be sure the vials stayed cold. "There is no drug information on it," John said. "There is no human being's name on it. There is no indication of what this is or how to use it. There is no doctor, no patient, no pharmacy name, nothing." White stickers on the vials had typewriter text that read "allergy one" and "allergy two."

"I don't know what it was," said John, "but my mind immediately went to . . . this might be EPO, because EPO needs to be refrigerated."

In the bedroom and bathroom that Rupp and Salazar shared, he found a slew of other random substances, including stool softener, vitamins, and an ergogenic drink mix the team was using. There were a few green bottles with commercial labels on them that read "Alpha Male," a product made by Biotest. John looked at the pills; they were green too. He found Rupp's wallet with $200 to $300 in it, then packed up the young runner's toy helicopter and Nintendo Wii as carefully as he could. In the dryer was Rupp's green-and-yellow Oregon running uniform.

On the bathroom counter, under a towel, John found an unopened clear plastic bag of at least a dozen hypodermic needles.

"When I saw that bag of syringes, I thought, *Jackpot*," John told me. "What the fuck is this? This is a Lance Armstrong story, right here. Have you ever known a track coach who uses needles with his athletes?" He and his wife were careful with the vials and wrapped them in dry ice at the Pack & Ship in Park City.

Later, Salazar called to say that they had arrived still cold, thanked the Stiners for their efforts, and let them know that he would arrange the details of their trip to work with the team at the Olympic Trials in Eugene.

The couple was excited to work with the most prestigious program in the sport, but John's curiosity was piqued. *What was in those vials*, he wondered, *and what's Alpha Male?*

Researching some of what he'd seen online, John quickly realized that for men with heart issues testosterone was a steroid to stay far away from. Reputable websites called it "contraindicated," he told me. (Chicago-based AbbVie Incorporated manufactures AndroGel and lists among its serious side effects "possible increased risk of heart attack or stroke.")

"I was loyal at that point, but my mind was unsure," said John. He noticed the book Salazar was reading was about the mafia. "I thought to myself, *Am I working for the Nike Mafia?*"

WITH THE US OLYMPIC MARATHON TEAM DECIDED, THE OLYMPIC TRIALS FOR THE REST of the track-and-field events began in Eugene on June 27, back at Hayward Field—events that would double as the 2008 USATF National Championships.

Nearly a year after Salazar was first felled by his heart stoppage, Nike hired the University of Wisconsin–Madison coach Jerry Schumacher as his successor. Schumacher's cross-country team had won the 2005 NCAA title and finished second on five different occasions.

"I'm looking to the future," Salazar told a reporter from the Eugene *Register-Guard*. "I have life insurance for my family, and now, with Jerry coming to Portland, I have coach insurance for Galen."

In Eugene for the Trials, Salazar instructed the Oregon Project runners to lie low and not talk to the media. Salazar and the team's sports psychologist, Darren Treasure, told them to put away their cell phones and handed out burner phones that only contained their personal contact information. Kara grudgingly powered down her first-generation iPhone. Later, she received a text on the burner that appeared to be from the *Oregonian*'s track-and-field reporter.

"Kara, this is Ken Goe, can I just talk to you for a few seconds for a quick quote?"

"Oh, I'm not talking to media right now," Kara texted back.

"Good job, you passed, this is Alberto!"

Adam, who was still working through myriad injuries, managed to put up impressive times in training and harbored hopes that he'd qualify for the Olympics in his two best events: the 5K and the 10K.

It was Kara who would first test her fitness in a final at the hallowed track at Hayward Field, however, lining up against the American record holder at the distance and fellow Nike athlete Shalane Flanagan. Kara placed second to Flanagan, who won by more than three seconds. NOP runner Amy Begley, who was almost dropped just a few laps into the race, came roaring back to take the third spot and complete a Nike sweep of the Trials, and a Nike-only Olympic Team for Beijing (Reebok athletes were fourth and fifth).

Adam advanced to the finals of the 5,000 meters by running 13 minutes and 56.25 seconds in his early heat. Now the pressure was on. He needed to run the Olympic A standard time, so he pushed the pace for several laps. But as the clock came into view midway through the race, he realized that he wasn't going to make it. He made the tough decision to jog off to the side as the race steamed on, wishing to conserve energy for the 10,000 meters.

Amid these highs and lows were the Stiners. Although there wasn't yet enough work for them to be hired full-time, there were rumors that the team would soon expand with Jerry Schumacher joining the Project. Then, the couple would have their hands full with work. There was even talk of Nike moving them to Portland to work with athletes.

Starstruck, the Stiners were now on the other side of the rope at the biggest race of the year. They were surrounded by America's best runners, both past and present. Carl Lewis and Michael Johnson were there walking around.

Salazar was, at times, stern with the Stiners. During the Trials he told John not to talk to the athletes during treatments. "He was really a freak, like . . . he turned," said John. "He acted all pissed for days."

On the day of the event, a sixteen-year-old Californian with a long ponytail bouncing behind her like a diminutive superhero's cape captured the hearts and minds of the Hayward Field crowd. In the 1,500-meter semifinal event, Jordan Hasay came from behind after the final bell, to finish fifth place, advance to the final, and set a new high school best time of 4 minutes and 14.50 seconds. As she posed with the massive time clock, the electrified crowd started to chant, "Come. To. Oregon." She finished tenth in the final, however, and would have to wait for her Olympic debut.

In the 5K event later that day, Kara, heeding Salazar's advice, went out conservatively and stayed in the pack until just before the twelve-minute mark of the race when Shalane Flanagan took the lead. As the US record holder in the event, Flanagan was the odds-on favorite to take gold. As she pressed the pace, Adidas athlete Jenn Rhines went with her, and Kara fell in behind her in third, instantly shattering the group and the Olympic dreams of those left in their wake. As late as the last lap, it looked like Kara might fall off the pace, but with about 150 meters to go, as the last turn straightened for the finish line, she flew by Flanagan, who couldn't respond. Rhines went with her and passed Flanagan too. Kara won the Trials and became the national champion in the 5K. One second later, Rhines finished in second, with Flanagan in third, to make up the US Olympic team headed to Beijing.

Still later on the same day, three Oregon Project men—Galen Rupp, Adam Goucher, and Josh Rohatinsky—lined up for the Olympic Trials 10K: Rupp had withdrawn from the 2008 collegiate track season—a process called "redshirting"—to focus on this race. On the track he was stolid and reserved. He possessed a sweetness, or at least a naïveté, that seemed positively contrary to the aggression required to outrun the most imposing athletes in America. All of the Nike athletes wore blue-and-black kits with the exception of Rupp, who lined up in his University of Oregon yellow and green, with a matching nose strip. He ruffled his poofy blond hair and waved to the hometown crowd, most of whom stood and applauded. Was he the most promising distance runner since Steve Prefontaine? The world was about to find out.

Tactically, Rupp seemed out of his league as the race progressed, and some commentators quipped that he still looked like a high schooler nervously trying to cover every move, no matter how early or inconsequential. But he was fit, there was no denying that, and he remained in the lead group, which also consisted of Abdi Abdirahman and Jorge Torres. When Rupp finally took the lead in front of a record crowd at Hayward Field, the 20,936 fans went insane with applause for the homegrown talent. He had taken the lead but hadn't pressed the pace. This allowed Abdirahman a respite before his final push for the finish line, passing Rupp on the last lap to win the Trials and the national championship. Rupp finished second, Torres in third—the three of them headed to the Olympics in Beijing. Adam, despite running a new personal best—by less than a second—finished in seventh place. His Olympic dreams dashed, once again.

"It was devastating," Adam said after the Trials. "It's hard to explain, to put into words, how hard we work, how intense the training is, how much it means to us and how hard it is to see others all reach their goals, and to be happy for them, but not be completely a part of it."

SHORTLY AFTER SCHUMACHER ARRIVED IN OREGON, ADAM APPROACHED HIM WITH A proposition. "I reached out to Jerry and asked if he'd coach me. He said, 'No.' He didn't want to piss Alberto off,'" said Adam.

"Salazar doesn't put up with that stuff. You either kiss Alberto's ass or you are off the team. It's as simple as that, and I wasn't going to kiss his ass."

Back in Beaverton between the Trials and the Olympics each member had a weekly chart to fill out with all the ancillary activities they had to accomplish. They were responsible for a certain amount of time exercising on the elliptical machine and the stationary bike, as well as stretching and strength training. The routines had names like Pilar, Waterloo, and Beton. At the end of the week, the athletes would hand in a sheet to Salazar listing exactly what they had done.

"It was supposed to make us feel like we had autonomy," said Kara, "but not really because he was standing over us telling us what to do most of the time."

Rupp was improving rapidly while Kara was becoming the running world's new darling, but Adam was the ornery athlete who challenged Salazar at every turn.

Treasure, who still wasn't licensed as a psychologist in Oregon, began attempting to pit the girls against each other with half-truths and manipulations. He told Kara that Begley was resentful and jealous of her success. He told Begley that Kara didn't want to share a room with her in the Olympic Village in Beijing. (Treasure did not respond to requests for comment.)

In China, since he didn't make the team, Adam had to leave Kara at the gate to the Olympic Village, which permits Olympians only. Isolated and overwhelmed by the pomp and circumstance of the largest sporting event in the world, she felt the mental fortitude she'd worked so hard to attain begin to slip. *I don't belong here*, she thought. *I'm not good enough to be here.*

Inside Beijing's Bird's Nest stadium, she had to fight back tears as her mind continued to reel with self-destructive thoughts. When the gun went off for the women's 10,000-meter final, the pace was torrid. Kara was caught off guard. Salazar had prepared her for more of a sit-and-kick race, but Tirunesh Dibaba and Elvan Abeylegesse flew around the track, just as they had at the 2007 IAAF World Championship, to finish one and two. At worlds in 2007, Kara had finished third, but today—even though she ran a personal best of 30 minutes and 55.16 seconds—she was the tenth athlete across the line in a race that was the second fastest women's 10,000-meter of all time. Dibaba ran a new Olympic record and lapped some of the field—including Oregon Project athlete Amy Begley—for the gold. Shalane Flanagan ran a new American record time for the 10,000 meters of 30 minutes and 22.22 seconds to finish with a Bronze medal.

"I didn't race well," Kara told the *Duluth News*. "I'm disappointed. I feel I let a lot of people down."

"We really made some bad choices in Beijing," Treasure admitted later. "Alberto had trained Kara for a specific race—a slower race—that just didn't end up happening. It was a lot quicker than we'd anticipated."

A few days later Salazar, still upset with Kara's tenth place finish, told her, "You're just not good at track."

"That's where my rift with Alberto started, right there," said Kara. "I'm at the Olympics and I have the 5K in three days, and he tells me I'm no good at track. So . . . what, should I just go home?"

According to Begley, after the race Salazar told her that she was too heavy, and that her previous success on the track was a fluke. "He said that to be on the elite level I needed to weigh less than I did," Begley said. "I weighed one hundred and fourteen pounds."

On August 22, an emotionally depleted Kara ran an uninspired 5,000 meters in the women's final event and finished in eighth place, this time beating Flanagan, but nowhere near Tirunesh Dibaba, who won her second gold medal, pulling off the double in the 5,000 and 10,000 meters.

In the men's 10,000 meters it was the three-time world champion, two-time Olympic champion, Ethiopian Kenenisa Bekele, besting the field with his finishing kick for the gold medal. Rupp finished thirteenth in the race, but his time of 27 minutes and 36.99 seconds was a new American Olympic-record time.

After failing in his debut in 2006, Ritzenhein was looking for marathon redemption at the Beijing Games. He was still a hot prospect, but the twenty-five-year-old suffered calf pains during the race and would be seen stopping on course to stretch them out, to no avail. He finished far out of medal contention in ninth place, in a race where half of the top ten places were occupied by Kenya and Ethiopia.

AFTER THE OLYMPICS, KARA'S FOCUS WAS NOW SQUARELY ON PREPARING FOR THE New York City Marathon. Volume was of the utmost importance to a marathoner, Salazar told Kara. He increased her mileage in the lead-up to her debut to a whopping 100 miles in a seven-day period, or almost four marathons per week. He also increased her weekly long runs, inching closer and closer to the 26.2 miles of the marathon. Kara loved that her coach also biked alongside her during these draining sessions, giving her form cues and keeping her on pace.

"No one else [on the team] was running the marathon at the time," said Kara, "and I loved that this was something that was all mine."

Her body seemed to respond, absorbing the new stress with few complaints or, at least, no full-blown injuries. The increased physical strain, however, brought her emotions closer to the surface. Though Kara now had some misgivings about Treasure's professional aptitude, she went back to work with him in rebuilding her mental game to be more resilient. Her new keyword was "confidence."

In New York for the marathon, Kara was outwardly the picture of a self-assured, professional athlete. She bounced around the city in the week leading up to the event, happily fulfilling media responsibilities and answering questions with aplomb. Inside, however, she was struggling to fight back the negative self-chatter, her mind ruminating on the fact that she had never run 26.2 miles before.

And Salazar, she thought, was just too much. At previous events, like the Olympics or the World Championships, Salazar had other athletes to obsess over. In New York, it was just Kara, and he stressed over every minute detail. Again, Salazar told her to shut out all distractions. Kara's mom, two sisters, brother-in-law, aunt, and niece had all flown in to watch the race and cheer her on. Two days out, Salazar limited her interactions with her family. Her coach's increased scrutiny acted to level up the stress, rather than decrease it.

The media exploited two storylines: a rematch with Paula Radcliffe, who was the returning champion going for a third New York City Marathon victory, and the fact that Kara's coach had won this very race in his marathon debut exactly twenty-eight years prior. Now it was Kara's turn. But back at the hotel, slumped in a chair, drained of energy and confidence, Kara told Treasure she didn't even know if she could finish the distance.

It was windy and dark on the morning of November 2 when Salazar and Treasure joined Kara on the shuttle bus to the starting line on Staten Island. "It's a little like the first day of school for your child," Salazar recalled after the race. "You're putting her on the bus with your fingers crossed."

Radcliffe, the reigning champion, assumed the lead at the gun. Kara tucked in behind the world record holder and mentally prepared herself to track her for the next couple of hours. Kara was so close, in fact, that when Radcliffe threw away a used energy gel packet, it hit Kara's arm on its way down.

Her mind was calm, and the pace felt surprisingly manageable. Her thoughts drifted to her father as they ran through the halfway mark in Queens, where he had lived. His untimely death just a week before Kara's fourth birthday had happened just across the river. Surprisingly, these thoughts brought an unexpected boost. *I'm going to make him proud today*, she thought.

Kara raced with what appeared to be the self-assurance of a marathon veteran, with the one exception of her water bottle pickups. Elite marathoners are allowed to leave special water bottles at aid stations throughout the course that include their own personalized nutrition. Kara's mother had decorated her bottles with her favorite skull-and-crossbones design (that would later become the Nike Oregon Project logo). But they had neglected to have Kara practice grabbing them at full speed and in full stride. She dropped the first one.

With twenty miles behind her and just six to go, she started to feel the new distance. Her stomach was upset. Her calves twinged with cramps. As she assessed her situation, Radcliffe began to pull away.

"I had to recoup," Kara said after the race. "I told myself, *It's a 10K. You can do this. Pull it together.*"

Confidence, she thought as she drove toward the finish line. *Confidence.*

When she crossed, the clock read 2 hours, 25 minutes, and 53 seconds. It was a new women's record for a debut marathon by a little more than a minute, and Kara became the first American to climb up on the podium since Anne Marie Lauck (née Letko) was third in 1994.

Adam, waiting near the finish line wearing a shirt that read "Mr. Kara Goucher" on it, embraced his wife after the race. Her circle of family and friends were ecstatic, but Salazar was disappointed.

Adam took a picture of coach and athlete sitting on a bench after the race, thinking it would be a great moment of posterity with mentor congratulating pupil, the next great American marathoner. "He showed me that photo after, and was like, 'This is so cute, Alberto talking to you about how awesome you did,'" Kara told me. "You know what he was telling me? Everything I did wrong."

Third place just wasn't good enough. Salazar changed his flight and flew home early. "He left New York because I didn't win," said Kara. "He changed his ticket and left the race right after. I was devastated."

The next morning, the Gouchers got a call from the hotel's front desk. They received a fax from Salazar. It was a graph of when, exactly, to have a baby and when not to, that read:

Option 1: You go through fertility treatment and have a baby.

Option 2: Adam takes a cold shower. Adam keeps taking cold showers.

"It was kind of funny," said Kara, "but it was also Alberto saying, 'You can't have a baby now.'"

13
LET'S RUN

NOW OFFICIALLY A MARATHONER, KARA SET HER SIGHTS ON THE STORIED RACE through Beantown on Patriots' Day. In March, in the lead-up to the 2009 Boston Marathon, Salazar and the Gouchers traveled to Portugal so that Kara could race the Lisbon Half, which would serve as an important barometer of her fitness.

She continued to work with Treasure on fortifying her mind against the internal assault of her thoughts. It was an ongoing process without a finish line, and there was no inoculation against what might be lingering online. Salazar warned her to stay off the running message boards, namely LetsRun.com: they could be a platform for cruelty, he explained, that for Kara, would be counterproductive to the work she was doing with Treasure.

Twin brothers Robert and Weldon Johnson started LetsRun in 2000, about a year before Salazar launched the Oregon Project. The BroJos, as they are known, created what would become *the* place on the web for the most obsessed runners and running aficionados to commune and talk shop (and sometimes talk smack). In the early years, the site comprised a mix of training advice, links to interesting running articles, the brothers' opinions on the track world, and the infamous message boards for the diehards (well-known runner and author Malcolm Gladwell has said that he starts his day reading the site, which has more than one million visitors per month). Presently, it's a place one can find both conspiracy theories and unfounded accusations alongside solid journalism and prescient predictions. Recent discussions on the boards include, "Why did Hall DNF? Was it cause she didn't have vaporflies?" "Anyone Who Wants Their Student Debt Cancelled is REPUGNANT," and "Why do many here hate on Kara Goucher?"

Salazar outwardly reviled LetsRun, and would remind the athletes to stay off it. But he had trouble following his own advice. According to Kara, whenever she would meet with Salazar in his Nike cubical, his browser seemed permanently stuck on the site.

He had another reason to keep Kara off the internet in the days before the Lisbon Half Marathon: he had said something that was being hotly debated online. Not realizing what the half marathon world record was, Salazar had told a reporter that Kara was more fit than her last attempt at the distance and therefore prepared to run under 66 minutes for 13.1 miles. Naturally, LetsRun reported it. The IAAF-ratified world record at the time was 66 minutes and 25 seconds, held by Lornah Kiplagat. If Kara ran under 66 minutes, as Salazar had said she was ready to do, it would be a new world's best time.

"So when we get to Lisbon," said Kara, "they are prepared for a record. All these reporters were asking me about the world record. They even have a giant check with my name on it and it says 'World Record' on the check. I was like, 'What the hell?'"

When she looked for answers from Salazar, he sheepishly admitted, "I must have said something." He was furious with LetsRun for being critical of his runner and possibly getting in her head.

On race day, Kara ran pretty close to world-record pace for nearly eleven miles, but eventually fell off the ambitious effort. She would go on to win, however, with a time of 1 hour, 8 minutes, and 30 seconds.

IT WAS A WINDY APRIL DAY IN 2009 WHEN KARA LINED UP WITH THE OTHER ELITE women in the field for the one hundred and twelfth running of the

Boston Marathon. She wore Nike NFL receiver–type gloves, a tight pink top with a race bib pinned to it that read "Goucher," and two press-on American flag tattoos on her right shoulder.

She felt great in the first miles of the race but gradually realized that the group seemed content to run a slower marathon than she had planned. She took the lead at the 10K mark. Twenty miles in, there were still twelve women in the lead pack and Kara still felt potent. Shifting into a new gear, Kara steadily increased the pace.

"You know how people do miraculous things when an opportunity is given to them?" said Kara. "I was like, 'I have to change this from the pretenders to the contenders.'" She pressed the group, flying down the street, leading the Boston Marathon.

In an egregious breach of etiquette, a race photographer riding along in the press truck yelled, "Kara you're running five-minutes pace, slow down." The comment sent a shot of recognition to Kara's brain. She had taken her mind off Salazar, and now she realized she was disobeying his advice. Doubt and fear overtook her.

While Kara eased her effort, Kenya's Salina Kosgei surged ahead before turning onto Hereford Street; Ethiopia's Dire Tune picked up her pace and covered the move. They charged through the streets of Boston together, with Kara drifting off the back. In the end, Salazar's strategy worked, just not for Kara. As the finish appeared, Kosgei and Tune sprinted for the line in the closest finish in the history of distance running's most fabled race. Kara faded out of focus, behind them by nine seconds. She collapsed into tears after crossing the finish line.

Salazar, who had no idea what had transpired on the course, was furious at Kara's race performance. Afterward he told her he didn't think he would be able to get over it. "That was my biggest crack with Alberto, when I didn't run Boston how he wanted," said Kara. "But I'm the one that ran, and I'm the one that put my entire soul out there. I felt so hurt. I'm living this life, one hundred percent dedicated and he's so mad at me after that."

To this day, she can't bring herself to watch the broadcast of the race, saying in our interview that this was the first time she'd ever talked about it without breaking down into tears and that she had to go to therapy over it.

To add to the disappointment, Kara knew she had let people down. Her brother-in-law worked at Nike and had let slip what they had planned for a media blitz upon her inevitable victory. In anticipation of a win, Nike had paid for the Gouchers to travel to New York City, where Kara was booked to be on the *Today Show.* They had a commercial ready and purchased full-page ads. They even planned to have the John Hancock building light up with news of the first US woman to win Boston since Lisa Larsen Weidenbach in 1985.

Instead, the couple flew to New York and sat with a few editors, which was now more formality than celebration. "The *Today Show* got canceled," she said, "because they don't want an interview with the person who got third at Boston."

A WEEK AFTER KARA'S RACE, RITZENHEIN WOULD TEST HIMSELF IN HIS THIRD MAJOR marathon, this time in London. He was once again almost five minutes behind the winner, Kenyan Samuel Wanjiru, finishing just out of the top ten in eleventh place. "London was really the last straw," said Ritzenhein after the race. "I thought, *I've put so much into this. Do I really want to be just mediocre?*"

In May, at the Oregon Twilight Track and Field Meet, Rupp said he was approached after his mile relay race by Chris Whetstine. According to Rupp, the well-known massage therapist rubbed his shoulders, ostensibly to say congratulations, but Rupp said he felt something wet on his back. As ludicrous as it sounded to most, there was still the conspiracy theory in the track-and-field community that Whetstine, in a vindictive act of sabotage, had caused Justin Gatlin's 2006 failed drug test by rubbing testosterone cream on his thighs after his race.

Salazar had been outspoken about how implausible a topical sabotage was back in 2006, but says he began to believe it was not only plausible, but probable by 2009.

When reached for comment about the alleged incident at the Twilight Track and Field Meet, Whetstine said that it never happened. "I wasn't even there," he said. "To my recollection, I've never been to a Twilight meet in my life."

Regardless, at their races, Salazar began insisting that the athletes never let their water bottles out of their sight or the sight of someone they "really, really trust"; otherwise they were to be locked in a cooler at all times. Any Oregon Project bottle found left unattended was dumped out. He instructed them not to give high fives to other athletes after races. He had Rupp keep his medications and supplements in a metal box when he traveled, so no one could spike them.

After the Twilight event, Salazar has claimed he put in a call to USADA's toll-free number, then emailed the CEO, Travis Tygart, to tell him what had happened, and that he was suspicious that Whetstine "could have possibly rubbed something onto Galen." (In an email response to this claim, USADA said that Salazar did send an email about Whetstine, but that they were suspicious about this claim for several reasons. They questioned the logic of Whetstine attempting to sabotage Rupp immediately after a track event, when "Rupp and his teammates were celebrating . . . in a public place when the teammates were being photographed after their record-breaking run." They then pointed out that after replying to his email, they never heard back from Salazar.)

In late June, the 2009 USA Outdoor Track and Field Championships took place at Hayward Field, the spiritual center of the Nike Corporation. "If you planted the Nike flag anywhere, you would have to, in my opinion, plant the Nike flag in the middle of Hayward Field," John Capriotti has said. "It is where Bowerman, the coach, and Phil Knight, the athlete, lived. It is also where Steve Prefontaine, an athlete who is very important to Nike, trained and raced and lived." With the crowd going berserk, Rupp launched himself into professional running by beating America's best runners—including Dathan Ritzenhein, who placed second—in the 10,000-meters event.

After the race, Ritzenhein made the tough decision to leave his coach of five years, Brad Hudson, and join his sponsor's Oregon-based project. By this point, success had come slow to Salazar and his quixotic program, but Nike was patient, and the coach was beginning to be lauded in the media as having built champions, from the ground up, of Galen Rupp, Kara Goucher, and Amy Begley.

Ritzenhein moved to Portland in early July with his wife and their toddler. Shortly after his arrival, Salazar showed the runner his basement pharmacy full of supplements. Some were good for fat burning, others for muscle building, Salazar explained, while others boosted testosterone. Bottles with names like Alpha Male and Testoboost lined the shelves.

After the public yet surreptitious shoulder rub at the Oregon Twilight Track and Field Meet, Salazar began working with Dr. Brown to test the efficacy of AndroGel. The men devised an experiment to use the topical testosterone on Salazar's sons, Alex and Tony (both of whom were full-time Nike employees at this time). To determine hormone levels pre-dosage, they would first have baseline urine samples tested. Then they would simulate race conditions by running 5K or 10K on the treadmill in a controlled 85-degree environment. After applying their father's testosterone cream to their skin, they would wait an hour before taking the second urine test. Comparing the pre- and post-dosage urine would determine if there was a spike in testosterone large enough to have caused a failed anti-doping testosterone-to-epitestosterone ratio drug test.

Despite the fact that both federal and state laws prohibit sharing prescription drugs, on June 30, Salazar rubbed AndroGel testosterone cream on his son Alex's bare back after he finished his treadmill run in an environmental chamber at the Nike laboratory. An hour later they took the second urine sample, and sent both to Aegis Labs for testing.

Tony had to be even more careful with the powerful steroid. He and his wife were either pregnant or trying to get pregnant at the time, and since testosterone creams can be transferred through skin contact, Dr. Brown was worried. The risks posed by possibly dosing a pregnant woman with the male androgenic steroid were real: an unnatural spike in a pregnant woman's testosterone could adversely affect a baby's development. If Tony were to leave testosterone residue on his clothing that she handled or towels and sheets they used, there was a chance she could absorb it transdermally.

Salazar first applied one pump of the cream, then two. Nike CEO Mark Parker had become interested in the experiment at some point and Dr. Brown began emailing him directly with details. On July 7, 2009, in an email to Parker he wrote, "We have preliminary data back on our experiments with a topical male hormone called AndroGel . . . We found that even though there was a slight rise in T/E ratios [testosterone to epitestosterone] it was below the level of four which would trigger great concern . . . We are next going to repeat it using three pumps."

In his email response, Parker wrote that it "will be interesting to determine the minimal amount of topical male hormone required to create a positive test."

After a strenuous basketball game on July 19, Tony was dosed with four pumps of the steroid cream, which only increased his T/E ratio from 0.8 before, to 1.4 after. A few days later, they tested Alex with four pumps of AndroGel. His tests showed an increased T/E ratio to 2.8, still well below the limit of the world anti-doping agency's drug tests.

After seeing the results, Dr. Brown wrote to Salazar, "Want to try 6 squirts?"

"I don't think it's worth it. The four squirts was an enormous amount that was easily noticed," Salazar wrote back, adding in a later email, "I'll sleep better now after drug tests at big meetings knowing someone didn't sabotage us."

Some of the details around the experiment remain murky, however. Salazar claims he used his own prescription for the tests. Dr. Brown testified that Salazar asked him to prescribe the AndroGel for testing purposes, but he refused. Instead, Dr. Brown told Salazar that it was up to him what he did with his own prescription, and that "I can't prevent you from doing anything."

To make the situation that much more confusing, Amy Begley and her husband, Andrew, told USADA that Dr. Brown sent Amy back to Portland, from an office visit, with a plain envelope that said "Alberto" on it, the same month Dr. Brown proposed running additional experiments. In subsequent conversations with Salazar, the Begleys remember the coach telling them that "the package that Amy had transported was the [testosterone] cream that he used on one of his sons to test it." In his most recent testimony, Salazar now claims to have "no recollection" of the Begleys bringing him testosterone.

That same month, Dr. Brown emailed Parker and Salazar to explain the increased risk of women being sabotaged. Testosterone's effects on female athletes, he explained, were enhanced—since a woman's natural levels are on average about ten times lower than a man's. He estimated that "probably as little as one or two squirts may well trigger a problem." In order to test this, he wrote, the men would need to create "a full fledged research protocol, secure volunteers, and get an institutional review to sign off on it." But Salazar didn't think it was worth it. Rupp was the priority, and they had ostensibly accomplished their goal of protecting him. Salazar and Brown had either figured out that the sabotage concerns were invalid due to the noticeable amount of cream that would have to be rubbed on an unwitting athlete surreptitiously, or they figured out how much testosterone cream could safely be used without triggering a failed anti-doping test.

He told the women on the team to start putting on long-sleeve shirts immediately after their races and to avoid contact with anyone.

FOR RITZENHEIN, THE NEW PROGRAM SEEMED TO SUIT THE TWENTY-SIX-YEAR-OLD well. In Berlin, Germany, for the 2009 World Championships in August, he finished sixth in the 10,000 meters and set a personal best of 27 minutes and 22.28 seconds, in a race where Kenenisa Bekele won with a championship record time of 26 minutes and 43:31 seconds. Zersenay Tadese of Eritrea earned the silver medal four seconds later and Moses Ndiema Masai of Kenya took the bronze. (Bekele would also win the 5,000-meter world title six days later.)

Anticipation was also high for the women's marathon scheduled for the last day of competition. Since moving up to the marathon distance, Kara had electrified the event, become a certified running superstar, and made the marathon must-watch television.

But during the race, she struggled with an ornery stomach that had her throwing up midstride. She finished tenth. In post-race interviews Kara seemed on the edge of tears. "I'm sorry, I'm gonna throw up," she said to reporters before stepping over to the trash can.

Adam stood behind her in case she went down. "I got you, babe."

"I'm seeing stars actually," she said before graciously admitting, "I wanted to deliver, but they were better. It was just not my day and they were better."

After going through the media scrum and drug-testing process, Kara sat with Salazar and Treasure at the hotel bar, still in her running outfit. According to her, the men carried on a discussion about Kara's talent and dedication as though she wasn't even there.

"Does she have it? Does she actually have what it takes?" Treasure asked her coach.

"No, no, she has what it takes," said Salazar.

Kara was speechless that they would openly question her dedication in front of her.

"At that point, I know that Alberto wanted me to run the New York City Marathon," said Kara. "But I was like, 'Yeah, I can't do this anymore, I feel like I'm hitting my head against the wall.'"

In late 2009, Salazar began focusing on fixing Ritzenhein. He thought the angle of the runner's pelvis was off, as was his arm carriage, which seemed excessive and a sign of wasted energy. Plus, his thumbs always pointed up, which Salazar saw as an unnecessary plyometric contraction that caused undue strain in the arm.

Internally, at odds with the suggestions, Ritzenhein swallowed his skepticism, saying, "Alberto told me, 'It's imperative that you believe completely in what we're going to do because it will be completely different from anything you've been taught.'"

Taking America's best through a full-scale biomechanical overhaul was an extremely hazardous endeavor. Many experts believe that athletes at this level, through years of training, have naturally found the most effective and efficient patterns for their individual physiology. But Salazar was nothing if not risky. He acknowledged that tinkering with Ritzenhein's form could cause a cascade of unforeseen downstream issues and possible injuries, but he was convinced that for Ritzenhein to bridge the gap to the next level, he'd have to take the risk. But Ritzenhein was already injury prone, having battled for years with stress fractures in his feet.

"To compete against the best, you've got to fix this," Salazar told Ritzenhein. "But there's a risk. We may injure you."

"I'm willing to take that risk," Ritzenhein replied.

THIS WAS RUPP'S FINAL YEAR AT THE UNIVERSITY OF OREGON. IT HADN'T ALWAYS been easy for the young Duck. Living like a professional athlete under the tutelage of Nike's high-powered coach meant he was often separated from and ostracized by the team. Although Nike had their guy Vin Lananna heading the program, Rupp's coach was undoubtedly Alberto Salazar, whom the runner trusted implicitly and had grown to love like a father.

Under Salazar's watchful eye, Rupp didn't partake in all of the team workouts as was normally required of a scholarship athlete. He slept in an oxygen-deprived environment. He took odd supplements and received regular shots at the health center for a phantom allergy condition few people had ever noticed him suffer from, for which he regularly also took the steroid prednisone, a drug so powerful it's used for cancer treatment and only prescribed for allergies in the most severe cases. (Athletes on prednisone report feeling invincible, owing to its pain-reducing, anti-inflammatory, and stimulating effects.) Rupp did workouts on a treadmill in his Eugene apartment with the heat turned up to 78 degrees. And as if the races weren't hard enough, Salazar would often have the runner do additional workouts, behind the stadium, immediately after dispatching the best young college runners in the country. It didn't make sense to the other coaches and athletes in the sport, who would frequently snicker about it, but the results were undeniable.

In his last year as a Duck, Rupp was spectacular. He won the individual NCAA Division I Championship for both the 5,000 and 10,000 meters at the outdoor track championships, helping his team secure another Pac-10 title. All told, he earned fourteen all-American honors and five individual championships, a relay championship, two NCAA cross-country team titles, and an indoor track NCAA team title. Rupp was also bestowed the inaugural Bowerman Award—along with Jenny Simpson (née Barringer) on the women's side—which serves as track and field's Athlete of the Year. Former Oregon Duck, teammate, and one of Salazar's best friends Rudy Chapa said Rupp's collegiate career was "going to be one of these things where years from now you look back and put him in a class, maybe not exactly with [Steve Prefontaine], but not far off that. If [Oregon fans] don't appreciate it now, then they will."

ELEVEN DAYS AFTER THE WORLD CHAMPIONSHIPS, RITZENHEIN CONTINUED TO ROLL, becoming the third fastest non-African of all time in the 5,000 meters. He'd only been on the team for three months when he ran 12 minutes and 56.27 seconds for third place at the Weltklasse Meeting in Zurich, Switzerland. Before the race, Salazar had preemptively warned Capriotti, who would be in attendance, not to get disappointed if their athlete ran much of the race in dead last. He thought Ritzenhein was fit enough to run around 13 minutes and 5 seconds. And as the race got under way Ritzenhein indeed found himself the last man in the group as they ran 4 minutes and 4 seconds through the first 1,600 meters. By the final 800 meters, however, Ritzenhein was closing on everyone, even the world record holder, Kenenisa

Bekele. He went from last place to third, taking down the thirteen-year-old US record by beating nine of the ten Kenyan runners in the race.

The American media immediately began heralding the performance as proof that US athletes, led by Salazar, were catching up to the Africans. And Salazar, who watched from his computer back in Oregon with Rupp, Amy Begley, and Simon Bairu, gave them the pull quote they needed: "In my mind, this shows that we're on the right track," said Salazar. "Americans are closing the gap. Now we have a guy who's right there with them . . . Brad Hudson made the cake. I'm just putting the frosting on it."

Ritzenhein wasn't done, however, and just six weeks later, in Birmingham, England, he finished third in the world at the Half Marathon Championship, running an astonishing 60 minutes for the 13.1 miles.

AT THE END OF 2009, THE SECOND OF THE BIG THREE WAS SUCCESSFULLY LURED TO Portland to train with Salazar. Alan Webb left his coach of ten years, Scott Raczko, to move to Oregon and train on the Nike campus. Webb played a large role in the renewed optimism around American running, but had hit a stagnant patch and decided on a big change.

Growing up in Virginia playing soccer and basketball, and swimming on the swim team, Webb wasn't the fastest kid on the playground, he told me, but he wasn't the slowest either. Like Magness, Webb excelled at the Presidential Fitness mile and set the school record for the distance in sixth grade and continued to do so through high school.

At thirteen, he tried his first road race at the Georgetown 10K. "I remember my dad saying, 'It's pretty long, don't go out too fast,'" Webb told me. "I had no idea how fast I was going, but when we got to a mile to go I was feeling pretty good and pushed it."

He won his age group, which provided an additional spark. He also won a raffle that day and received a gift certificate for shoes. At the local running store, he hemmed and hawed between Nike's

Zoom Airs and the Air Max. He was in middle school, so he told me the choice came down to whichever pair looked the coolest: he chose the Nike Air Max.

After high school, Webb committed to the University of Michigan. The summer after his high school graduation, he took a trip to Oregon with Raczko for a tour of the Nike headquarters. He met Phil Knight in the Joan Benoit building. "It was surreal, he and Bowerman started Nike," said Webb. "He was friendly and curious and seemed like somebody who really cared about his employees."

Later, Webb lined up against some of the best milers on the planet—including Hicham El Guerrouj, the world record holder at the distance—at the Prefontaine Classic. He laced up a custom pair of Zoom Kennedys that his Nike contacts had given him in the U of M colorway of blue and maize.

In those shoes and a blue South Lakes High School running kit Webb shattered the coveted thirty-six-year-old record for an American high schooler. With the Hayward Field crowd roaring with excitement, the young gun clocked 3 minutes and 53.43 seconds, surpassing the legendary Jim Ryun's high school best of 3 minutes and 55.9 seconds. Webb shared the victory lap with Hicham El Gerrouj.

The record changed Webb's life. He was flown out to New York City to be on *The David Letterman Show.*

"This must be the most exciting time in your life so far," said Letterman during the interview. "You're on top of the world, right?"

"I'm on the *Letterman Show*!" an excited Webb responded, astonished with where running had taken him.

After just one year of competing for the University of Michigan, he famously left college and relinquished his scholarship to turn professional in the summer of 2002. He transferred his studies to nearby George Mason University so he could return home to his high school mentor and running coach, Scott Raczko.

After a hearty pitch from Reebok, Webb eventually signed a contract with Nike in which they agreed to pay for his college education as well as a reported yearly salary of $250,000 for the next six years, plus bonuses for wins at major competitions. The deal made Webb officially ineligible to run in college meets but is widely viewed as the beginning of a new era of big contracts for runners, not yet commensurate with other high-profile sports, but a huge step in that direction. As part of the deal, Raczko was also signed by Nike for $25,000 annually.

In 2004, at twenty-one, Webb represented the US at the XXVIII Olympiad in Athens, Greece. There has never been an Olympic mile event, so Webb competed at 1,500 meters, the distance closest to the mile. He had won the US Olympic Trials in the distance, but didn't manage his energy levels that day in Athens. He lined up in the prelims with a pair of Nike Air Zoom Milers and fitness to burn, but the race became a sit-and-kick to a degree he wasn't properly prepared for. "It was a mess," he told me. "I ran terrible and didn't make it out of the first round."

But by 2007 he was back on track and proving himself to be one of the world's best runners. He won the national championship in the 1,500 meters, out-sprinting Bernard Lagat on the final half of the straightaway before the finish line. In July, after winning the 1,500 meters at the IAAF Golden League meet, he decided to turn his attention back to the mile record. Just two weeks later, in Brasschaat, Belgium, he ran a brilliant race in which he evenly increased his effort as the meters ticked by. He broke Steve Scott's twenty-five-year-old American record of 3 minutes and 47.69 seconds with a time of 3 minutes and 46.91 seconds.

"Nothing was left at the end," he said of what would stand as the fastest mile run in the world that year. "That was every last bit. I was totally tapped out." His aim now, the only aim there could be, was to set a new world record for the mile and solidify himself as the greatest American distance runner of his generation.

Going into the 2007 IAAF World Championships in Osaka, Webb had the top time in the world—a personal best of 3 minutes and 30.54 seconds—and was the favorite to win the 1,500 meters. But coming into the final turn with the leaders, he faded hard and finished a disappointing eighth. Asked by reporters if there were any lessons he could take away from the effort, a frustrated Webb said, "Nothing. I learned nothing. It was a waste of my time. I should have just stayed home. I had no idea what was going to happen. I was going to run as hard as I can and see where it fell. It fell mighty short." For the rest of 2008, Webb was "plagued by inconsistency, injury, overtraining."

Around that time, Steve Magness, apropos of nothing, called Webb's coach, Scott Raczko. After graduating summa cum laude from the University of Houston in May 2008, Magness had stagnated, and was now considering graduate school for exercise science. George Mason was on his list, so he reached out to see if he could help out and possibly train with Raczko and Webb. On a subsequent visit to the campus, Magness met up with the two men, talked training, and hit it off. By January 2009 he had moved to Virginia, was enrolled at George Mason, and was training shoulder to shoulder with Alan Webb while studying exercise physiology.

Graduate school allowed him to train hard and make tangible strides toward a career after running. His brother was a political science professor at George Mason, so he stayed in his extra room, and trained with Raczko's group. He'd been a top-tier runner his whole life, but he'd never seen anything like Alan Webb. The man seemed to have a preternatural ability to handle hard training, recover quickly, then come back for more.

"Alan could train like no other, he did some impressive stuff," said Magness. "I was used to training like one hundred and twenty miles a week, so I thought, *I can handle whatever*, but what got me with Alan was just the density of work he could handle. If there was any box to check, Alan was going to check that box."

Now one of America's biggest talents was under the care of the world's most powerful endurance coach, Alberto Salazar, training daily on his sponsor's luxurious campus. He told the press it was hard to leave Raczko, but he needed a change. Salazar, homing in on Webb's considerable upper body musculature, said, "He won't touch a weight the first six months. He has to lean out."

Webb moved into the infamous Nike altitude house, whose days were numbered at that point. Shalane Flanagan also moved to Portland in 2009 to work with her new coach Jerry Schumacher at the Nike facilities. Although Schumacher was brought in as Salazar's successor at the helm of the Oregon Project, he found working with the head coach unsustainable and began laying the groundwork toward splitting off with his own stable of athletes.

"We very much stick to our neck of the woods," said Flanagan. "We kind of quarantined ourselves. Once Jerry Schumacher broke off with Alberto in 2009, we've been very separated." Instead, Schumacher took over a competitive arm of the Bowerman Athletic Club, which came to be known as the Bowerman Track Club.

Salazar took it in stride, and by year's end would have a new credential on his mantle—USATF's Nike Coach of the Year Award. Rupp's NCAA titles, Ritzenhein's American record in the 10,000 meters, and Kara's third place at the Boston Marathon were all lauded by the organization as outstanding performance metrics.

While riding high on their prodigal coach, Nike was beginning to feel the effects of a controversial book published in May by journalist Christopher McDougall called *Born to Run: A Hidden Tribe, Superathletes, and the Greatest Race the World Has Never Seen.* In the book McDougall unpacks an argument implicating modern-day running shoes in the astronomical injury rates among runners. Instead of patterning running shoes after how a bare foot splays and the Achilles structure absorbs and returns energy, he points out that built-up running shoes heavy with high heels, lateral posts, and unnecessary accoutrements of the marketing team's dreams are based on flawed design theories. Bowerman's original conceit, that elevating the heel would lessen the stress on the Achilles tendon and allow for a longer stride and faster running, was now openly debated in light of eight million years of hominid evolution, and it was not faring well.

McDougall so masterfully weaved the critique into an ultrarunning adventure story that his book quickly became a breakout success and reached farther and wider than the insular running world.

It sold more than one million copies the first six months—on its way to more than three million total—and in doing so sparked the minimalist shoe revolution, overturning the trough from which Nike was so greedily eating. But it wasn't just Nike: every brand that had followed their lead was forced to reassess the fundamental ideas behind their shoe designs or continue to lose ground to the minimalist shoe brands.

14
I PAY YOU TO RUN

THE GOUCHERS KNEW THEY WANTED TO HAVE A FAMILY, BUT WITH KARA'S CAREER ascending there was always another race around the corner. Taking a year off in her prime could, at worst, mean an end to her ability to make a living through running, and at best, cause her to miss a year's worth of prominent races that could launch her into the stratosphere. As many athletes learn, however, prime athletic years are also prime fertility years.

The couple pored over Kara's contract to see if there were any guidelines for women to navigate this tricky time of their careers. It seemed as though she would be fine, but they had heard other Nike professional athletes complain in the past about having their salaries "reduced" for spending periods of time away from racing, in particular for being pregnant. Furthermore, they found no mention of maternity leave, reduction of pay, or suspension of pay for an athlete who became pregnant. The contract did include exact specifics about how often she had to race, including an explicit clause stating that if she didn't compete in a sanctioned race for 120 days, that her salary would be reduced.

Nike made it clear that they wanted to make decisions for their athletes. Most athletes who sign on with the brand envision themselves becoming a "lifer," or someone who is financially supported by the sportswear goliath for the rest of their lives. There are, however, only three athletes who have signed lifetime deals with Nike: soccer star Cristiano Ronaldo, and basketball greats Michael Jordan and LeBron James. There was a less official way to continue to cash Nike checks after you retired and it seemed to rely on staying active in the sport and, of course, in Nike's good graces. But Adam and Kara could only think of one runner who had managed this—Joan Benoit Samuelson, who was making $20,000 a year from the brand (and whose daughter now works for Nike). Running stars like Craig Virgin and Frank Shorter at some point had gone elsewhere with their talents during their competitive years, an act of treason to the swoosh executives that resulted in there being no chance of reconciliation.

As 2009 came to an end, the Gouchers approached Salazar about the prospect of having a child. They had discussed it before with their coach, and they now saw this as their last opportunity, but they wanted to make sure it wouldn't affect Kara's professional standing with Nike. They were building a new house and didn't want to commit money to it if there was a chance Nike could decide to reduce Kara's salary or stop paying her altogether.

Salazar told the couple he'd speak about it to John Capriotti, head of sports marketing. Later, he relayed what Capriotti said: "Don't even go there. As long as she stays relevant, as long as we still see her out and about and she's making appearances, she's fine."

The Gouchers understood this to mean that as long as she remained in the public spotlight as a Nike athlete, they would honor their agreement and continue to pay her the same monthly salary she had been receiving. They reviewed her contract again, just to make sure there was no mention, let alone an admonition, of childbirth.

Kara had some trouble getting pregnant and endured an arduous fertility treatment before becoming pregnant. Nike's initial response seemed to be one of jubilation. Capriotti even took the couple out for a celebratory dinner. "We thought we were family," said Kara.

Though her pregnancy was considered high risk, Nike took full advantage of her time away from racing by sending her to major photo shoots.

When Paula Radcliffe and her husband arrived in Oregon for a training spell at the Nike facilities, the two women hit it off. Kara asked her for advice on how to train through a pregnancy, since she had personally watched Radcliffe come back from childbirth and win her third New York City Marathon just ten months postpartum; she wanted to know what her secret was, if there was in fact a secret. A few days later, Radcliffe too learned she was expecting.

The women proved a great pair, often getting each other out the door to train when the realities of being pregnant made it less than appealing.

"No matter how hard we train and no matter how many marathons we've run, nothing compares to pregnancy," Radcliffe told the *New York Times*. "And that was really surprising to find out. That's why it was so nice to have Kara to train with, especially in the first trimester when we were really feeling that exhaustion. We were able to get each other out of the house even when we really didn't feel like it."

Nike told Kara and Adam to keep the pregnancy a secret. No Facebook announcements. No Twitter posts. This was normal inside the Oregon Project, and Salazar often forbade the athletes from exposing much of anything publicly, but there was another reason. Nike had been orchestrating a public announcement with the *New York Times* to be published on Mother's Day, May 9, 2010. If word got out beforehand, the big reveal would be spoiled.

The *New York Times* piece titled "A Friendship Built for Long Distance" was received well by the public and the large population of female athletes. Pregnant professional runners were rarely covered by large media outlets even though women make up the majority of runners of US road races (60 percent), and there was a dearth of information about training while pregnant.

At Nike's urging, Kara continued to do appearances and interviews with major media players. And she was asked to do photoshoots; one in particular for *Self* magazine had her hanging out on a dangerous cliff to get the shot. Truth be told, she would have rather had her feet up, resting, safe at home. Usually, female athletes recede into the background when they are pregnant, but Kara was out and proudly supported by Nike. Her popularity spiked.

Arriving home, exhausted after media appearances, the Gouchers were surprised to find a message from their financial adviser. "Your Nike quarterly payment didn't come in," he said. The couple reached out to their agent, who reached out to Capriotti, but he was out of the country. More than a week later they finally heard back. Kara had been suspended for a "medical condition" and would not receive pay. The reason? Because she wasn't running. Apparently, Nike saw pregnancy as a medical condition worthy of financial punishment.

Gobsmacked, Adam and Kara took their case to Salazar, reminding their coach what Capriotti had said, that "as long as we still see her out and about and she's making appearances" she'd continue to be paid under her Nike contract. After checking with Capriotti, Salazar told the couple, "Listen, they're not going to budge on it. Your two options are to let it be or get a lawyer and fight it, but that's an option you don't want to do. You'd be sinking your ship, basically. In the end you'll get what you get out of it and never work for Nike again."

"No, no, no, we have to be able to talk to someone," said Kara. "We can figure this out."

This led to a standoff with Nike's bombastic vice president of global sports marketing, John Slusher, whom they met for lunch in the Boston Deli inside the Joan Benoit Samuelson building on campus. Slusher began the meeting by preening about how he was the person who reduced Tiger Woods (referring to a reduction in salary, based on an athlete not upholding contractual clauses) before telling them that his wife and daughter were big fans of Kara's.

"Let's get down to it," said Slusher, talking through a mouthful of salad and spraying it on occasion. "I pay you to run, and you aren't running." He told the couple there was nothing he could do for them. Not only would Nike not pay them for the missed months, but, he said, "You owe us that time back too."

Nike PR had told Kara that she had become the most requested athlete for media appearances and interviews, so Adam countered. "You don't pay Kara to just run," he said. "You pay her to be Kara."

"No, we pay her to run."

"But Cap told Alberto that Nike was going to pay as long as she stayed out there and relevant, and she has. So, was Alberto lying?"

"I've talked to Cap about that," said Slusher, "and he doesn't remember saying that and there's nothing in writing. Cap is one of my best friends. If he remembered, he'd tell me."

Owing the time back meant that Nike was going to extend the end of Kara's contract the same amount of time that she hadn't raced. But Slusher threw them a curveball when he said, regardless of the time spent away from racing, her extension could be between twelve to seventeen months, depending on the budget.

ON APRIL 9, 2010, SALAZAR FORWARDED A BBC ARTICLE TITLED "SUNBATHING UPS men's testosterone" to Ritzenhein, Webb, and Nike researcher Dr. Loren Myhre:

> Hi Alan and Dathan,
>
> We may be on to something here! Google Vitamin D and effect on testosterone. There's lots of articles that talk about their relationship. Both of you have had low testosterone and low Vit. D and stress fractures. Galen has been on Vit. D for the last two months, his testosterone levels two days ago are the highest they've been since highschool and he's training very hard right now. Dathan, do you know what your latest Vit D levels were?
>
> Loren, could you check on Dathan's last blood tests and see if we got a Vit D and testosterone level on him? Thanks! - Alberto

The article served to reinforce another theory that Salazar was testing, in which taking high doses of vitamin D would increase testosterone. Ritzenhein responded that he'd never gotten a stress fracture in the middle of the summer, which Salazar saw as further confirmation, and enough for him to tell the runner to forgo any additional blood work and begin immediately taking prescription strength Vitamin D.

> HI Dathan, Good, I think we're on to something here! Forget the blood tests this monday, let's have you start on Vit. D now, you've had enough tests that show you in the lower range of normal for Vit. D and with very low testosterone levels, so we'll know if you get an increase in testosterone from the Vit. D, and even if you don't, you still need to have higher levels of Vit. D. I've left a bottle with one Vit. D pill in it just inside our front door. (never locked) I'm getting my prescription for them refilled and will get you more, but you only need to take one per week. Talk to you later- Alberto

Vitamin D dosages at these levels carry health risks—and it's one of the reasons they are available only through prescriptions—but Salazar inserted himself as medical adviser wherever he saw fit and Ritzenhein started taking Salazar's prescription vitamin D the next day. Later, Salazar called a physician that worked with the team, Dr. Robert Cook, to have him "call in a long-term prescription for Vit D," apparently allowing Ritzenhein to skip the consultation and blood work altogether.

In May, at the Payton Jordan Invitational Track and Field Meet, Nike's two most famous coaches went head-to-head, protégé versus protégé, when Salazar added Galen to the race, the day before the event, to line up against Jerry Schumacher's athlete Chris Solinsky. No one thought Solinsky had a chance: it was his first attempt at a 10,000-meter event, and Rupp was the reigning national champion. Running pundits had debated which American-born athlete would finally achieve the sub-twenty-seven-minute 10K, but they'd expected it to be Ritzenhein or Rupp, not Solinsky, who was considered too big for a distance man at 6 feet and 0.83 inches tall and approximately 165 pounds. Eritrean-born American Meb Keflezighi had held the American record for the past nine years of 27 minutes and 13.98 seconds.

The race was now billed as Rupp's record attempt. Nike executives had posters printed featuring Rupp running that read "Go Galen!" as well as giant cutouts of his head. Salazar even procured two Kenyan athletes to run as rabbits, athletes who run at a specific pace, breaking wind for the others, then usually drop out. The rabbits, Mathew Kisorio and Simon Ndirangu, did their work perfectly then pulled off the track at 6,400 meters. Rupp took the lead. "Just a couple of seconds slow!" Salazar screamed to him as he ran past.

Solinsky ran in Rupp's shadow until 900 meters to go when he passed him with authority on the outside and quickly gapped the young Oregonian. Once Rupp lost contact with Solinsky, the race was over, as Schumacher's runner extended his lead all the way to the finish line, not only winning the race but also becoming the first American-born runner to break 27 minutes for the distance. Rupp did go under Keflezighi's record time, by more than 11 seconds, but before he finished, that time had been made obsolete.

Some had argued that without the pacers and Rupp's efforts leading the race—basically everything that Salazar orchestrated—Solinsky would not have had the opportunity to speed past them all in the last 900 meters for the coveted mark. And though Rupp solemnly congratulated Solinsky after the race, animosity festered between their coaches.

Although Rupp had been overshadowed by Solinsky for much of 2010, he had set new personal best times in every one of his distances; 3 minutes and 56.22 seconds in the mile, 7 minutes and 42.40 seconds at 3,000 meters, 13 minutes and 7.35 seconds at 5,000 meters, and 27 minutes and 10.74 seconds for 10,000 meters.

Alan Webb, the American record holder in the mile, was having less success under Salazar and spent much of his first year with the coach just getting back to the point where he could begin to seriously train again. Shortly after arriving he had decided to get surgery on a nagging Achilles tendon, then suffered a stress fracture in March. He spent a lot of time on the AlterG and underwater treadmills before easing back into actual running in May, then speed training in July.

Salazar told reporters that Webb would run faster than ever under his tutelage. He would break Bernard Lagat's American record in the 1,500 and he would improve his own mile time. To accomplish this, he cited the changes he was making to Webb's original "waddling" running form.

"I would have to say that even my athletes agree as a group—Galen Rupp and Dathan Ritzenhein—we all know he is the most talented in the whole group," said Salazar. "But of course, staying healthy is another necessary thing. But Alan is so talented and he really hasn't been gone that long and I believe the systematic approach that we are doing is going to pay off for him."

After the Fifth Avenue Mile in New York, which served as the end to Webb's shortened comeback year, Salazar acknowledged the whispers about his athlete. "A lot of people think he's finished and washed up and he'll never run well again . . . Yesterday, we had a tremendous workout where he ran fifty-three seconds for his last four hundred, so we can see the acceleration curve is going up tremendously. So yes, he is very talented. Only God knows the future, but I fully expect that Alan Webb will break American records in the fifteen hundred and the mile. Alan Webb will run faster from the eight hundred to the ten thousand meters than he has before."

KARA GAVE BIRTH TO A BABY BOY NAMED COLTON MIRKO ON SEPTEMBER 25, 2010; Radcliffe had her boy, Raphael, days later. After childbirth, some female athletes never manage to return to their pre-pregnancy running levels. Kara, working hard to get back into race shape, hit a breaking point just two months after Colton was born: while running on the family treadmill, her brain clouded by sleep deprivation, she ruminated over life's big questions. *What do I really want? Do I want to start over and do something else?*

As if supplying an answer to her unease, Adam arrived home and relayed some news from the team doctor. "She thinks you need more sleep," he said.

More sleep? When? Kara thought.

Newborns aren't known for increasing parents' sleep quantity or quality, and the Gouchers were no longer enjoying full, uninterrupted rest. Overwhelmed with exhaustion, Kara hit the big red emergency stop button on the treadmill and began to cry.

The notion of staging a comeback on Patriots' Day, 2011, at the Boston Marathon had seemed like a glorious way to return to elite-level racing. Now, in the middle of a ninety-mile training week that she didn't think she could finish and suffering from infant-induced sleep deprivation, Kara was having second thoughts.

"I can't do this," Kara told Adam through tears. "I can't physically do this anymore."

"Maybe you just need to take some time," Adam said.

But there wasn't more time, not if she wanted to compete in Boston in less than five months. More important, Kara needed to race in a USATF-sanctioned event so that she could comply with the contract and begin to receive her Nike salary again. Adam was stuck in a sort of athletic purgatory, not disabled, but injured enough that he couldn't quite perform at the elite level expected. This left Kara as the sole breadwinner of the family. The couple calculated that by the April 18 race date, Nike would have withheld approximately $162,500 of Kara's salary. Looking at these numbers made Adam furious and Kara exhausted.

Soon after her treadmill meltdown, Kara went for a run with Joan Benoit Samuelson, who was in Oregon for a Nike visit. After listening to Kara's situation, the marathon gold medalist advised Kara to fight the unfair suspension of the Gouchers' livelihood.

MEANWHILE, STEVE MAGNESS HAD BEGUN A TENTATIVE TRANSITION FROM ATHLETE TO coach by working with a family friend, a young student who was competing in the US High School Nike Cross Nationals in Oregon. He reached out to Alan Webb to see if he could stay with the professional Nike runner while he was in town for the race. Webb was happy to host Magness, and told him, "*Of course, come on out!*"

While there, Magness joined Webb for his daily training on the Nike campus. It was here that he met Salazar for the first time. On a typical overcast day in Beaverton, Magness watched Paula Radcliffe run strides on Nike's Michael Johnson track. The group was videotaping her so they could run it back in slow motion to see if there was anything biomechanically that they could improve upon. When a question seemed to stump the group, Magness couldn't help himself, and provided particularly discerning commentary about Radcliffe's push-off and how it affected her landing. The highly paid and Nike-adorned group of men looked at each other, like, *Who the hell is this guy?* before realizing Magness was probably right.

A month later, he got *the call*. "Hi, Steve, this is Alberto Salazar. I read this blog you wrote analyzing Kenenisa Bekele's biomechanics from the World Championships and I was really impressed. Would you be interested in interviewing for a coaching position with the Oregon Project?"

After Salazar spoke to Webb, who had nothing but kind things to say about Magness, Nike flew the young science student out to see the campus and the condos, and partake in a round of interviews.

Salazar raved about the resources, the cryosauna recovery chamber, the underwater treadmills, the blood testing, but it wasn't until Magness saw the campus for himself that he was convinced. Nike spared no expense. It was a sport-science geek's dream.

Not even fresh out of college, he sat down for his first professional interviews with Treasure, Salazar, and Capriotti in the Mia Hamm building, which housed the sports marketing department. Though the meetings were surprisingly informal affairs—more like conversations than interrogations—he gussied up a seventy-page distance running literature review that he had written for graduate school and presented it to the men during his interviews.

Magness was struck by Capriotti's honesty. "Alberto's come a long way," he told him. "Initially he wasn't a very good coach. I had to get him better athletes, and now he is." Though Magness was confident in his sports science knowledge he thought his lack of experience coaching would put him at a distinct disadvantage.

Making sure he did his due diligence, Magness spoke with Webb's former coach, Scott Raczko, about the opportunity. Raczko relayed what happened on his visit with Webb that didn't sit well with him. He told Magness that Salazar tried to convince him to get Webb to "take five or six pills of iron a day." Raczko thought that was crazy and told him as much. Salazar laid out a theory that ferritin levels need to be at the very top end of the range.

Raczko also seemed to despise Salazar for perceived skullduggery. On his Nike visit with Webb, Salazar convinced them that the athlete should begin to use Treasure to improve his psychological tactics around sports. What Treasure did instead was convince Webb to dump Raczko to be coached by Salazar.

"I love Scott as a person and as an athlete," said Magness, "but I had in the back of my mind the fact that he might have some bitterness toward Salazar, which might be understandable."

In hindsight, Magness now believes that rather than working against his prospects with Nike, his naïveté was one of the key reasons he got the job with the Oregon Project. Salazar couldn't seem to get along with older, more experienced coaches who called it like they saw it and wouldn't put up with nonsense or perceived cheating. Coaches had been quietly and unceremoniously dismissed from the Oregon Project. A malleable young coach like Magness, however, who wasn't yet set in his ways, could be easily manipulated.

DANNY MACKEY WAS FINDING NIKE'S CORPORATE CULTURE FRUSTRATINGLY HARD TO navigate by 2010. His job as a researcher had seemed like a dream when there was still the prospect of moving up into a more reasonable salary range. He seemed to have topped out at $48,000, however, even though he knew the researcher one position made at least $60,000 a year and his work was at least at that level.

He was still interested in a possible coaching career and had heard that the Nike assistant coaches made from $75,000 to $175,000, with Salazar and Schumacher making north of $300,000.

"I was doing good work, and I had two patents already," Mackey told me, "but I was having trouble moving up."

He found an open position under John Capriotti, in sports marketing, which paid $75,000. "Cap is the Godfather of the sport," said Mackey, "so, I was a little intimidated interviewing with him." But the two Illinoisans found common ground in their mutual obsession with the track. According to Mackey, Capriotti was an extremely smart, quick thinker who surprised him with his instant recall of even the smallest Illinois track meet.

The grassroots sports marketing manager position went to someone else, however, so Mackey began his search anew, but this time he would consider other companies.

"Adidas came along and offered me double what I was making at Nike," said Mackey. "I felt like, 'The market has spoken, that's my real value.'"

His last day was October 30, 2010, when he left for a job just down the street working as a footwear developer for Adidas. Mackey knew he was likely sealing his fate with Nike, because taking a position with Adidas would be seen as an unforgivable act of treason from inside the berm.

"When I left, my boss was like, 'You know, you'll probably never be able to come back here,'" he said. "It's very tribal."

ON JANUARY 1, 2011, A TWENTY-SIX-YEAR-OLD MAGNESS MOVED TO PORTLAND TO begin work, accompanied by his mother and sister who helped him unpack and settle into his new home. After a day of moving and heavy lifting, Magness met Salazar and headed to the Duniway Park Track for the day's workout. While there Salazar said, "You may have heard about me and this doping stuff. It's all bullshit. People are jealous and none of that occurs. Everything we do is aboveboard." Unsure of how to respond, Magness just listened and nodded.

Nike gave him a cubicle next to Salazar in the Mia Hamm building. Right away, some of the things the Oregon Project's head coach told Magness didn't seem to add up.

"Early on, Salazar told me [former assistant coach] Vern Gambetta was crazy," he said. "And I knew him a little bit. He seemed nice to me."

Magness had been writing a blog called *Science of Running*, as well as contributing training articles to the magazine *Running Times*. Gambetta was impressed and reached out to talk shop with the science student and keen coach. Magness had also gone to school with Gambetta's daughter. To him the man seemed intelligent and far from crazy.

Around this time in early 2011, it was not uncommon to see a diminutive Somali man hurtling around the Nike campus track at more than fifteen miles per hour in Adidas shoes. British athlete Mo Farah was secretly training with Salazar while still contractually obligated to run in the German brand's three stripes, causing consternation and dirty looks from the swoosh minions in Oregon. Salazar had to get it cleared through upper management for Farah to be on campus and use the facilities in the enemy's trefoil logo, something that had never before been allowed.

It was a world championship year, with the Olympics on the horizon, so the twenty-seven-year-old Farah began working with Salazar in secret. Farah raced well on the international scene, but was good, not great, and had never medaled in an international championship event. He moved his growing family across the Atlantic for a chance to work with Salazar, whom he revered. "The matchstick man from south-west London," as he's been called, was also handsome and affable, and seemed like the type of athlete who might shine in Nike advertisements, especially with the London Olympics approaching.

But the Oregon Project was created with the expressed goal of putting top American athletes back on international podiums, so Salazar heard the criticism. "When somebody said, 'Well, why are you wasting time with this British athlete?' Well, because he's going to make our athletes better. So I brought him over for that reason. And also because the London Olympics were coming up, and Nike said, 'Hey, it would be great if you could get this guy to run well.' It's good for Nike. So sometimes, it's a business, I'll make an exception."

As contract negotiations were occurring behind the scenes, Farah became acquainted with the accoutrements of the Nike Project as UK Athletics did their due diligence on Salazar. They found that there were ample rumors about the man, including his involvement with banned athlete Mary Decker Slaney. They asked Salazar, point-blank, about his association with Decker Slaney during her 1996 failed anti-doping test.

"That's the question I asked before [joining Salazar's training camp in 2011]," said Farah, "and Alberto said he wasn't coaching her at the time."

Despite their misgivings, Ian Stewart, who was head of Britain's endurance running, approved the coaching change. "We had a long discussion and I think it's a great decision for Mo. Alberto has attention to detail. The place is specifically an endurance place. He's very, very driven on where we need to go."

On January 10, Salazar allowed the *Oregonian* to publish the news that Farah had joined Nike and the Oregon Project team. "Mo is wearing our gear," he told them. "I don't have to sneak him in the back door at Nike anymore. He was the first exception ever there. People would be staring at him, and I had to get out the word that it had been OK'd at the highest level for him to train in Adidas shoes."

Nike coaches are not allowed to work with athletes who are sponsored by any other shoe company. The only exception to the rule is if a newly hired coach had previously been working with an athlete from another brand before the coach signed with Nike. Even then, the allowance was for only one athlete, no more, and all other non-swoosh relationships were to be severed. Salazar and Farah hit it off to such an extent that Salazar said he would have argued to subvert the rule if Farah's contract negotiations broke down and the Brit stayed with Adidas. But that didn't happen.

"I've made a big decision to move forward in my career," said Farah. "Last year was a great year for me and, if I'm ever going to get close to a medal in a world champs or Olympics in 2012, something needed to change a little bit. I believe he can just make that one-to-two percent difference to get close to a medal."

Magness arrived in time to help Farah and his family move from their temporary Nike condo into a more permanent home. Farah notched right into the program, running one-hundred-mile weeks, seeing Darren Treasure, and using the underwater treadmill every other day for extra mileage. And he seemed to click with Rupp, which was essential for longevity in the Project.

WHEN KARA RETURNED TO TRAINING AFTER GIVING BIRTH SHE NOTICED A DRAMATIC shift in the program. "It was all about medals and results," she said. "Big-time results."

To stop the clock and start getting paid again by Nike, Kara entered the Rock 'n' Roll Arizona Half Marathon as a tune-up race just three and a half months after giving birth to Colton. She was still nursing and didn't feel prepared to run hard, but every day she didn't race the Gouchers didn't get paid, and finances were tight. "I was wearing a diaper to practice for like four months after giving birth," said Kara. "I was still bleeding. I wasn't healed, and my hip was hurting all the time." She was also nervous about flying to races with her newborn.

Four days before they were to leave for Phoenix, Colton was hospitalized. He had developed a staph infection in his jaw, and went into corrective surgery.

Snow, freezing rain, and fog enveloped the Nike campus track that day. Salazar was notoriously squeamish about training in bad conditions. He called Kara to talk about final race preparations and to remind her about the planned workout at eight at night in the halls of the Mia Hamm building. Kara told him she didn't want to leave Colton and she didn't think she would be able to race the half marathon. Salazar insisted that she not only show up for the night sessions, but also that she would race in four days' time in Phoenix.

Adam, who was resolute about staying right there by Colton's side throughout, told Kara she could go. He'd be there.

When she arrived, Salazar was there measuring out four hundred meters of hallway with the distance wheel. Kara ran eight-hundred-meter repeats, sprinting to the end of the hallway, turning around at the cones, and sprinting back past Salazar, who stood to the side with his stopwatch.

This training session, in particular, heightened the Gouchers' feeling that Nike was taking advantage of them. And when they talked through the situation, they came to believe that Nike had no legal justification to withhold Kara's salary when she was clearly working on campus daily. Adam had been trying to find a lawyer and possibly fight the brand for the lost income, but even that was proving difficult due to Nike's influence. It seemed as though each and every lawyer that they contacted would eventually call back and tell the couple that their firm worked with Nike in some other capacity, or were on retainer, and that it would be a conflict of interest for them to take on the Gouchers' case.

ON RACE DAY, JANUARY 16, 2011, KARA HAD NO IDEA WHAT TO EXPECT FROM HER body. *Would her muscles, joints, and tendons be more or less resilient to the pounding of the road? Would she be able to pick up where she left off, instantly returning with ease to her elite-level speed? Would she feel that extra gear sometimes touted by new moms after childbirth?* she wondered. She felt underprepared and ruminated about Colton lying in his hospital bed, so many sizes too big for him. *Was he okay? Was she a bad mom for doing this?*

With all of that on her mind, she lined up in Phoenix for 13.1 miles of fast pavement. She wanted to prove to her followers, and the mom-runners that had rallied behind her, that she was still the same athlete, but more important, she needed to begin getting paid again by her multibillion-dollar sponsor. For this, she was willing to be embarrassed with a bad performance.

But Colton's condition weighed heavy. "My mind was not there at all," she said. "I forgot my warm-ups, my drink mix. I couldn't believe it. I thought, *Who am I? I am so scattered.* I wasn't thrilled with the performance. I had no fight, nothing."

Mexico's Madaí Pérez won the race, with Kara 2 minutes and 13 seconds behind her in second place. And finally, she would begin receiving the paycheck she had worked so hard for so long to get.

Colton would be fine and pull through it all with nothing more than a little scar on his jaw.

FOR THE GOUCHERS, NIKE'S INSCRUTABLE FOUNDER, PHIL KNIGHT, HAD PROVEN TO BE a bit of a mystery in their time living in Oregon and working on campus. Though they had personally met him a few times over the years, he wasn't involved in any of their day-to-day activities. Occasionally, Salazar would talk about going to meet with Phil or would step out of the room to take calls from the seventy-three-year-old founder. But in early 2011, Knight was scheduled to emerge from his office and appear on the *Oprah Winfrey Show* and the television crew needed footage from around campus.

"That was the most I ever talked to him," said Kara. "They wanted some B-roll, and so they put dots on and pretended to test me." Kara was struck by how affable Knight was throughout the process. "He seemed to know a lot about me."

On Oprah's show, the eponymous host told the fellow billionaire CEO that she was just getting back into running. Knight, red-eyed and wearing a Livestrong bracelet beside his Nike watch, gave Oprah a pair of custom LunarGlide+ 2 running shoes, which she immediately put on before declaring, "Now this is a shoe!" Streams of assistants then poured out from behind the stage with boxes of the new model for each and every audience member. (How they knew everyone's size we'll chalk up to TV magic.)

It was around this time that Salazar began the process of putting together his biography with the *Runner's World* contributing writer John Brant. The men would often sit in conversation, lubricated by wine or beer, and go through the life of one of America's greatest runners. Writer and subject first met in 1983 at the magazine's offices in Mountain View, California. Brant later called it a "divine visitation," one in which he could sense Salazar's shyness, but also "his taut disdain for any person or circumstance diverting him from the pursuit of his destiny."

Now, all these years later, the writer and the coach sat down to write a book together. Salazar told Brant all about an exciting new addition to the Oregon Project, an assistant coach named Steve Magness.

Magness was everything Salazar was not, most notably a fact-based thinker and a scientist with deep knowledge of human physiology. But he was also green.

Magness almost immediately began to distrust Salazar, beginning with the coach's frequent inappropriate comments. During Kara's return to competitive speeds, in the months after giving birth, Salazar would frequently comment on her body. "Have you noticed how much bigger Kara's boobs are?" Salazar said on more than one occasion, or "Steve, do you think Kara's boobs are bigger?"

"I don't know," Magness would reply, squirming, thinking to himself, *She has a kid and a husband.*

"I think they look really good," Salazar replied.

This was not hidden from Kara or Adam, who have both attested to hearing Salazar make sexual comments about Kara. "After my son he was obsessed with the size of my breasts," Kara told the BBC. "And he would be talking about it openly in front of other people and making comments that were sexual in nature. It's just inappropriate." When asked if she ever spoke up to defend herself against her coach she said, "No, because that was the culture. You do not stand up to Alberto. If you do, you're a negative person, you're out."

ELSEWHERE ON CAMPUS, A SECRET GROUP OF SPORTS SCIENTISTS, DESIGNERS, AND researchers was tasked with nothing short of starting a revolution in training methods. Deliverables from Nike's SPARQ performance center fell squarely on its director Paul Winsper's shoulders. Launched in 2008, the effort is an acronym for Speed, Power, Agility, Reaction, and Quickness, and was a laboratory for developing new techniques in high-performance training that would result in new products—such as speed ladders, agility rings, resistance parachutes, and, of course, shoes—that Nike could then sell to the public. Finding innovative ways to improve athletes' speed and agility meant that Winsper was at the cutting edge of training and performance research, in which he often heard about promising new supplements.

The SPARQ performance center was housed between a daycare center and a security firm on the Nike campus, in a nondescript garage with tinted windows and nothing but an "A" on the always locked front door. The facility was so secretive that in 2013 *Fast Company* magazine called it "the twenty-first-century equivalent of the advanced facility Russian boxer Ivan Drago trained at in Rocky IV." (An unfortunate comparison if you recall that Drago was using cutting-edge technology, including nondescript injections the viewer understood to be illegal performance-enhancing drugs, while Rocky chopped down trees and carried logs through the snow.) Winsper told the journalist, "We don't want anybody to know about this," and one of his Nike handlers jokingly said to the writer, "After the article runs, we're going to have to kill you."

In January 2011, Winsper told Salazar about an exciting new amino acid supplement that contained L-carnitine. Dr. Francis Stephens, an associate professor at the University of Exeter in the United Kingdom, in conjunction with his research group at the School of Biomedical Sciences at the University of Nottingham, had published studies that showed possible benefits for endurance athletes. Stephens's research showed that if you could increase the amount of the amino acid stored in an athlete's muscle, you could improve their fat metabolization, exploiting one of the two energetic pathways used in endurance sports, and sparing the other. Athletes predominantly use carbohydrates (ingested glucose and stored glycogen) and fat for energy during exercise; as the effort level increases so too does the percentage of carbohydrate utilization. Carbohydrate storage is limited, however, while fat storage is abundant even in the leanest racer. Every endurance athlete knows the feeling of running out of their onboard carbohydrate stores, affectionately known as "bonking," a term used to describe hypoglycemia, resulting in a loss of energy and a feeling that you cannot go on. One of the main differences between elite runners and the rank-and-file is that they already have a superior ability to use fat as fuel at higher intensities. L-carnitine promised to make that difference even more pronounced. If an athlete can more efficiently metabolize fat, then his onboard carbohydrate stores are spared, resulting in an increase in endurance on the track or the road.

The fundamental problem with L-carnitine as a supplement, as the research group found, was that it was hard to get into the muscle. It couldn't simply be ingested in the morning to improve performance later that day, like, say, caffeine. The group determined that insulin, a hormone that is released by the pancreas into the bloodstream to regulate blood sugar, could increase the absorption of L-carnitine into an athlete's muscles. They figured out that there were essentially two effective ways to load L-carnitine: injection or ingestion. By needle or by drink. The former worked immediately, while the latter would take months to be effective.

Winsper told Salazar, "You might want to think about [the new sports drink] for your athletes that could really be, you know, a benefit." He introduced the coach to George Clouston, who was developing a new sports drink company in the United Kingdom (that would later be called NutraMet) based on the promising research that L-carnitine was going to be the next big endurance supplement.

On January 25, 2011, Salazar emailed Clouston to inquire about the "new supplement and other supplements you have that might benefit my runners," explaining that his Nike Oregon Project was known for "our cutting-edge sports science and medicine protocols," and that he had two athletes he wanted to load with L-carnitine in the coming weeks, before their upcoming races— Mo Farah and Galen Rupp.

Clouston responded that to "load the muscle with carnitine, the clinical trials demonstrated that the performance benefits were obtained when athletes consumed two doses per day over a twenty-four-week period. It takes this time to load the muscle with carnitine. Clearly the athletes you mentioned would not gain any performance benefits in the six weeks leading up to their next races."

Salazar forwarded the information to Magness and asked him to look into the research paper. "He asked me to tell him if it was legitimate or not," Magness told me. "The research was really well done from a scientific standpoint and the head researcher was a guy that was one of the first researchers on creatine as a supplement, there were no obvious scientific holes. So I told Salazar, 'It's a supplement. They tested something like forty people and are extrapolating this to elite athletes.'"

15

DO YOU HAVE ANYTHING TO CONFESS?

Feb 4, 2011: Email from Alberto Salazar to Dr. Jeffrey Brown.
Subject: Galen Blood Test

Hi Dr.Brown, Galen had great results from his blood tests yesterday. Hightest [sic] HGB levels ever, total testosterone of 617 just two days after a hard workout and after months of 130 miles per week. Ferrtin [sic] of 230, TSH of .07 and Free T3 of 500. Vit D of 66. Do you want him to cut back on the Cytomel now and go back to the one compounded pill in the morning that he was on prior to taking prednisone? He will start dropping his mileage now so we don't want him getting hyper.

Thanks! – Alberto

BLOOD WORKUPS FROM DR. BROWN'S OFFICE OR THE NIKE LAB WENT TO SALAZAR first. He would forward them on, of course, with his suggestions, but over time, as USADA describes in their Interim Report, this broken chain of privacy was inundated by the coach. Shortly after the blood work referenced above, Rupp was off to Europe to race in a couple of indoor 5K events, one in Dusseldorf, Germany, and another, five days later, in Birmingham, England.

Salazar wanted Rupp to take the drug prednisone, the corticosteroid that he claimed was for the runner's asthma. The drug was banned in competition by WADA in 2007 due to its powerful performance-enhancing benefits, so Salazar and Magness worked to get him a therapeutic use exemption. They were denied, but Magness told USADA that Salazar had Rupp take the drug anyway (Salazar denies this claim).

Rupp then flew ahead to Germany, while Salazar had Magness fly a container of Rupp's urine to Minnesota and drop it off to get tested at the Mayo Clinic. Inklings of doubt percolated in Magness's mind as he walked through the airport with the container. It seemed to Magness that Salazar was obviously getting Rupp tested to see if he would fail a drug test for prednisone. And if his sample failed, Rupp would presumably just bow out of the race and feign illness to avoid being tested altogether.

After the Mayo Clinic drop-off, in which he was relieved of his piss-delivery duties, Magness continued on to coach Rupp overseas, where he met the runner in Dusseldorf. He was just getting to know the young athlete and although they spent most of the trip together, he admits they didn't ever connect in any meaningful way. But it wasn't for lack of trying. Magness brought up innumerable topics to see if he could arouse the young man in conversation, but Rupp didn't seem interested in talking. That is, until Magness stumbled on basketball—Rupp loved to talk basketball.

More than a decade into his relationship with Nike, Rupp had become the anodyne champion that no one wanted. The teen who once had to ask his coach what he should order at restaurants had become a tractable adult. In interviews he would praise Salazar, God, and Nike, while managing to avoid saying much of anything insightful, interesting, or, heaven forbid, off-script. Lacking in Prefontaine-like joie de vivre, the milquetoast young runner was seemingly worthless to Nike's marketing department.

At some point early on in the European trip, Rupp started to complain of cold symptoms. Magness scoured the local pharmacies for medicine he could take that wasn't on the USADA banned list.

He gave over-the-counter medication to Rupp, but Salazar didn't think it would be strong enough. "Expect a package," he told his assistant coach. When the package arrived at the hotel, Magness opened it and found a magazine and a novel with a ship on the cover, written by Clive Cussler. It looked normal enough, but when he opened the magazine, pills fell out. When he opened the book, he discovered a homemade secret compartment cut into its interior, an inch deep into the middle of the pages. Inside were more pills and some nasal spray taped into the bottom. Rupp commented that it was like something out of the movie *The Shawshank Redemption.*

"I just gave it to Galen," said Magness. "Looking back, I think it was probably prescription drugs, that Rupp probably didn't have a prescription for, and he wanted to get it through customs without being questioned."

Shortly thereafter, Salazar became worried that while in Europe—essentially at sea level—Rupp would lose the benefits of his altitude acclimatization. Based on the science on red blood cells and altitude training, Magness didn't see how this would be possible, but was unable to convince the head coach.

Salazar spent a few thousand dollars to ship out an altitude tent for Rupp to use for two of the four days between the events. Though Magness remained incredulous, they set it up in Rupp's hotel room. "It just makes zero sense scientifically," said Magness. "In that short of time, nothing bad can happen, red blood cells aren't going to dissipate and they are not going to fall apart." (Magness later heard Salazar had done something similar with Ritzenhein, flying a cryosauna out to New York and having it set up in the basement of the runner's hotel before the race.)

Farah had joined the Oregon Project team a mere six weeks prior, and his training times were already indicating a serious leveling up for the Brit. When he lined up in Birmingham, England, next to Rupp, he was officially a Nike athlete. UK team uniforms were provided by Adidas, however, so Farah covered the three-stripe logo on his jersey with his number eight bib before the race began.

The two started at the back of the pack of athletes, with Rupp tracking Farah, and at times subserviently stepping into the outside lane to let him pass on the inside. Farah made a decisive push, sprinting away from Rupp with precise tactical execution, moving to the front with 420 meters to go, on sub-four-minute pace. He kicked a second time and dropped Rupp at 200 meters for the win and for a new European 5,000-meter indoor record. Farah's time, 13 minutes and 10.6 seconds, was ten seconds faster than the twenty-nine-year-old British record previously held by Nick Rose. Rupp too set a new mark, with an indoor American record of 13 minutes and 11.44 seconds. After the event, Salazar took both athletes through a 45-minute high-tempo training session—introducing Farah to the hard training after hard racing method.

"I think it has made a difference mentally and physically," Farah said of his time working with Salazar to that point.

IN EARLY MARCH 2011, LANCE ARMSTRONG FLEW OUT FOR A NIKE VISIT. HE HAD RE-tired from professional cycling for a second time after a disappointing twenty-third-place finish at the 2010 Tour de France. (While he was pedaling in the race, a federal grand jury was assembling in Los Angeles to hear evidence against him in a possible criminal case.)

In recent years Nike's commitment to Armstrong was beginning to approach their support of Michael Jordan. The brand had pledged at least $7.5 million a year to his nonprofit organization, Livestrong, under a five-year deal, with Armstrong's personal take-home at a reported $2.5 million per year.

This was despite the fact that credible evidence was hitting the newsstands at a shocking frequency that pointed to Armstrong's years of deceit. Former teammate Floyd Landis had broken the omertà around professional cycling back in April 2010, when he sent an email to the then CEO of USA Cycling, Steve Johnson, that detailed year by year all the ways in which his team, the United States Postal Service led by Armstrong, had used illegal performance-enhancing drugs and methods to dominate the cycling world and win the Tour de France, the sport's most prestigious event. In fact, as Landis told me, it was Armstrong who gave him his first PED, a stack of 2.5 milligram testosterone patches.

The email would prove to be the beginning of the end for Nike's highest-profile athlete. Fearing that the Teflon-esque Armstrong would once again emerge from the accusations unscathed, weeks later Landis also filed a whistleblower lawsuit under the federal False Claims Act. The suit claimed that Armstrong and his team had defrauded the government by taking the US Postal Service sponsorship money while knowingly cheating in their cycling events.

Though Armstrong cited the same reasons for retiring as he did in his first attempt at it—spending more time with his family and his nonprofit organization—he couldn't sit still. Just a couple of weeks after the second announcement, he was on the Nike campus to learn what he could from Alberto Salazar. The cyclist was newly focused on a return to his first sporting love, the triathlon (specifically the much longer Ironman triathlon distance of a 2.4-mile swim, 112-mile bike, and a 26.2-mile marathon). To be competitive in the multisport event, the swaggering thirty-nine-year-old cancer survivor knew he would need to become a better runner.

On campus, Armstrong was ingratiating and affable, presenting as the ultimate guy's guy, casually swearing and joking with everyone. Salazar was starstruck by the seven-time Tour de France winner and considered him one of the toughest athletes of all time. Armstrong seemed thirsty for admiration and he got plenty of it on the Nike campus. In fact, he was surrounded by toadies despite the fact that there was now credible proof to back up the years of rumors that Armstrong's career was one massive fraud. Magness watched in bewilderment as his Nike colleagues were mesmerized by the former champion.

As far as Magness was concerned, Armstrong was the enemy, "the guy who did things the wrong way." As the minions fawned over Armstrong, the assistant coach couldn't bring himself to do the same mental gymnastics it took to admire such an athlete. To Salazar, Armstrong was an honorable and true champion, though at this point, even he would privately admit that he knew Armstrong was a drugs cheat—the mountain of circumstantial evidence had become just too massive to deny.

The two men had kept in touch since Salazar had paced Armstrong in the New York City Marathon in 2006. Salazar was now supplying a steady stream of tips on what Armstrong needed to change in order to become an elite runner. To do this the coach took him through biomechanical testing in the Nike lab and videotaped his running form for analysis.

Armstrong, whose marathon PR was now 2 hours and 46 minutes, told Salazar that he needed to be able to drop that time to 2 hours and 30 minutes. If he could run a marathon that fast by itself, then Armstrong figured he'd be able to run 2 hours and 50 to 55 minutes after the swim and the bike in an Ironman triathlon. And if he could do that, he reckoned, he could win Ironman races.

Salazar told reporters that Armstrong's 2-hour-and-46-minute marathon PR was built on very little running, that they had inflated the distance of his long run when talking to reporters before that race, and that the cyclist's aerobic prowess was so superior that he had a lot more room for improvement. What Armstrong needed to become a 2-hour-and-30-minute marathoner, the coach thought, was to drop some of his newly acquired upper-body weight, improve his flexibility, and be careful not to let his prodigious aerobic abilities cause him to run paces that, for now, were too fast for his physiology.

AS KARA'S NEXT RACE APPROACHED, SALAZAR GREW INCREASINGLY WORRIED ABOUT the extra weight he thought she was carrying. In an effort to help her lose it faster, she says he told her to take the thyroid drug Cytomel. She didn't have a prescription for it, was never advised to take it by a physician, and was already on a different thyroid medication prescribed by Dr. Brown—all of which Salazar knew—yet he persisted. "He kept saying, 'Ask Galen. Ask Galen,' and I never did, so he just brought it to me."

Eventually, Salazar handed her a pill bottle full of the drug (which Kara still has) with the identifying details torn off the label. In its place Salazar had handwritten "Cytomel" on the side. Kara took the bottle to appease her coach but had no plans on ingesting the pills for fear of what they might do to her thyroid. She knew that too much synthetic thyroid hormone could be just as debilitating as too little.

Salazar had no idea where her thyroid hormone levels were; he just had a vague concept that if you ramp up thyroid production with drugs, athletes seem to gain energy and lose weight. When she asked Dr. Brown about taking this new drug, he said, "You do not need that. Don't take it," before chiding Salazar, "Alberto needs to stop trying to be the doctor."

In the days before leaving for her next tune-up race, Kara received an email from Salazar in which he reminded her that she needed to be strong. "You may be about to seize the dreams you've always had of winning Boston," he wrote, "but you must now focus on the good things you have in your life, in order of importance:

1. Your husband and son and their good health
2. The rest of your family, mom, sisters, etc. and their good health
3. Your running is going great now
4. Your true friends love you and are rallying around you

"So looking at all of the above would you like to trade a one-year Nike payment suspension for one of the above?" he wrote. Salazar then empathizes with Kara by mentioning how his own poor investment decisions made him angry and distressed for years, before suggesting she continually thank God for her blessings, and signing off with an admission of familial love for her.

In New York for the half marathon, Kara sent a last-ditch email to Nike CEO Mark Parker, just as Salazar had advised. In it, she said, "Everyone thinks that you guys were supporting me and I wasn't getting paid. I feel like you are asking me to lie."

Kara lined up the next day, a cold Sunday, March 20, for the half marathon with her teammates Galen Rupp and Mo Farah, who were making their debuts at the distance. Things had gone sideways over the last few months of the project for Ritzenhein, who was now hampered by injuries, and Webb, who was struggling with erratic performances.

The Oregon Project athletes sported their team's new kits at the race, all black shorts and shirts with the new flaming skull NOP logo on the right chest, inspired by Kara's favorite skull-and-crossbones design. Rupp also wore a black mask wrapped around his face that Salazar made him wear to filter out allergens that might slow him down (providing endless fodder for the LetsRun message boards). "I knew it would look goofy but I agreed to do it because it was silly to take a chance (with asthma problems)," Rupp said.

During the seventh mile Rupp ran too closely to Kenyan Peter Kamais and the two became entangled before crashing to the pavement. But Rupp got up quickly, newly adrenalized, and finished in third place.

"I was most excited for Kara's race," Salazar told the *Oregonian*. "For her it was huge. She's busted out of a plateau recently and she's at a whole new level now. She believes now."

Kara placed third in the women's race, in 1 hour, 9 minutes, and 3 seconds, just 11 seconds behind a pair of Kenyan athletes, Caroline Rotich in first and Edna Kiplagat in second. It was an impressive return to serious racing, hard proof that she was ready to take on the world's best at the Olympics. The top three women all broke the previous course record set the year before by Mara Yamauchi of Great Britain.

After the race Kara opened Parker's response to her email. The chief executive, whose total compensation package around this time was close to $15 million (this would triple in 2016, to more than $47.6 million, when he took over the chairman role vacated by Phil Knight), congratulated her for an impressive run before writing that there was nothing he could do about her lost income. He trusted these men to make the right choice and he couldn't go against their judgment.

"I was pretty crestfallen when I got that because I felt like I had a great relationship with him," said Kara. "He's a father and I really admired his wife and her running and I just thought, *He'll get it.* I think I was a little green with a lot of my relationships looking back. I thought these are people who will be in my life forever, but they were in my life because I was running well."

BACK IN OREGON, SALAZAR FINALLY GAVE MAGNESS SOME DATA TO DIG INTO IN THE form of altitude reports prepared by the Nike lab, and the assistant coach was excited to put his hard-earned knowledge to good use. The small binder full of paperwork included data for three athletes, Adam, Kara, and Galen. It listed when they used altitude tents and how it affected their hemoglobin values over time. The notes were mostly about races but also included some annotations on medications and supplements each athlete had been on during these periods. Most of the data had been recorded by a senior scientist at the Nike Sport Research Lab, Dr. Loren Myhre. One of the line items for Rupp read, "presently on prednisone and testosterone medication," in a date range that corresponded to when the runner was in high school.

Shocked and frightened, Magness left his Nike desk and, as casually as he could manage, scampered into the stairwell. There, secluded but panicked, he took a picture of the medical printout, just in case he was ever forced to speak about what he'd seen. The prednisone was not a surprise—he knew Rupp took that—but testosterone, that was the favorite drug of almost every chemical athlete he'd ever heard of; it was a substance that no thinking coach or support staff would bring around an elite-level runner for fear of guilt by association at best, and actual contamination at worst.

He called his parents for advice on what to do next. His father told him to take it directly to Salazar and ask him about it, which is what Magness did. Salazar reviewed the document and then told Magness it was a mistake. He blamed Dr. Myhre. "He said Loren was crazy, that's why it was wrong," said Magness, "he never said anything about it being a supplement." Myhre was suffering from

ALS at the time (he passed away in 2012 from the disease) but the contentious notations were from 2002, a year that Myhre received an award from Nike for the quality of his work.

Salazar told him to take it down to Brad Wilkins, a researcher at the Nike Sport Research Lab. Wilkins told Magness that it was probably a mistake, and that he would check with Loren Myhre. That was the last he ever heard of the issue, and no one ever followed up.

Magness managed to convince himself it was likely an error, though logged it in the back of his mind as yet another red flag. He was becoming wary of what he was witnessing on the Oregon Project.

ON MARCH 15, THE GOUCHERS FINALLY HEARD BACK FROM JOHN SLUSHER ABOUT Kara's salary suspension for her medical condition. The Nike executive vice president of global sports marketing emailed them: "While I fully appreciate that you probably don't like or agree with my decision that a 1 year suspension of your contract is the appropriate course for Nike, I hope you realize that I am trying to do my best to be fair—to both you and our company—in this business decision." Twelve months was indeed better than seventeen months, but losing their only source of income while Kara was working harder than ever felt unfair to the Gouchers. "You win the Boston Marathon and this is all going to go away," Salazar kept telling her.

For the 2011 Boston Marathon, Salazar had Kara focus on training the downhills in preparation. "Steve, on the LetsRun message boards people were talking about training for Boston and they said the biggest thing you have to worry about is the downhills," Salazar told Magness. "In my head I'm thinking, *Dude you won Boston why are you reading LetsRun to understand this?*" said Magness.

This served to add to Magness's growing disenchantment with his role as scientist on the Nike Oregon Project. He grabbed a private moment when Salazar wasn't around to get some council from Treasure.

"I don't know what I'm supposed to do here from the sports-science side," he said.

"Steve, you cannot do sports science in this group," said Treasure. "Everybody thinks Alberto is very scientific and cutting edge, but Alberto will read something, get attached to it, and think this is it and the program will be on that thing tenfold."

Salazar invited running journalist John Brant on one of Kara's training runs in her preparation for the marathon, and the three of them drove toward the Duniway Park Track in southwest Portland where Magness was working with Rupp.

The coach and the athlete barely spoke while driving in Salazar's BMW. Before they reached the track, Kara got out of the car and started to run, just as the clouds released a light rain. Salazar thought the downhill section of road to the track was a perfect simulation for the downhill near the end of the Boston Marathon.

The session was one of her last important workouts before the race, now just a week away. As the car drifted past Kara, her ponytail bouncing in the rain, Salazar told Brant, "Kara needs to be patient. In her earlier marathons, despite all that we talked about, she lost focus and went too hard too early. I told her to think of the marathon as if it were a ten-K. In a ten-K, would you start racing hard halfway through? Of course not. So why would you do that in a marathon? Trust your training, trust your speed, wait until that final mile."

The Boston Marathon was Kara's A race and an event she had dreamed of winning, a dream that helped her get through the tough training it took to return to the top of the sport.

"I had a hang-up with speed. I was able to run throughout my pregnancy and so I felt strong," said Kara. "My long runs came back really easily but my speed was so hard. I'd go to the track and be like, oh man, I can't break thirty-five seconds for two hundred. Finally, I had to let it go. I remember the day, being at Duniway running two hundreds and hills with Alberto, and it was like, I'm just going to go for broke on this two hundred and if I blow up, who cares? I ran like 34.9 and that honestly turned it around. I was like, okay, I'm getting somewhere. I had to quit worrying about what my competitors were doing and I had to focus on myself. A year earlier 34.9 would not have been good, but at the time I had to take it and cherish it and kind of use that moving forward."

Back at Salazar's house, Brant had trouble keeping Salazar focused on what they were there to do: work on the book. The coach was too excited about Kara's prospects at Boston and Rupp's impressive progress, not to mention what the addition of Magness could mean to the program.

"My biggest dream, what I really want to accomplish with the Oregon Project," Salazar said, "is to build a body of sport science and training knowledge that runners and coaches everywhere can draw on for years to come."

Prompted by Salazar's talk of his legacy, Brant asked him, "Do you have anything to confess? Is there something that would come back to haunt you?"

"Absolutely not," said Salazar. "Next year at the London Olympics, people can root for Galen, they can root for Kara, they can root for any Oregon Project athlete, and feel confident that they're clean. Every day we train on the Nike campus. Phil Knight can look down from his office and watch us work. Do you think I would do anything to embarrass Phil, or betray the trust he's put in me? Do you think I would do anything to harm Galen or Kara or any other runner in the Project?"

IN THE DAYS BEFORE THE BOSTON MARATHON, MAGNESS RECEIVED AN URGENT CALL from the team massage therapist, Al Kupczak. He told Magness that Salazar was having Kara run eight downhill 400-meter repeats on the course, "just to ingrain it," he said. "You need to talk him out of it."

It was just two days before the race, which was way too close to do such a damaging workout. Her muscles simply wouldn't be able to recover before the starting gun, Kupczak explained. Magness called Salazar and pleaded with the coach to alter the workout to one or two 100-meter strides. "We just want to be safe," Magness said. "I think the training's done." This time, thankfully, Salazar relented.

At the media press conference before the 2011 Boston Marathon, Kara fielded questions about which runners had a chance to finally break the American winless streak on Patriots' Day. No athlete born in the United States, man or woman, had won the famed event since Lisa Larsen Weidenbach's victory in 1985.

"This pressure is different. We're desperate for an American winner," Kara told reporters. "I'm trying, but Blake [Russell] and Desiree [Davila Linden] have just as good of a chance." The interview, as was typical, moved on to questions about her famous Nike coach and what it's like to work with him. "I love him, he's like family to me," Kara said. "It's a really special relationship and I feel really lucky to have it." She was now just six and a half months post-childbirth and although the compressed training window made preparations extremely challenging, Kara wanted to get another marathon race under her belt before the big one—the 2012 Olympics.

Post-race headlines would be carried by Kenyan Geoffrey Mutai, who won the men's race in a blazing 2 hours, 3 minutes, and 2 seconds—a time that would have been a world record if not for the fact that the Boston Marathon course is not IAAF ratified as record eligible, due to its being more than 50 percent downhill. With Mutai's victory, Kenyan men had now won eighteen of the past twenty-one runnings of the Boston Marathon, a race the *New York Times* called "the cradle of 26.2-mile glory." Ryan Hall broke the previous course record time as well, on his way to the fastest ever American time of 2 hours, 4 minutes, and 58 seconds and a fourth-place finish.

The women's race was also won by a Kenyan; Caroline Kilel was the first to break the tape, in a time of 2 hours, 22 minutes, and 38 seconds. In a breakthrough race, Davila Linden finished in second, just 2 seconds back. Kara placed fifth in a time of 2 hours and 24 minutes—the fastest she'd ever covered 26.2 miles in her life.

After the race, Salazar and Treasure had trouble celebrating the positives. According to Kara's mother and sister, Salazar said, "Don't tell Kara, but she is still too heavy. She needs to lose her baby weight if she wants to be fast again."

"No celebration on her tremendous run," Adam said, "just judgment on her body."

Racing successfully stopped the clock on her salary suspension and started their payments from Nike again, but the twelve-month suspension meant that the Gouchers would lose her entire salary. Adam still wanted to fight it, but Kara was tired.

"I wrote a book. I did twenty-two appearances. I was on the cover of *Runner's World*, it's not like I was sitting around," said Kara. "Had I known that there was no value in that, and that they were going to say that's not worth any money to us, I would have spent time with my family. I would have enjoyed my pregnancy and I wouldn't have trained so hard during it."

Just days after the marathon, Adam finally found a lawyer, with no Nike affiliations, who was willing to help them. Kara went to Salazar and told him, "Look, I'm not going to accept the twelve months. I found a lawyer."

When Salazar told Capriotti, the Nike executive flew into a fit of rage. After he calmed down, he agreed to counter with a six-month suspension . . . but there was a catch: Kara had to agree to add another year on her contract. Salazar and Treasure urged her to take the deal, reminding her how blessed she was. *You have to take this. You have to move on. This is holding you back as an athlete. This is holding you back as a mother*, they said. "In the end I took it because I just needed it to end," said Kara. "I had a seven-month-old. I can't even tell you how stressed out I was. I just needed it to be over. I just needed to get paid and for it to be over, I couldn't take it anymore."

THE ONGOING FEDERAL INVESTIGATION INTO DRUGS IN CYCLING HAD ROOTED OUT AT least two other teammates who were now turning on Armstrong and disregarding the omertà, cycling's culture of secrecy, to tell the truth. On May 19, 2011, *60 Minutes* aired clips from an interview with cyclist Tyler Hamilton, a former US Postal Service teammate of Lance Armstrong's. He was coming clean and admitted that he had seen the seven-time Tour de France champion inject EPO in preparation for his first Tour victory.

Armstrong was defiant. On the day of Hamilton's confession, he went on the attack, tweeting: "20+ year career. 500 drug controls worldwide, in and out of competition. Never a failed test. I rest my case."

Hamilton's confessions meant that he'd be vacating his gold medal from the 2004 Olympics, a race in which Viatcheslav Ekimov came in second place. A couple of hours later, Armstrong tweeted: "Congratulations to @eki_ekimov on his 3rd Olympic Gold Medal!!"

That Sunday, at training camp in Park City, the Oregon Project gathered together to watch the entire interview on CBS's *60 Minutes* television program. As Hamilton stuttered his way through the conversation, he exposed on national television that he and Armstrong had used performance-enhancing drugs, including EPO, testosterone, and blood transfusions on multiple occasions.

At dinner in Park City, the interview was all any of them could talk about. After all, Armstrong had just visited the Nike offices and worked out with Salazar, who was helping coach him. "I mean, obviously Lance doped, everybody knows he doped," said Salazar. "But Tyler's just trying to sell books." Salazar seemed to think Hamilton was the dishonorable one in the scenario. The group exchanged stunned expressions.

"I was super disappointed in that," said Kara of Salazar's comments that night. "When you are a clean athlete, you just don't want any associations. The rumors are uncomfortable, but they are one thing, but someone who is definitely doping his ass off and then showing up at practice, is another."

BY THE FALL OF 2010, WEBB WAS MOVING OUT OF THE NIKE ALTITUDE HOUSE HE HAD shared with Rupp. It was clear that Salazar's "systematic approach" wasn't going to catapult him to new records as the coach had once predicted. By March, Webb was looking for a new opportunity. Nike had lost confidence that the twenty-eight-year-old athlete could return to greatness, and they signed him to a much-reduced contract.

Webb told me that he'd heard a lot of rumors about Salazar while he was on the Nike Oregon Project team, but never saw anything illegal, or even in the gray area, and he was never asked to do anything untoward. "I treated them as rumors, you know," said Webb, "because my first-person experience was kind of counter to that." When I asked clarifying follow-up questions, he became exasperated, angrily reminding me that he didn't like being asked about Salazar.

While Webb was in the process of moving out, Jackie Areson, a University of Tennessee standout, came in, upending her life to become a professional Nike athlete. Areson was born in Hong Kong and lived on a boat until she was ten years old before attending high school in Delray Beach, Florida. She ran the 1,500 meters with a respectable PR of 4 minutes and 18 seconds.

Areson's college team had worked with a nutritionist who also worked with the Oregon Project. She offered to speak to Salazar about Areson's professional prospects after she won the NCAA Division I 1,500-meter outdoor championships.

Nike flew Areson out to visit the Beaverton campus for a few days in March, where she met with Salazar. She admits to not knowing much about him at the time. "I'm surprised with myself that I didn't do my crazy-scientist-research thing on Salazar," Areson told me. "I think mostly that was because it was made very clear from the beginning that Steve [Magness] would be my coach." The plan since Salazar's heart attack had been to find a replacement for the ailing Cuban. Areson was Magness's chance to show he could coach an athlete on his own.

With Areson, Salazar preemptively brought up drugs (and his reputation) more than once, telling her, "I don't know what you've heard about me, but I've never been involved with drugs." She hadn't heard anything, and didn't think much about the comment in the moment.

During contract negotiations, Areson made the strategic mistake of telling Nike too much. They knew she wasn't talking to other sponsors and that she was leaving her current coach, which she now says Nike used to lowball her salary offer. They also made the deal contingent on her joining the Oregon Project, which the team began doing around this time.

Areson was fine with that, having left her college coach; she signed a contract for $30,000 a year, with an additional $7,500 travel and medical allowance, and moved out to Beaverton in May 2011. She took over the rent on a Nike town house recently vacated by Mo Farah and his family. Like the Nike altitude house, every threshold in the rental was sealed so the oxygen content could be lowered.

After a lifetime of traveling and moving, Areson was ready to settle into a place where she could focus on becoming one of the best in the world. There was no drastic overhaul of her training or tinkering with her form, but Areson said Magness was meticulous, making gradual tweaks to her plan to avoid causing any injury issues. Her personality and sensibilities meshed well with the assistant coach, but Salazar seemed manic to Areson. "Alberto was a little loopy, kinda crazy, and very energized."

The Oregon Project was no democracy, she said. Salazar ruled as dictator, always having the last word and always getting his way. Meanwhile, Magness was the opposite of the Project's head coach: where Salazar flew by the seat of his pants and changed training plans on a whim, Magness had a long-term vision and a detailed plan to get there. This worked well for Areson who was prone to overtraining and becoming injured as a result. Coach and athlete would meet for coffee and talk for hours about her training and the drama within the team, and they quickly started to see incremental success as well. After just a couple of months, she set a new 1,500-meter personal best time of 4 minutes and 12 seconds.

Shortly after Areson's first meeting with the team's sports psychologist, Darren Treasure, Magness and Salazar received a debrief of the session. What Treasure and Salazar didn't realize was that Magness now felt more of an allegiance with Areson than he did the Nike boys club; and he was shocked that Treasure completely ignored one of the fundamental principles of psychiatry: client confidentiality.

Treasure told Salazar and Magness things he and Areson hadn't even talked about and claimed she had certain issues that didn't at all track to the conversation Areson remembered having with the psychologist. He seemed to be making things up and editorializing as he went. "He was extremely inappropriate and unprofessional," Areson told me. "We had also heard that Darren did this to Adam and Kara, to try and manipulate their relationship. So I was like, 'No, I'm not doing this.' I refused to see him again."

"That's a deal breaker," Salazar told her. "You are out of the group if you don't see him."

"Fine. I don't really care, I'm not seeing him," she replied.

Treasure, for his part, had all the right credentials, a bachelor's degree in sports science and history, as well as a doctorate in philosophy and a master's of science, but after all these years still hadn't become licensed in Oregon as a psychologist. On the Nike Oregon Project website, Nike was careful with the language, calling Treasure "one of sport's most recognized 'mental performance coaches.'"

In his fight with Areson, Salazar had a problem with his harsh stance that every athlete had to see Treasure or be forced off the team. Everyone already knew that Amy Begley had previously put her foot down when it came to the "psychologist," and Salazar had allowed her to see someone else on Nike's dime, so they had a precedent to cite. And eventually, Salazar relented and agreed to have the team pay for Areson to work with someone else.

16
INFUSED

WITH THE RETURN OF KARA GOUCHER, THE CONTINUED UPSWING OF GALEN RUPP, and the addition of Mo Farah to the Oregon Project, the squad was more competitive than ever, and poised to prove it against the world's best athletes. Before their final exam, at the 2012 Olympic Games, the team set its sights on training harder and more precisely than ever for the 2011 IAAF World Championships in Athletics in Daegu, Korea.

To integrate Farah into the program, Salazar took him through a typical overhaul, assessing every aspect of the athlete's life and training protocol, from running form and speed work to supplements and substances ingested. Salazar had the Brit running up to 130 miles a week with a few twenty-minute ancillary sessions on the underwater treadmill thrown in to augment his volume. He increased Farah's time in the gym, from none to four times a week, twice for general strength and two sessions focused on core exercises. All the athletes kept "in touch with speed throughout the year" as Salazar put it, with weekly 200-meter speed-laps at race pace. Altitude was manipulated day and night, even at training camps, where the team used tents cranked up to an equivalent oxygen content of 10,000 to 12,000 feet above sea level.

This level of training produces a physiological load that few athletes can handle for even one week, let alone the months that make up a training block. To mitigate the deteriorating effects of such insults to the human organism, Dr. Brown put Farah on numerous new substances. There were prescription doses of vitamin D, which the entire team took orally in clinical doses, two 50,000 IU pills twice a week. The Recommended Dietary Allowance (RDA) of vitamin D for adults is 600 IU a day, or 4,200 per week. But Salazar had a theory, which he had tested on his own children, that massive doses of vitamin D would increase testosterone.

The team, and now Farah, were also on calcitonin—a hormone that helps regulate calcium levels in the body and which prevents the breakdown of bone. It is administered by nasal spray or injection and is used in men and women with osteoporosis to help reduce bone loss and to slow the rate of bone thinning. Running is a high impact, repetitive sport, and stress fractures are a scourge among its serious participants. Salazar thought that calcitonin could prevent his athletes from suffering from the painful tiny bone cracks.

Lastly, Salazar had Farah take ferrous sulfate, an oral iron supplement he thought would prevent iron deficiency. Whether the athletes tested low for iron or not, they took the supplement, which he hoped would help with oxygen transport during hard running efforts.

Farah indeed performed spectacularly, and in early June at the Prefontaine Classic in Eugene, he smashed the English and European 10,000-meter record against a stacked field, running 26 minutes and 46 seconds.

After watching the races, Kara called Adam. She told him she couldn't believe what she had seen from the athletes she had trained with.

"Do you think they are cheating?" Adam asked his wife.

"Yes," she said.

Beginning on June 23, the USATF National Championships were held at Hayward Field on the University of Oregon campus. In the 10,000-meter event, Rupp lined up in his all-black Nike Oregon Project uniform, featuring the NOP skull opposite the Nike logo, and his black allergy mask, once again absorbing the heckles and countless Hannibal Lecter jokes from the crowd. The pace was easy. Rupp stayed toward the front, but was reluctant to take the lead. He threw off his mask just past the halfway mark. Though Schumacher's athlete Matthew Tegenkamp stuck close to Rupp, he wouldn't have the power to overtake him at the line. Rupp won the race, with Tegenkamp in second.

Kara, who had to petition her way into the 10,000-meter event due to her time off for pregnancy, proved she belonged by finishing second in a race dominated by twenty-nine-year-old Shalane Flanagan. Flanagan took it out hard, running alone for all twenty-five laps, demonstrating without a doubt who the best 10,000-meter runner in America was and setting a new Hayward Field record. Flanagan lapped many of the other women in the race, including the Oregon Project's Amy Begley, who finished sixth in the nation.

Salazar had had it with Begley. After the race he berated her for her poor showing, told her she had "the biggest butt on the start line," and promptly kicked her off the team. She had entered the race at 112 pounds and a body composition similar to what it was in the 2009 world championships, a race where she finished sixth in the world, but none of that mattered to Salazar, who callously ended her career as a professional runner.

The following day, Rupp—the masked man in black—finished third in the 5,000 meters, with Bernard Lagat and Chris Solinsky going one-two.

"It's like in war," Salazar said after the race. "The soldier has to learn how to fight and do everything—be physically fit, be a one-man army. But then you try and equip him with every bit of top science—everything you can—to keep him alive. That's what we do. We use science, every bit that we can, on top of old-school training. We are going to train as hard as anybody else, and then we're going to train more by adding things that don't get us injured. And we're going to train smarter than anybody else."

THE TEAM THEN MOVED ON TO FONT-ROMEU-ODEILLO-VIA, FRANCE, FOR A TRAINING camp. UK Athletics sent British sports medicine physician Dr. John Rogers to observe. The organization wanted to gain some insight into Farah's dramatic improvement and his coach's training methods.

Within weeks, Dr. Rogers had grown concerned about the myriad sports medicine practices that Salazar employed. He had learned through conversations with Salazar that the coach had many off-label and unconventional uses for prescription medications.

For example, Salazar told Rogers about the drug thyroxine, one of a few prescription thyroid-stimulating drugs used for an underactive thyroid gland that Salazar thought boosted both energy and testosterone levels. According to partial records reviewed by USADA, Brown had, by now, prescribed Adam and Kara Goucher, Galen Rupp, Dathan Ritzenhein, Chris Solinsky, Amy Begley, Arianna Lambie, and Joaquin Chapa various thyroid medications.

Lambie's prescription was possibly the most egregious of them all. Her levels during five tests between September 2008 and June 2009 showed TSH levels at 1.048, 0.864, 0.670, 1.098, and 1.430. Brown knew the normal range for TSH levels was between 0.45 and 4.12, but he saw himself as a groundbreaking physician who helped athletes win gold medals. He told Lambie she didn't have a thyroid condition, but explained that athletes in particular require treatment at much lower levels, before he prescribed her a high dose of the drug Levoxyl.

After about two weeks on the drug, Lambie started to feel terrible and exhibited signs of hyperthyroidism (an overactive thyroid gland). She sought a second opinion on her condition with a different endocrinologist, who immediately began weaning her off the medication.

With little regard for her long-term well-being, she says Salazar and Brown changed the birth control she was on in an effort to manipulate her blood levels and improve her performance. "I was using one that would give me one period a month," she told the BBC, "and they wanted me to try one that was one period every three months."

Lambie claims Salazar even brazenly tested an unknown supplement on her, then had her tested to see if she would fail a doping test. "I was given a supplement to take, then gave Alberto urine that he could test to find out if it would, presumably, test positive on some result," said Lambie. "But I don't know what the substance was." (In a statement to the BBC, Salazar said, "At no time did I give any supplement to any Oregon Project athlete for the purpose of determining whether that supplement would result in a positive test for a banned substance.")

After just eighteen months, overtrained and burned out, Lambie was dropped by Nike and left the program. Another once promising career cut short.

On July 16, Dr. Rogers emailed colleagues expressing his apprehension. None of the substances that he listed in his email were on the WADA anti-doping banned list, but their use seemed to lack rigorous scientific logic and disregarded the athlete's long-term health in favor of immediate success.

Putting an athlete on thyroxine, for example, runs the risk of causing horrible side effects, including "irregular heart rhythms, insomnia, and loss of bone density," according to Harvard Medical School. There is also fear that taking exogenous doses of the hormone will down-regulate the body's own ability to create it.

Since Salazar wasn't Farah's physician, he didn't know that the runner had hypercalciuria, a rare but potentially serious condition that occurs when an individual's kidneys produce higher levels of calcium than normal. Taking calcitonin, another drug he was on that was meant to increase calcium, could exacerbate issues with brain, heart, and renal function as well as create kidney stones. More shocking still is that too much calcium has shown to weaken bones, not strengthen them, and for Farah would likely have the exact opposite effect on him than intended.

Prescription doses of vitamin D—yet another chemical substance he was told to take—were also potentially dangerous for Farah, because vitamin D toxicity causes a buildup of calcium in your blood, which could exacerbate the entire cascade of issues for the runner because of his hypercalciuria.

Salazar told Rogers that most of the off-label or out-of-range uses of these substances were meant to increase the athletes' testosterone levels. As is evident in Salazar's own career, overtraining can cause low levels of the vital hormone. When the human organism is running too hard, it creates a state of constant stress, which puts the body's autoimmune system into fight-or-flight response mode, which, in turn, produces hormones like cortisol and adrenaline. Cortisol is an antagonist to testosterone, so when cortisol levels are elevated, the body lowers its testosterone production. Evolutionarily, an extremely stressed person consolidates resources to survive, not to thrive or reproduce. In women, overtraining often results in amenorrhea, a condition by which their menstrual cycles simply stop. In men, testosterone drops to such a degree that they lose their libido as well as the energy with which to train.

The high stress of running 130-mile weeks, sleeping at extreme altitudes, and traveling to foreign places constantly pushed Salazar's athletes into a physiological hole. And when the runners inevitably broke down, rather than reducing the load and increasing the focus on rest, their coach used prescription drugs to get them back into normal hormonal ranges. Boosting testosterone levels and thyroid hormones was key to this scheme. But the human body is a complex system, and hormones don't exist in a silo. This is what makes endocrinology—the branch of medicine that deals with the endocrine system and its specific secretion of hormones—a complicated and evolving field of practice.

Although Farah was performing exceptionally well, the whole thing was a mess as far as Rogers could tell. He immediately told Farah to stop all four substances.

Rogers, despite being concerned by Salazar's tactics, concluded his email with a compliment: "His attention to detail, medical and sport science knowledge was considerably better than any other endurance coach I know. After discussion he seemed to have a low threshold for asking for advice on any minor medical or MSK issues and was very keen to have UKA input with Mo. I felt reassured after speaking with him that he was very frank and open about the methods they were using."

IN AUGUST, SALAZAR WAS ASKED HOW CHRIS SOLINSKY, THE AMERICAN RECORD holder over 10,000 meters, had gotten so good all of a sudden. "It's the thyroid," he said, before claiming that four other Schumacher athletes were on the medication. It was a feather that he was sticking in his cap, as though his meddling and the medications were the determining factor, not Solinsky's hard work, or Schumacher's training guidance.

Though both teams worked out on the Nike campus, they kept to themselves for the most part. They didn't actively avoid each other, per se, but the athletes didn't hang out and the teams rarely ended up on the track at the same time. But during the excitement around Dr. Brown, there was internal pressure to send athletes to see him in Houston. After their visits, however, Schumacher found it unbelievable that they all returned with similar, newly diagnosed conditions and prescriptions to remedy them. He grew suspicious and told his runners to see at least two other doctors who were outside the Nike ecosystem for second and third opinions.

Kara and Salazar flew together to Daegu, South Korea, for the 2011 IAAF World Championships in Athletics. According to a confidential source, it was on this flight that a drunk Salazar tried to kiss Kara on the mouth. Kara, who was working with the United States Center for SafeSport during my reporting of this, told me, "I'm not saying that it didn't happen, but I don't want to talk about it right now."

Adam wasn't scheduled to arrive in Daegu for another five days, but he could tell something was bothering Kara. Every time they Skyped she was in tears, but he chalked it up to the pressure of racing at the top level.

On the way to their events, Salazar handed out solid blue pills, which he said was B12. But Kara knew B12, having taken it in the past, and this didn't look like the vitamin to her. She also watched as Salazar manipulated WADA rules to get Rupp intravenous fluid before races, instructing the athlete to tell medical staff that he was dehydrated and unable to drink. The entire trip proved "glass shattering" for Kara. She was injured (and would find out later she had a stress fracture in her foot), and because her family couldn't afford the reduction in salary that would result if she didn't race, she endured a painful event and an embarrassing result.

Kara's 10,000-meter final took place on the first day of events in South Korea, during the 13th World Athletics Championships. An American contingent set the early tempo, with Shalane Flanagan and Jennifer Rhines joining Kara in an attempt to push the pace. But it would be a Kenyan sweep of the podium, as the reigning 5,000-meter world champion, Vivian Cheruiyot, took the victory. Flanagan was the first American across the line in seventh place, with Jennifer Rhines next in ninth, then Kara, who finished in thirteenth.

Farah and Rupp raced next. Before the first of their two races in Daegu, neither man had ever medaled in a global championship. Rupp had never placed higher than eighth, and Farah no better than sixth. This was their moment. The twenty-eight-year-old Farah lined up with his twenty-four-year-old Oregon Project teammate for the 10,000-meter final. Farah launched a surge with 500 meters to go, but he couldn't hold the lead and was dramatically passed in the last 100 meters by Ethiopian Ibrahim Jeilan, who won by just 0.26 seconds. Flanked by Ethiopians, Farah became the first English athlete to ever earn a world medal in the event. Despite his superior hydration status, Rupp was not quite ready to level up to the international competition and finished in seventh place.

The final in the 5,000-meter event took place a week later, and the Oregon Project's top men once again lined up against each other. This time, however, Farah would not be denied. He led through the final lap as a host of Ethiopians, including Imane Merga and Dejen Gebremeskel, were closing on him in the final stretch but ran out of real estate to overtake him. Bernard Lagat, swinging out wide of the melee, came from behind in lane three to eke out the silver medal on the final straight. Rupp finished an upsetting ninth place.

Later, Kara and Rupp went for a training run together. Rupp seemed distraught, and while nearly in tears, he admitted to Kara that he was thinking of leaving Salazar. "He's too controlling and involved in my life," he told her.

During the cab ride to the airport after the events, Adam was contemplating how, exactly, to tell Kara that he wanted her to leave the Nike Oregon Project. In deference to her career and her mental state he hadn't said what he'd been feeling for a long time—that Treasure and Salazar were pitting her against him by undermining his input and constantly questioning if he was the right partner for her. But in the cab, before he had a chance to bring it up, Kara said, "I'm leaving Alberto. I'm leaving this. I can't do it anymore. I can't be here."

The homecoming proved to be a truth serum for Kara, who finally explained to Adam what had happened on the plane to Daegu with Salazar. Shortly after she told him, Adam's best friend Tim Catalano arrived at the Gouchers' house for an extended visit. He and Adam had a book tour planned for their recently published *Running the Edge*. Adam and Kara quickly filled him in on the situation.

"Kara was upset. Adam was upset," said Catalano. "Adam just kept saying he wanted to go to Alberto's house, like, right now." Tim helped keep Adam from tearing off across town to confront his former coach.

"She really loved Alberto as a father figure, and that, that was it though," said Catalano. "How do you come back from that? You don't come back from that."

Working with her agent, Dan Lilot, Kara began quietly reaching out to other coaches. She was shocked when she started to hear back that some of them assumed she was a doper and wanted nothing to do with her, while others simply wanted to wait until after the 2012 Olympics.

In the fall of 2011, the couple contacted Jerry Schumacher, who came over to the house to discuss the prospect of coaching Kara. "I didn't know what to expect," she said of the meeting. "I was injured and quite a bit overweight. I was really depressed and he just made me super, super excited. When he left, Adam and I were like, 'That's the guy.'"

Schumacher went to Nike with the proposal. They wanted what was best for Kara, but thought the move might "kill Alberto." The group explored other options, but two weeks later, Schumacher came back to Kara and said, "I'll rattle cages if you're willing to rattle cages."

"Let's go rattle cages!" she replied.

After hearing of her decision, Salazar called and gave his blessing; he wouldn't be upset if she went elsewhere. "You can do whatever you want," he calmly told her.

With just three months to get ready for Olympic Trials, Kara joined Schumacher's group, the Bowerman Track Club. She went out for her first run with her new coach's star pupil, Shalane Flanagan, and quickly realized she had her work cut out for her. Flanagan was running a twenty-mile-long run. Kara only lasted for eight of them.

BEHIND THE SCENES SALAZAR'S VITRIOL FOR SCHUMACHER'S GROUP FESTERED, HOWever. The picture he had painted for his assistant coach, Steve Magness, was that "Jerry's kids" controlled Jerry. The team was old-school and anti-science to their own detriment. Conversely, Salazar considered the Oregon Project progressive and boasted about using research and high-tech gadgetry. Schumacher wouldn't even use Treasure with his athletes.

In August, Chris Solinsky, who had been suffering through hamstring issues, tripped over his dog coming down the stairs and completely tore two of his hamstring tendons off the bone. A hamstring avulsion, as it's called, is a gruesome and painful injury that can end a runner's career.

Salazar reveled in the news of misfortune among Schumacher's athletes. Solinsky's fate was no different. "Alberto was happy about it," Magness told me. "He was like, 'See, I knew this would happen to them.'"

Though clearly aligned with the Oregon Project, Magness couldn't conjure the schoolyard vitriol for Schumacher's team, whom Salazar had been calling "our mortal enemies." Magness also knew Solinsky a little bit, having run against him in high school, and had always thought he was a nice guy. "I felt bad for Solinsky, he's not a bad dude, he's a good guy," said Magness, "and we're cheering that he's injured, so now it looks like Alberto's a better coach because Galen is going to beat Solinsky. He was obsessed with beating Jerry and being better than Jerry."

AFTER A SOLID TEN MONTHS OF TRAINING WITHOUT CATASTROPHE, ADAM'S THOUGHTS of retirement started to dissipate and the idea of running an Olympic marathon qualifying time for the 2012 London Games began to seem possible. And by August he had clawed his way back to elite-level fitness.

He finished the Rock 'n' Roll Philadelphia Half Marathon course in twentieth place, with a respectable time of 64 minutes and 53 seconds. More important, it qualified him to line up at the US Olympic Marathon Trials in Houston the following April. But that realization was short-lived as Adam came to terms with the knee pain that plagued his every step during the race. A subsequent MRI showed a torn meniscus and substantial damage to cartilage in the joint. His professional running career was over.

In November, just before undergoing another knee surgery, he announced online that he was retiring from professional running. "I am retiring from elite racing but I am not done running. Not by a long shot!" Adam wrote. "I am looking forward to new starting lines and joining the millions of runners who find inspiring reasons to run that do not include Olympic berths or even personal bests. Those days might be behind me but I feel like my running career is just beginning. I am a runner. That will never change.

"This coming Friday (November 11, 2011) I will go under the knife again to fix what is broken . . . There will be no Olympic Trials marathon in January. That is one starting line I will not make."

WITH THE OLYMPIC MARATHON TRIALS APPROACHING, SALAZAR WAS MANIACALLY focused on ensuring that Dathan Ritzenhein got pumped full of L-carnitine, the magic supplement. NutraMET, the UK-based manufacturer of the first L-carnitine drink, agreed to make an early special batch exclusively for the Oregon Project athletes before it was commercially available, "so your athletes will still gain an advantage," they told Salazar.

Upon this news the coach sent an email to the marathon athletes on his team that the "greatest sports endurance supplement is on the way" and that the first batch was "getting tested at an independent lab to make sure there's nothing bad in it and then on the plane.

"You will each start on ot [sic] immediately as it takes months to build up." When the pallets of the L-carnitine sports drink arrived on September 28, 2011, Salazar emailed Dathan Ritzenhein, Galen Rupp, Alvina Begay, Lindsay Allen-Horn, and Magness:

> Hi Everyone, I'm bringing a box of the new sports drink we got from the UK to Nike tomorrow. I've got enough for six months for each of you. It takes up to four months to take effect, so for the marathoners you need to start now. It definately [sic] will help a 10k runner. Possibly a steepler and 5k runner. Steve, is it worth giving to milers? All of you need to get it from me tomorrow. I'll be at the track at 9:30? For Jackie, Lindsay and alvinas [sic] workout. Steve is it ready to go at9:30 ? [sic] – Alberto

Though not yet available to the wider population of athletes whom NOP runners would be competing against, there was nothing illegal about the drink (L-carnitine has never been on the WADA banned list). But it was clear to the athletes by now that this team was all about marginal gains; if there was even a chance that it would improve their running speeds, they were expected to drink their medicine.

Everyone took their cases of the drink home with instructions on how much to consume each day. Without consulting Salazar, Magness told Areson to cut the amount in half. "There was so much sugar in it," said Areson, "Steve was like, 'This seems unnecessary, we don't want you to gain weight, that seems like the opposite of what we want.'"

Areson's dedication showed. Her weight was down, and her times were up. But that didn't seem to be good enough for Salazar, who she said was obsessed with the numbers on the scale. He had a callous disrespect for women, she said, and no one escaped his critique. "He said I had the biggest butt he'd ever seen, even though I was one hundred and five pounds," Areson said. "It just became comical, we were like, 'he's just a lunatic.' He just thought every woman should look like a ripped Galen and Mo." At meets, she said, he'd pick a person of the day and comment about them ad nauseam. "I think he's a psychopath or a sociopath," said Areson. "I think he's one of the worst people I've ever come across in my life."

Salazar worried that with the Marathon Trials happening so soon that Ritzenhein, in particular, didn't have enough time to load the L-carnitine. "In their article it talks about getting the same results in a few days with infusions," he emailed Magness. "Please check into those asap with Dt. [sic] Brown to see if he can do it and of course if it's Wada legal. For everyone else we have time for the supplement to work, for dathan we may not. This has to be a top priority for you this week. Jackie, Ciarán, even Galen and mo take backseat to getting dathan ready. I don't care if you come to work, just get this figured out asap."

Later the same day Magness responded, it's "no good" because "it has to be infused with Insulin to work like in the studies. Insulin IV is banned by WADA." He was alluding to the fact that the world anti-doping rules are strict and clear around being infused or injected. In a six-hour period, an athlete may not receive more than 50 milliliters (an amount equivalent to 3.4 US tablespoons) of any substance, whether it is on the banned list or not.

Ritzenhein and Begay were both scheduled to race in the US Olympic Marathon Trials on January 14, 2012. Ritzenhein had struggled with injuries for the entirety of 2011—his IAAF profile page tells the tale, not a single race is listed for the world championship year. (In March, he had surgery to remove his left Achilles tendon sheath and a neuroma in his right foot. His comeback was then slowed by a severe allergic reaction to the stitches, causing an infection which led to another surgery, further delaying his comeback to top form.)

Magness worked with a couple of different doctors in the research group to figure out if there was any way to increase the rate of L-carnitine loading, explaining to the men that one of their runners only had about fifteen weeks until the Marathon Trials. On October 19, 2011, Dr. Paul Greenhaff emailed Magness back with the infusion protocol, but admitted, "that should work—having never done it I can't be sure."

"Salazar became super obsessed with it," said Magness. "He pinned a lot of hope on the supplement and everything took so long that he started freaking out." At Salazar's instruction, Magness reached out to Dr. Kristina Harp, an internist in Portland, Oregon, who acted as one of NOP's team doctors at this time. "If Salazar needed an injection of iron or something for an athlete, he'd use her," Magness told me. "Every once in a while, UK Athletics would request that Mo get injections, like B12 or whatever random vitamin they wanted, and they'd send Mo to see Harp to get it done." Magness didn't know Harp, and she never responded to his questions about infusing the Oregon Project athletes with L-carnitine, but he assumed she called Salazar and told him no directly.

Either way, he was instructed to move on to working with Dr. Jeffrey Brown, in Houston, Texas, to figure out the logistics of the infusion. By sheer coincidence, or prominence, Magness had worked with Brown as his doctor in high school, though he hadn't seen him in over a year by this time. "I was a patient of Dr. Brown's since I was like fifteen years old. He wasn't yet nationally known back then, but he was *the* local dude," said Magness. "So I knew him, and he was a good doctor to me, I thought. By then he had become involved with USATF and he was well respected. He was the guy to figure things out."

Brown was skeptical that the L-carnitine infusion would even work. Additionally, he told Magness that he thought there were potential risks in an L-carnitine infusion for someone with thyroid issues, which narrowed the pool to effectively zero on the Nike team. So Salazar told Magness to set Areson up as the guinea pig, but Magness wouldn't allow it.

"What if we just try it with Dathan?" Salazar wrote in an email. "We have nothing to lose, if it works it will get his Lcarnitine [sic] levels up quicker. If it doesn't there's no harm."

Brown agreed to help the team figure it out, though he emailed them that he was leery of giving the "insulin and glucose clamping" to Ritzenhein, since he too had an ongoing thyroid issue. Disregarding the fact that Magness also had hypothyroidism, Salazar came to believe that his assistant coach would be the perfect test dummy, since he was still fast and often ran with the athletes during training sessions. Magness was registered as an athlete and a coach with USA Track and Field and enjoyed entering the odd race here and there. And it had become part of his job with Nike to stay fit and ready to race and pace if need be. It turns out that it's hard to find athletes who can keep up with the fastest runners on earth, and although Magness was no longer technically at their speed, he was close enough to help out on occasion. The team would usually pay pacers somewhere between $100 and $500 per session, but with Magness on board, and fit, they would instead pay him with the funds as cash bonuses.

In an email to Magness dated November 15, Salazar wrote that they would do "pre L-carnitine exercise tests prior to Thanksgiving, then you fly there [to see Dr. Brown in Houston, Texas], get the L-carnitine infusion, come home and retest . . ."

Brown procured the infusion bags of L-carnitine from Corner Compounding Pharmacy in Sugar Land, Texas. Before he left for Houston, Magness underwent a grueling treadmill test on the Nike campus designed to measure an athlete's maximum oxygen uptake, or VO_2 max. This is a value that can be indicative of an endurance athlete's performance capacity, and when coupled with a measurement of a person's fat metabolism efficiency at high intensities, can begin to predict somewhat reliable performance estimates. These are the lab numbers generally used to measure training progress.

"I was freaking out but I was putting a lot of trust in Dr. Brown," said Magness. "Since he was my doctor I pinned my expectations on him as to whether I should do it or not. And he said it was safe."

Four days after Thanksgiving, Brown stuck a needle directly into a vein in Magness's arm and administered a continuous gravity drip infusion of 1,000 milliliters of solution that contained the requisite sugar, in the form of dextrose, and 60 millimoles of L-carnitine. That amount of L-carnitine alone was over the allowed limit by WADA, but the bag that hung over Magness as he sat in Brown's office on that November day was approximately twenty times the volume that is considered legal in a six-hour period by anti-doping agencies the world over. Not only was Magness support personnel, but he had raced about a month prior and was scheduled to compete in the December 10 Cross Country Championships in Seattle, Washington.

Back home in Oregon and brimming with L-carnitine, Magness retook the VO_2 max and substrate utilization test. On December 1, he emailed a spreadsheet to Salazar summarizing the astonishing improvements:

> Both adaptations would result in very significant performance enhancement that is almost unbelievable with a supplement. The changes in VO2max (7.6%) is within the range that research has

> shown is the change that occurs with blood doping (5-9% according to Glehill et al., 1982).

Salazar, giddy beyond belief, told Magness his work on this was worth a huge raise, then stressed that it was to be kept an absolute secret. "We can't tell anybody about this," he warned Magness. Salazar thought L-carnitine would improve a marathon time by about a minute and a half. He emailed Nike CEO Mark Parker and the Nike executive who helped him launch the Oregon Project back in 2001, Tom Clarke, with the results.

He then sent an email to Lance Armstrong that read:

> Lance, call me asap! We have tested it and it's amazing! You are the only athlete I'm going to tell the actual numbers to other than Galen Rupp. It's too incredible. All completely legal and natural. You will finish the Iron Man in about 16 minutes less while taking this. – Alberto.

The last step in Salazar's ad hoc L-carnitine testing was to see if these improvements could be seen in the real world, through raw speed on the ground. For this, Magness set out with Ritzenhein for one of his long-tempo runs: fifteen miles at a 4 minutes and 50 seconds per mile pace. A stupendously fast "tempo" that few humans can maintain for even a single mile, let alone fifteen. They started with laps at the Nike track, before venturing off for a loop around campus. Rupp and Farah joined them for part of the run, as did Salazar, who cruised alongside on his bicycle. With the US Marathon Olympic Trials a little over a month away, this was a key session for the twenty-nine-year-old Ritzenhein.

It was six days after Magness's infusion and three days after they had gotten the promising lab results back. The group was already drinking the L-carnitine sports drink and knew Magness had had an infusion of the substance, so as the miles ticked by and he was still running this aggressive pace, the conversation centered on just how good this supplement was. Magness ran 11.5 miles of Ritzenhein's workout that day and logged a new 10-mile PR.

"It was very annoying," Ritzenhein would later tell USADA, "and Galen was quite upset about it afterward, that he just kept hanging in there." Salazar, only half-joking, told Magness he looked so good that he should think about making his marathon debut at the Trials.

Infused-carnitine, it seemed, was the real deal.

Days later, Ritzenhein was instructed to travel to Houston to receive the same infusion. Brown told him that it would take the same amount of time that Magness's did, "about four to five hours." Salazar mentioned that Nike was now looking into purchasing NutraMET so they could secure the competitive advantage for their athletes (not to mention that the drink was expensive, costing about $1,000 per person for a six-month supply).

Magness was now in charge of setup and administration of the treadmill tests to measure economy, $V0_2$ max, and substrate utilization data for the athletes before and after infusions.

Things get a bit murky from here, however, because Brown neglected to list the infusion quantities on the medical records for most of the athletes. For reasons he could not sufficiently explain, even under oath during his USADA deposition, he had not properly annotated the details for most of the NOP athletes' infusions.

After spending so much time injured and not able to race, Ritzenhein was no longer being paid his salary by Nike. And it was clear that everything was riding on his ability to make the 2012 US Olympic marathon team. To make the likelihood of that easier, Salazar stressed to the runner, he needed to get the infusion. But the procedure made Ritzenhein extremely uncomfortable. He flittered back and forth between going through with it or not. He knew that if he lost favor with Salazar by disobeying him, he could be cut from the team entirely.

Ritzenhein confided in his wife, Kalin, that he was nervous about the infusion. They talked about it for days before Ritzenhein went to see Salazar and, for the first time, pushed back on what the coach wanted. "Are you sure this is legal?" said Ritzenhein. "This doesn't sound legal."

The runner now believes his comments to the coach prompted him to contact USADA to gain some modicum of assurance that what they were doing was within the anti-doping rules. Salazar sent a confused email in which the coach claimed Magness's infusion was a "clinical trial." The coach then forwarded the email on to Ritzenhein and Magness with a note:

> Hi Dathan, we are cutting edge but we take no chances on a screw up. Everything is above board and cleared thru USADA. They know me very well because I always get an okay before doing anything!

"I remember reading that," said Magness, "and thinking, *Clinical trial? What?*" USADA science director Dr. Matthew Fedoruk called Salazar and then followed up with an email in which he stressed: "Infusions or injections are permitted if the infused/injected substance is not on the Prohibited List, and the volume of intravenous fluid administered does not exceed 50 mL per 6-hour period."

Sufficiently coerced, Ritzenhein flew from a high-altitude training camp in Albuquerque to Houston for his infusion with Brown. Over the next nine months, one by one, all of the athletes on the Oregon Project would fly to Houston—with the exception of Farah, who would have his done in England—and sit on Brown's office couch to get infused with an unknown amount of L-carnitine.

17 LOOPHOLE SALAZAR

SHORTLY AFTER HIS INFUSION, RITZENHEIN BEGAN FEELING SICK. HIS BLOOD WORK showed a spike in TSH, and he was steadily losing faith that Salazar could properly orchestrate the panoply of substances he was subjecting his body to. When questioned, the coach would tell the athletes, *I can't coach you if you don't do this.* They knew Salazar was the Oregon Project, he answered to no one, and crossing him would result in an end to their professional careers.

When Ritzenhein complained, the coach discounted his concerns and told him that his side effects weren't due to the L-carnitine. However, a review of Nike emails sent by Salazar just days later showed the coach either changed his mind or was telling different athletes different things. A few days after telling Ritzenhein it wasn't the L-carnitine that was driving up his TSH and making him feel terrible, he wrote to Rupp on December 20, 2011:

> Hi Galen , take a full extra levoxyl tonight and start on cytomel right away. We're not going to wait for another TSH test. You've been sick before and never had a TSH go to 3.0. It's got to be the Lcarnitine. This is great news! no one else will know this possible side effect of the drink. -Alberto.

> Ps – vie [I've] got cytomel. If you don't have it, call me and I'll drive it over. I have to drop Maria off at a party anyways.

All the NOP athletes, it seemed, were on prescription medication with dubious justifications. And Salazar's tactics were beginning to spread like a disease among Nike athletes, even inside Schumacher's camp. "The year that I left Nike I was running with Chris Solinsky and Matthew Tegenkamp," Kara told me. "They were like, 'Just so you know, USADA has been sniffing around so you might want to lower your Advair dose.' I was like, 'I'm not on Advair.'" Kara said they seemed shocked, as they were both on a high dose of the asthma drug.

Early in the new year, 2012, Salazar sent an email to Ritzenhein, Rupp, Begay, Treasure, and Magness, cc'ing Alex Salazar, that read:

> HI Dathan, Alvina ,and Galen, For your interest. When asked about an Infusion, you are to say no. LCarnitine and Iron in the way we have it done is classified as an injection. So no TUE's and no declaration needed, not online and not when asked about infusions when getting drug tested in or out of competition. Thanks.- Alberto

This didn't seem right to Magness, who was internally questioning the legality of what they had done. If Salazar had truly gotten everything cleared through USADA, it wouldn't matter if the athletes declared the infusions during drug tests or not. When he brought it up with Brown and Salazar, however, they again claimed to have gotten USADA approval. "It now seemed like Alberto was intentionally obscuring what had happened," said Magness. "And, things started mounting."

In early 2012, Magness's desk was moved into a cubicle that he shared with Alex Salazar, Alberto's son. Alex has an MBA and was the business manager for the Oregon Project (he was also a registered coach with USATF). One day, Alex sent Magness the Oregon Project team budget to print, which the assistant coach couldn't help but peruse. "It was something like a million dollars, not including athletes' contracts," said Magness. "It was a lot, and we'd always go over budget, but Alberto didn't care." A former Nike coach had told Magness that the joke around the office was that Salazar was Phil Knight's bastard son. It was readily apparent that they were close, and the coach seemed untouchable, able to go over budget with impunity and essentially unable to do anything wrong in Knight's eyes.

In friendly office conversations, Alex shared many of his crazy-dad stories with Magness. Like the time Salazar gave Alex "gas station dick pills" to see if they actually increased testosterone. Or another where Alex took megadoses of vitamin D, again to see if it would increase his testosterone. And then, of course, the time his dad rubbed actual testosterone cream on him to see just how much it would increase his testosterone.

Alan Webb had told Magness that Salazar had once given him his daughter Maria Salazar's prescription medicine. Salazar hadn't removed the details from the label, which identified that it was meant for Maria on the side. Magness now thinks it was most likely a anti-inflammatory drug (Webb later confirmed this), but this memory made him consider all the times Salazar had handed out the prescription drug Celebrex at the track as if it were candy, or the time he made Magness drop everything in the middle of working with an athlete to drive from Park City down to Salt Lake City and pick up a prescription medication for Matthew Centrowitz Jr. that, for some reason, he needed immediately. So yes, to Magness it all seemed crazy, haphazard, and quite frankly illegal.

THE US MARATHON TRIALS IN HOUSTON ON JANUARY 14, 2012, WOULD ONCE AGAIN make the heartless distinction between third and fourth, the misery or blessed relief of being one of the three athletes in the country to represent America at the Olympics. Rupp flirted with the idea of running his first marathon that day but ultimately withdrew from the field prior to race day. Ritzenhein, who had been the top American marathoner at the 2008 Olympics, fell off the pace around eighteen miles into the event and watched the top three—Keflezighi, Hall, and Abdirahman—disappear from view. Though he rallied near the end and made up much of the time on Abdirahman, he would ultimately be 8 seconds too slow to make the Olympic squad. Keflezighi became the oldest man to win the Trials, running 2 hours, 9 minutes, and 8 seconds at thirty-seven years old, and taking home a prize purse of $80,000. Ryan Hall was next 22 seconds later, with Abdi Abdirahman snagging the last spot on the Olympic marathon team.

Ritzenhein crossed the line stupefied, into the midst of the three athletes before him who were celebrating their Olympic dreams coming true. He bent over in exhaustion, then squatted down to gather himself and began to weep—Olympic rings tattoo visible on his shaved left calf. The broadcast cameras momentarily fixed on him, capturing his teary walk through the finishing chute. Nike reduced his salary from the $200,000 he had made the year before, down to $100,000 for 2012.

Kara's training for the Marathon Trials had gone well, all things considered. She had rapidly come into form while working with Schumacher and chasing her training partner, Shalane Flanagan.

Flanagan shined. It was only her second attempt at the marathon distance, but she ran like a veteran, patiently tracking the front group until the twenty-first mile, where she took the lead. But Hansons-Brooks runner Desiree Davila Linden wasn't going to let the win go that easy and stalked Flanagan all the way to the line, finishing 18 seconds back, in second place. Flanagan's time of 2 hours, 25 minutes, and 38 seconds was a new American Olympic Trials record.

In her second marathon in nine months Kara claimed her first spot on an Olympic marathon team by finishing third in 2 hours, 26 minutes, and 6 seconds. Flanagan, Davila Linden, and Kara were headed to the 2012 London Olympics to represent their country. "I never would have made that team in 2012 without Jerry and Shalane," Kara said. "I owe so much of that to her."

"We've had a great journey together," Flanagan said after the race, "and I think it's going to get even better now that we get to train full-time for London."

ON FRIDAY, FEBRUARY 3, JUSTICE DEPARTMENT PROSECUTORS ANNOUNCED THAT THEY were closing a criminal investigation of Armstrong. No charges of performance-enhancing drug use would be filed, they said, without explaining the reasons for the decision. Despite the mountains of circumstantial evidence, including eyewitnesses, it seemed as though Armstrong would never suffer a reckoning.

"It is the right decision and I commend them for reaching it," said Armstrong. "I look forward to continuing my life as a father, a competitor, and an advocate in the fight against cancer without this distraction." A disaster, it seemed, had been averted by Armstrong, and Nike.

Rupp, who was now twenty-five years old and had just gotten married to his college sweetheart the previous September, headed to the USATF Classic in Fayetteville, Arkansas. He lined up with the intention of breaking Bernard Lagat's 2-mile indoor American record. Before the event, Salazar was having trouble finding someone to pace him at the 4-minute-per-mile speeds it would take to set a new record (just under 15 miles per hour). "I don't want you to pace Galen's race," Salazar told Magness, "because USADA knows you had an infusion."

"That was all he said about it," Magness told me. "I remember feeling that shot of adrenaline, like 'Oh shit, what the hell does that mean?' I think he was worried that if I assisted Galen in the race then the record wouldn't count." It was Salazar's first acknowledgment to Magness that despite what he said, the infusion hadn't been USADA approved, and might have actually been a serious rules violation.

Rupp ran mile splits of 4 minutes and 7 seconds and 4 minutes and 2 seconds for a new American indoor record with a total time of 8 minutes and 9.72 seconds. He told a reporter that his biggest takeaway from the race was to not lose sight of the competition while chasing after a record. When you set a record but come in third place in the race, it tends to dull the shine of the achievement.

"In the end that's what it's all about, winning," said Rupp, "and beating people." In the same interview, he talked about his relationship with Salazar. "We have complete trust in each other," he said. "That doesn't mean that we don't fight and yell, but at the end of the day it is like a family, ya know? . . . I've always had complete one hundred percent trust in what he's done with me, his training, and all of that."

As the Oregon Project headed to New Mexico for a training camp leading up to the Indoor National Championships in late February, Magness and Salazar continued to wear on each other. While there, one of the athletes, Dawn Grunnagle, had started to complain that she wasn't feeling quite right and had begun to cough. Salazar would always overreact in these scenarios, worried that Rupp would get sick too. He called Brown and demanded a Z-Pak—the brand name for the drug Zithromax, a prescription antibiotic— for Grunnagle. However, this time, Brown said no. The drug had side effects like nausea, stomach pain, constipation, diarrhea, and the lethargy that often accompanies a disrupted microbiota, but the real threat in overprescribing the drug is antibiotic resistance. Magness, who was in the room during the call, could tell that Brown, on the other end of the line, was saying no. Salazar was growing more and more agitated, his voice rising until he was screaming at the doctor all the reasons why he needed the prescription. He hung up in frustration.

"Dawn's not even sick and he's trying to get her antibiotics," said Magness, "because if she got Galen sick, that's the end of the world."

Thank god Dr. Brown is saying no, Magness thought. Not ten minutes later the doctor called back to relent and told Salazar the pharmacy where he could pick up the drugs for his athlete. "Alberto can get anything he wanted," said Magness, "if he yelled loud enough."

Albuquerque can be windy during the winter months, and since Salazar hated to have Rupp run in adverse conditions, he would often use Magness to make workouts less arduous for the athlete. On this trip, he had Rupp run a long workout in one direction, going with the wind. And rather than having him turn around to fight the wind on the way back, Salazar had Magness follow in a car and shuttle him back instead, so Rupp could resume running with the gusts. He had, at times, done something similar with the runner's hill training. Magness would drive him down or up the hill, depending on the goal of the session.

Even though Farah was clearly the star athlete on the team—more talented and charismatic than Rupp—more than one Oregon Project member told me that Salazar would joke that the Project should be renamed the Galen Rupp Project, because he was all that mattered. Farah was a foreigner, after all, and Nike and their Oregon Project's sole purpose had begun as an effort to return American runners to global podiums.

Magness and Areson would often escape after dinner to hide in the rental car and talk about how crazy everything was. There was the normal level of Salazar capriciousness, but during this period, it was widely known that one of the athletes was cheating on their spouse with another team member. "We'd just try to wrap our minds around this, what we were in," said Areson. She said Salazar distributed prescription painkillers "like it was candy." And she made a point to avoid letting the coach massage her "because I knew he had the testosterone cream and didn't trust him." Both Magness and Areson had come to the conclusion that they had to leave.

The team performed well at the Indoor Nationals and then moved on to a training camp in Park City as they prepared for the Olympic Trials, which also served as the US National Championships for the year. New to the team, Lindsay Allen-Horn was finding it hard to feel comfortable on the Oregon Project. She invited her friend, professional steeplechaser Shayla Houlihan, to join them so they could train together. Houlihan was living in Salt Lake City, sponsored by Brooks, and working remotely with her coach Kyle Kepler toward the upcoming Trials. Salazar graciously gave the okay for her to stay with the team in the Park City condos.

It was an eye-opening experience for Houlihan, who corroborated what Magness had called "Bizzaro world." She concluded that she needed to stop running professionally after seeing the lengths, and expense, to which the Nike team was going. "I didn't have access to any of that on my low income," said Houlihan. "These are the people I'm trying to compete against and they have access to all this stuff, and I don't." (Houlihan is now a professional coach for the Under Armour team in Flagstaff, Arizona.)

Salazar would often wonder out loud what time it was, with the understanding that at some point it was time to drink. Treasure would remind him, "Not until four o'clock, Alberto." Magness wasn't much of a drinker and would often slink away to his room to go to bed while on work trips. At the IAAF World Indoor Championships in Istanbul, Turkey, for instance, Salazar became insistent that Magness join them in the bar for a drink.

"You gotta join us now," Salazar told him. "You gotta join us."

When Magness awkwardly declined again, Salazar started yelling at him. "If you want to be part of the group," he said, "this is where the group is."

Then, a week or two before their next New Mexico training camp, Magness realized that Alex hadn't booked him or his athletes hotel rooms for the Occidental Invitational track meet. Magness took it upon himself to make their reservations. By the time Salazar realized the rest of the team didn't have a place to stay, it was too late to get the type of accommodations he liked, and he had to stay at a lower-end hotel instead. He blamed Magness, and the two men argued about it.

Unjustly blamed, Magness was now determined to leave, though his parents and close friends urged him to stay through the Olympics. He knew how it would look if he left during an Olympic year and he worried about his reputation. But he couldn't do it anymore: Salazar seemed unstable, the Project seemed illegal, and the men who ran it were just too overly disrespectful to the women on the team.

During the next training camp in Park City, Utah, Magness took a deep breath and drove up the hill from his condo to the Black Bear

Lodge where Salazar was staying to tell him that he was moving on and leaving the Oregon Project. "I was ungodly nervous," said Magness. "He's an intimidating dude and he could ruin my career."

The conversation was awkward, but Salazar took it well. He told Magness that he thought the Oregon Project experience had been good for the young coach. Then the two agreed that Magness would work through the Olympic Trials and then they would go their separate ways.

IN JUNE, ARMSTRONG AND FIVE FORMER CYCLING TEAM ASSOCIATES, INCLUDING THE doctor who provided the performance-enhancing substances, were formally accused of doping and drug trafficking by USADA. The anti-doping agency had picked up the case after federal investigators had failed to bring charges.

Later that same month, the US 2012 Olympic Trials for Track and Field were once again held at Hayward Field. As usual, the Nike presence ran deep at the Trials. The swoosh was omnipresent, and Nike had their own VIP section, so luminaries like John Capriotti could watch the athletes in peace. A keen observer noticed Mark Block—a former track-and-field agent who was serving a ten-year ban from the sport for his role in the BALCO doping scandal—also enjoying the action with the aristocrats of track and field in the Nike VIP skybox instead of with the masses in the stands. Someone snapped a photo of him leaving the area that made its way to LetsRun.com, which wrote, "We thought that the tide had turned once and for all and felt that dopers and its supporters would no longer be welcomed in the sport with open arms or with a wink-wink. Maybe we were wrong."

In an attempt to counter the overwhelming Nike presence in Hayward Field, Washington-based Brooks Sports, who sponsored marathoner Desiree Davila Linden and 10,000-meter runner Amy Hastings Cragg, flew a plane over the stadium. Trailing behind it was a banner with Brooks's popular ad slogan, "Run Happy." USA Track and Field and their title sponsor, Nike, were apoplectic. Citing fuzzy rules, USA Track and Field warned Brooks the next morning to stop all such advertisements. But they hadn't seen the admonition in time to stop the next day's flight and the banner once again flew over Hayward Field as the stadium's Bill Bowerman statue looked on. For this transgression, the Brooks executives in attendance had their credentials pulled and were summarily escorted out of the stadium.

This was Nike's track. This was Nike's sport. This was Nike's world.

And this being Nike's golden opportunity, Galen Rupp did not disappoint, proving himself more dominant than ever, by winning the 5,000-meter final in stunning fashion and beating Lagat by less than two-tenths of a second. He also took the national title and an Olympic berth in the 10,000 meters while setting a new 10K Trials record of 27 minutes and 25.33 seconds for the distance. Ritzenhein, who had refocused his training around speed after his marathon Trials failure, was passed in the final moments of the race by Matt Tegenkamp, but managed to hold on for third place. He would be headed to the London Olympics after all. On the rain-laden track Ritzenhein fell to his knees, then made the sign of the cross. "It was a huge weight off the back," he later told *Runner's World*. "I was saying thank you for prayers that were answered. I was overcome with emotion, knowing that I had come back from so much and could finally move forward."

When Areson told Salazar that she was leaving the team, he said, "You're wearing the Oregon Project uniform at the Olympic Trials, or else." She ran her last race in the Nike Oregon Project kit that day, a disappointing sit-and-kick, in which she failed to make the final.

BEFORE THE LONDON OLYMPICS, FARAH WAS ASKED WHOM THE BIGGEST THREATS would be in the 10K. "It's definitely the Kenyans and Ethiopians," he said. "There's always someone new coming through, so you never know who will be your biggest threat, but I'll just give it my best shot."

The London stadium, filled to capacity with English fans, buzzed with anticipation as Farah and Rupp lined up for the now famous 10,000-meter race. Both men stayed near the front through a slow first half until it was time to make moves. Farah took the lead with 700 meters left and never relinquished it. Rupp, with his Nike track spikes flashing yellow, steamed past Ethiopia's Tariku Bekele to take second in the final meters in a thrilling race, in front of Farah's hometown London crowd. The two men appeared to be putting the world on notice that the East African dominance was a thing of the past. English and American athletes were now on the same level in the middle-distance events. To say the stadium shook with applause after Farah won gold and Rupp silver would be an understatement.

Ritzenhein, meanwhile, let a late race surge get away from him, and that was that. The twenty-nine-year-old spent his finishing reserves just trying to catch up to the front, and crossed the line in thirteenth place.

According to Nielsen, a record 219.4 million Americans watched the 2012 Olympic Games, among them Steve Magness, Mary Cain, Josh Rohatinsky, and Pete Julian—to differing effect.

Magness was shocked and saddened.

Cain was inspired. Salazar's athletes had conquered the world. The high school phenom was considered "a once-in-a-generation talent" by running pundits, and she dreamed of someday winning an Olympic gold medal in Nike shoes.

Rohatinsky was incredulous. The 2006 NCAA Division I champ who ran for Salazar from July 2007 to July 2009 watched in utter disbelief. Like Magness and the Gouchers, he found it hard to square Rupp's performance with the athlete he'd trained with all those years ago.

Julian was blown away. "I was watching it on TV and it was such a wonderful day," he told me. "I got swept up in it, for sure." Soon after the race, Salazar called him while still in the stadium. Track's most famous coach was now requesting Julian join the team he had just watched win gold and silver at the Olympics. "Pete, you gotta come out," Salazar said.

"He had been working on me for like a month at that point," Julian told me, "and it wasn't that I didn't trust what was going on, but I was a Pac-12 coach and I loved my team. I loved what I was doing. I loved it."

The next day, Kara and Shalane ran at the front of the marathon pack that spanned both sides of the road as the group streamed past Big Ben, Parliament, then Buckingham Palace. It was a rainy day and an undulating course, consisting of one short and three long loop sections passing some of London's best-known landmarks. The two Americans were subsumed by the group just past the 9-mile mark, as Ethiopian Tiki Gelana carried on for the win. Flanagan was the first American in tenth place, with Kara 16 seconds behind her in eleventh.

On August 11, Rupp and Farah lined up for the final in the 5,000 meters. There was a charge in the air, like an approaching thunderstorm. Farah put on a clinic, hanging in the back of the pack for the early part of the race before gradually moving up to the front. Though Rupp surged up to Farah's shoulder just before the final lap bell, he couldn't maintain the pace to stay at the front. Farah, running as if lightning were nipping at his heels, streamed off for the win as the London crowd went wild for the double-gold medal winner in the Union Jack (the noise in the stadium was so loud that the sound waves, estimated at more than 140 decibels, distorted the finishing photo). Rupp came through in seventh, more than four seconds later. Farah had entered rarefied territory, becoming the first British athlete to win gold at the 5,000-meter event and joining a group of just six other men who have won the 10K and 5K double-gold medal at the same Games. Farah got on his knees, bowed to the north, then flipped over and did some crunches, as if to say, *That was nothing, I've got more in me.*

AFTER THE GAMES, ON FRIDAY, AUGUST 24, 2012, AT 5:46 P.M., SALAZAR'S LAWYER, Roy B. Thompson, sent brothers Robert and Weldon Johnson, the twin owners of LetsRun.com, an email with the subject: "Alberto Salazar." Attached to the email was a letter:

> Rather than filing a legal action against you . . . we request the following actions:
>
> 1. The immediate production to this office of any and all ISP addresses and any and all identification information provided by those users listed in Exhibit A; and
> 2. The immediate removal of all defamatory and untrue statements regarding Mr. Salazar from your site.

Exhibit A listed 117 usernames, including:

> Darth Salazar
> Lance,aka Galen [sic]
> thyroid
> loophole Salazar
> Mary Slaney's Dirty Coach
> Al Sal A Drug Czar
> Boy, I wonder

Weldon wrote back: "LetsRun.com will not provide to you the IP addresses or identification information of our users without a valid court order," adding, "Being a public figure, Mr. Salazar should come to expect a fair amount of criticism from being in the public sphere." He then cited a joint project of the Electronic Frontier Foundation and clinics at Harvard and Stanford Law Schools who have legal resources on internet freedom, pointing out the fact that internet forums are not liable for the posts therein. Thompson thanked Weldon "for the Harvard and Stanford cites," and was never heard from again.

During this period, Salazar was working on getting Julian to move to Portland and join the Oregon Project as the new assistant coach. To do so, Julian would have to abandon his duties at Washington State University in Pullman, Washington, just as his coaching efforts were beginning to bear fruit. But he didn't care for the Pullman weather and the job meant he had to live away from his wife, who lived and worked in Boulder. The timing couldn't have been worse for Julian, whose school year was just getting started. "My kids were already on campus for the fall cross-country season," he told me. "It was terrible. That was one of the hardest decisions I've ever made."

Before he called Salazar with an answer, Julian dialed his friend Adam Goucher. "Adam, is there any reason I shouldn't take this job? Tell me right now if the Oregon Project is doping."

According to Pete, Adam told him that they weren't doping, but that Adam didn't like Salazar. "'I hate that dude,' he said. 'But I give you my blessing and I think it's a wonderful opportunity for you.'"

"It was on Adam's word," Julian told me, "why I actually took the job."

Adam remembers the conversation differently, however. "I told him, 'Don't go there.' And he kept asking me, 'Why?'"

"They're cheating," he told Julian.

"People say that about everyone. People say that about you, Adam."

"Yeah, but you know me, man. Do you think I cheat?"

"No."

"I understand if you really feel like you need to do this for your family, then fine, it is what it is," said Adam. "But, dude, you gotta promise me that when you see something, you say something, you expose it."

Either way, Julian packed up and moved to Portland, a city he was pretty familiar with, having gone to the University of Portland for his undergraduate degree. He said he and Adam haven't talked since then, though he was on the receiving end of a nasty text message from Adam after one race where his Nike athletes performed well, which Julian declined to share with me.

When he arrived in September 2012 to work for the Oregon Project, most of the team was away on break after the Olympics. One of the first athletes he met was Tara Erdmann Welling, who had joined the team despite being thoroughly warned off by both Magness and Areson (through an email correspondence with a fact-checker, Welling claimed this is untrue).

One of Julian's first official tasks was to meet Welling in Houston and accompany her to her appointment with Dr. Brown. Welling, whose mother had recently been diagnosed with cancer for a second time, and who was dealing with an eating disorder while also trying to train at a world-class level, initially told USADA that she'd never seen Dr. Brown. An email between Salazar and Rupp reviewed by the anti-doping agency suggested otherwise, however (Welling would eventually cooperate with the investigation).

Her medical records show she received four injections of 10 cc resulting in a total L-carnitine level increase of "an extraordinary, almost 11,000%," though Welling claims not to know whether she received an injection.

"I can't speak for Tara," said Julian, "but we were there." When asked about the injections, he doesn't know if he was present in the office or if he sat in the waiting room. "There is no way I could ever tell you what was in the bag, or if there was a bag, or what," he told me. According to Julian, Welling would be the last Nike Oregon Project athlete ever sent to see Dr. Brown.

LANCE ARMSTRONG'S HOUSE OF CARDS FOLDED IN ON ITSELF FOR GOOD IN OCTOBER when USADA published a dossier called the "Reasoned Decision," that indicted the cyclist as the ringleader in the "the most sophisticated, professionalized and successful doping program that sport has ever seen." They imposed a sanction of a lifetime ineligibility and disqualified his competitive results dating all the way back to 1998.

Nike, who had recently defended Armstrong, finally took action after publication of the USADA report. At the time, they had ninety-eight products bearing the Livestrong name next to the Nike swoosh.

Armstrong remained defiant and continued to deny that he had ever doped, even claiming that the entire USADA investigation was one-sided and unfair. In their nine years together, Nike and Armstrong made the yellow Livestrong silicone bracelet an internationally recognized symbol in the fight against cancer, while raising more than $100 million for the nonprofit that would soon be rudderless.

Nike endorsement contracts generally have a broadly worded "Morals clause," which allows the brand to terminate the relationship should an athlete-endorser do something that would tarnish their reputation and, in turn, the reputation of the brand. But Nike had a track record for standing by their most important athletes, even as the news steadily ruined their personas, like when the NBA star Kobe Bryant was charged with felony sexual assault in 2003, and when Tiger Woods's façade crumbled around him after news of his many extramarital affairs became public during Thanksgiving 2009 and he lost half a dozen other sponsors. Nike would eventually help Woods, their highest-paid athlete, rebuild his reputation. Phil Knight greatly admired Woods and later wrote in his biography that he "will not stand for a bad word spoken about Tiger" in his presence.

For Woods and Bryant, their transgressions could easily have fallen under the morals clause, but they were technically off-the-field offenses. Armstrong had cheated *in* the races, or on-the-field, where he was supposed to dominate because he trained harder (and wore Nike gear), not because he had a better doctor and ingested more chemicals.

Nike had to act.

Armstrong stepped down from his nonprofit organization, Livestrong. "This organization, its mission and its supporters are incredibly dear to my heart," Armstrong said in a statement. "Today therefore, to spare the foundation any negative effects as a result of controversy surrounding my cycling career, I will conclude my chairmanship."

Nike had already removed Joe Paterno's name from their childcare facility for his part in the Jerry Sandusky child sexual abuse scandal, and it was only a matter of time before they removed

Armstrong's name from the employee fitness center. There was no denying the evidence.

On October 17, 2012, Nike released a statement that read: "Due to the seemingly insurmountable evidence that Lance Armstrong participated in doping and misled Nike for more than a decade, it is with great sadness that we have terminated our contract with him. Nike does not condone the use of illegal performance-enhancing drugs in any manner. Nike plans to continue support of the Livestrong initiatives created to unite, inspire and empower people affected by cancer."

Nike was the linchpin to Armstrong's sponsorship support; once they abandoned the Texan, the rest followed. In one day, he lost seven sponsors and an estimated $75 million. Days later, the Union Cycliste Internationale (UCI) stripped him of his record seven Tour de France victories. He was banned for life from competing or having any official role with any Olympic sport or any sport that follows the World Anti-Doping Code.

Nike's feigned shock and their claim that Armstrong had deceived them all these years, as Juliet Macur puts it in her book, *Cycle of Lies*, "was as if one of the world's most sophisticated sports companies knew nothing of doping's history in cycling, though Tour winner after Tour winner had admitted to doping."

When she asked Armstrong who, exactly, knew about the doping, the journalist had this exchange with the cyclist:

"Who else knew?" Macur asked.

"Everybody," said Armstrong.

"Everybody?"

"They knew enough not to ask."

"Bill Stapleton?" *Silence.*

"Nike?" *Nothing.*

"The board of directors of Livestrong?" *Not a word.*

"I ain't no fucking rat," Armstrong finally said, "like these other pussies."

18
YOU'RE A NOBODY

LANCE ARMSTRONG'S FATEFUL DECISION TO ALIENATE THOSE WHO HAD FORMERLY been in his inner circle led to his downfall, and proved prophetic when it came to Salazar. He had been discarding athletes and coaches with dispassion for so many years that there was a growing swell of discontent in the community. Everyone seemed to have a Salazar story, or had at least heard their fair share. He was often gruff and aloof toward his competitors at events, displaying that Nike air of confidence, or arrogance, which didn't help his image. And as the rumors continued to swirl about his tactics, it wasn't long before journalists began to get caught in the eddy.

On March 31, 2012, Salazar emailed the team—Mo Farah, Galen Rupp, Dathan Ritzenhein, Lindsay Allen-Horn, Dawn Grunnagle, and Matt Centrowitz Jr.—to make sure they were all on the prescription drug calcitonin, which he thought would inoculate them from the bone injuries that often plague runners.

> Hi Everyone, you should make sure you get on the Calcitonen [sic] nasal spray to prevent stress fractures. Matt, send me your pharmacy number so Dr.Cook can call it in, also make sure you are all on Vitamin D. Thx - Alberto

Salazar wanted his athletes on the nasal spray version of the drug regardless of whether they had seen Cook or Brown for stress fractures. Cook prescribed it for Centrowitz. Brown prescribed it for Rupp.

Oregon Project athlete Dorian Ulrey decided to research the drug, and what he found worried him. One such finding showed an increased risk of cancer. Ulrey presented this to Julian—who had not yet joined the project when Salazar began instructing the athletes to take the drug—and Julian immediately emailed the team on November 29, 2012:

> Hi Team,
>
> We sent out an email earlier today requesting that you begin (or continue on) a calcitonin nasal spray for bone health. Dorian wisely pointed out that there has been some very recent research showing that long term use of calcitonin may slightly increase cancer risk over time. Although the FDA has not restricted calcitonin at this time, it does appear that it is revisiting the long term safety of this drug. If you are currently taking calcitonin, we recommend that you immediately stop taking this prescription until the FDA brings more clarity to the matter. In the meantime, we will look to explore some alternatives to managing good bone health. PJ

Ritzenhein emailed Julian back, "Is this some kind of joke? I have been taking this for the last four years." Rupp must have missed the admonishing message from Julian and none of the coaches told him personally that there were now cancer concerns associated with the drug. He filled his next prescription on December 21, 2012.

KARA AND ADAM WERE CONSIDERING GOING TO USADA WITH WHAT THEY SAW AND experienced on the Nike Oregon Project, but there was much at stake: Kara was still a Nike athlete, and at the time provided her family's only source of income. She also still felt deeply conflicted about turning on the team she once—no, that she still—loved.

Then, a man she had never seen before appeared on her television and a way forward presented itself. That man was USADA's chief executive officer, Travis Tygart.

Armstrong and his team were in damage control mode, working every public relations angle they could. Along the way they had pitted themselves against Tygart and USADA, who they said were just trying to justify their funding by head-hunting the biggest name in sports: Lance Armstrong. By then, the public had turned its collective back on Armstrong and his brand, even altering their Livestrong bracelets by editing out the *v* so they read "Lie Strong."

Attempting to stanch the bleeding, Armstrong flew to Chicago to sit down with Oprah and explain himself on cable television. Despite the animosity, he still had supporters and America loves a redemption story, so maybe if he just fell on the sword, he could begin to mend his broken image. But the interview went sideways, as Armstrong was unable to summon the appropriate level of contrition (among his many missteps he made a fat joke).

When Tygart did his own round of interviews, to counter the Armstrong spin, Kara was watching. She told me his erudition, confidence, and apparent honesty struck her as trustworthy.

"It's hard for an athlete to figure out who to trust, but it was obvious that this guy cares about us," said Kara. "I told Adam, 'If you get him, I'll go to USADA."

Her husband had made connections through his years of visiting the Olympic Training Center, so he contacted the former head of sports medicine at the US Olympic Committee and was soon in touch with Tygart.

In Colorado Springs, at the USADA offices, Kara sat in front of several lawyers and was duly intimidated. The interview process, which she assumed would take thirty minutes, dragged on for hours. She cried a lot during the meeting and felt as though she was betraying those she cared about.

"I didn't know what would happen after that first day. I was incoherent," said Kara. "And I didn't know if they really believed me or if they thought it was something that was important enough.

But it became crystal clear that an investigation was already going on, and that we were suspects."

She left the meeting physically and emotionally exhausted and dazed by the process. Fearing for her contract and livelihood, the couple didn't tell anyone about what she'd done.

MAGNESS SENT THE MOST CONSEQUENTIAL EMAIL OF HIS LIFE TO USADA IN EARLY December 2012, but didn't hear back from the anti-doping agency for about two nerve-racking months. In February 2013, USADA reached out and wanted to hear more about what he'd seen while he was employed by Nike.

He handed over his computer and his cell phone, then agreed to supply them with his medical records from Brown's office. The anti-doping agency's tech department made mirror images of his computer hard drive and cell phone.

"If we are going to do this, we're going to do it the right way," said Magness. "I'm not going to pick and choose what emails to send, I'm going all the way. Which from a legal standpoint and a saving-my-ass perspective I probably shouldn't have done, but from an ethical standpoint I was like, 'Here ya go.'"

One of the main questions USADA had was exactly how much Magness infused on that day on Brown's couch. "I don't know," said Magness to the group in front of him. "I sat there [in Brown's office] for three hours, but I don't know. I've never had an infusion in my life, but it was obvious it was more than the three tablespoons."

As the disgraced heroes piled up, Nike was being roundly criticized in the media for their apparent inability to vet athletes and judge the market. They had lost a step, the *New York Times* wrote, and were having "a harder time standing out amid the clutter, bringing out fewer ads that are widely deemed hot, or cool."

In March 2013, the company put out a marketing campaign celebrating Tiger Woods's victory at the Arnold Palmer Invitational and his unlikely return to the number one world golf ranking (which he had ceded in October 2010). One ad featured Woods, crouching with golf club in hand to examine a prospective putt. Over the photo was the phrase "WINNING TAKES CARE OF EVERYTHING," which given his recent marital transgressions seemed crass and nonsensical.

There was an uproar online, where the Twitter and Facebook ads garnered thousands of comments, such as, "Nike's new ad is hilariously blatant and their feigned ignorance of its double meaning is even better," and, "I understand Nike is a business, but the Tiger Woods ad is beyond irresponsible."

The Woods campaign was just the latest example of Nike pushing their win-at-all-costs ethos, as if athletic success could absolve Woods of his marital sins and egregious lapses in judgment.

As journalists began looking into the Nike Oregon Project it became apparent that using physicians was central to the secretive team's methodology. In April the *Wall Street Journal* published a story about the Oregon Project's doctor under the headline "U.S. Track's Unconventional Physician." The subtitle summarizes the article nicely: "Dr. Brown treats runners for a disorder not known to afflict them. His patients' medal count: 15 Olympic golds." The piece publicly called into question Brown's tactics. Was he diagnosing hypothyroidism in young athletes to improve their athletic performance without a true medical need and contrary to good medical practice? To many inside the sport, and the regulators, it sure seemed that way.

In addition to the censure of Salazar, American athletes Carl Lewis, Ryan Hall, Galen Rupp, Amy Begley, Bob Kennedy, and Patrick Smyth were all called out in the piece as being treated by Brown or other unnamed physicians for hypothyroidism. Smyth was used in the piece to explain how it worked: "When a physician near his California home found no evidence of thyroid dysfunction, Smyth flew to Houston to see Brown, who conducted some blood tests and diagnosed him with the condition."

In the article, Brown admits that he'd never heard of other endocrinologists treating athletes for hypothyroidism in this manner. The journalists, Sara Germano and Kevin Clark, then go on to quote a few other physicians who were openly skeptical of Brown's methods. "To see large numbers of young, athletic males being treated for thyroid deficiency would be certainly considered unusual, if not a bit suspicious," said Mayo Clinic endocrinologist Ian Hay. Salazar is then quoted in the piece as saying he considered Brown the best sports endocrinologist in the world.

For his part, Brown compares himself to the popular TV doctor at the time. "I'm like that guy House on TV," he said. "I'm like a detective." But as much as it must have felt good for him to be recognized, the article brought a great deal of criticism and unwanted attention to the secretive Nike running program.

By now, Areson and Magness had fled Portland and moved to Houston. The tight connection they had developed under duress in the Nike Oregon Project had blossomed into a romantic relationship, and in Houston, the two began dating.

While there, Areson carried on seeing Dr. Brown who was no longer a flight away—it just made sense—but she noticed a change in his attitude toward his former partner-in-prescriptions. He used to explain Salazar's tactics as him being "a crazy mad scientist coach," while claiming that "You gotta be crazy to be great." Now, Areson said, he was no longer speaking positively. "I went back to Steve and told him, 'You are never going to believe this, but Dr. Brown didn't have anything nice to say about Alberto this time.'"

In the media, Salazar was working to counter the narrative that followed the *Wall Street Journal* piece, that Oregon Project athletes were using off-label benefits of prescription drugs as performance enhancers. He told the *Guardian*'s Sean Ingle that Farah's giant leap from middling pro to Olympic gold medalist could be attributed to "the professional approach" taken by the team. He added that Farah's training before joining his project "was haphazard. He was all over the place. He did no weight training. He would jog and do five minutes of drills with no stretching afterward. And technically, Mo tended to over-stride toward the end of races. That's why he lost at the 2011 world championships in Daegu."

Also, by this time, Mark Daly, a Scottish investigative journalist for the BBC, had begun looking into doping in athletics. His team began by focusing on some of the famous British athletes that had possibly evaded detection in the 1980s. During his reporting he was urged by coaches and athletes to investigate an ongoing case of misconduct instead of the planned historical one—and a single name kept coming up again and again. "They pointed to one of the most prominent figures in the history of the sport: Alberto Salazar," said Daly, "coach of Britain's Mo Farah."

He heard about the famous coach's unorthodox methods and unnecessary prescriptions, as well as his use of banned substances and methods. Many runners would joke that being a fast runner was only one of the requisites for being selected to the Oregon Project, you also had to have prescriptions for asthma and thyroid medications. They called the coach "Albuterol Salazar" behind his back, for the asthma drug that all of his athletes seemed to be on.

Farah was at his peak popularity after his double-gold medal performance at the London Olympics, and Salazar was hailed as the mastermind behind the success and the mentor who brought the Brit from sporting obscurity to success on the world stage. As if putting a fine point upon his athletes' dominance, Rupp finished in second place—the first and second Olympic medal produced by the Nike Oregon Project in its more than eleven-year existence.

Daly's team was introduced to Steve Magness, an insider with a story to tell. The risks of speaking out were high for Magness, but he had grown frustrated with the lack of progress that USADA seemed to be making. Magness sat down with them in front of the cameras and told Daly, "I'm essentially the David taking on the Goliath of the biggest company and some of the biggest, one of the biggest, names in the sport, which is absolutely terrifying because they [Nike] control the sport."

The BBC, working with the American nonprofit organization and investigative website ProPublica, gathered interviews from coaches, athletes, and support personnel about the Oregon Project. The evidence was damning. They reached out to Danny Mackey to see if he'd be willing to participate in their documentary. "I wanted to go with my face because I'm an upfront person," Mackey told me. "But, I talked to a well-respected, prominent professional coach who told me, 'Look, I've been in rooms where Capriotti's talking to an athlete and has offered to pay them to trip another athlete in a race. They might not jump you. You might work for Brooks forever, but they will fuck with your athletes. It's not worth it, man. They aren't going down anyway, so don't put your name on it.'"

Mackey did interviews about his experience for the BBC and ProPublica, but did them anonymously, and was shown as a blacked-out face in the BBC television special.

Kara, too, had reservations about speaking to Daly and the BBC. She initially declined. When his team arrived in Boulder, they recorded interviews with Adam alone, but before they flew home, the crew stopped by the Gouchers' house to meet Kara.

"That night I started to get this nagging feeling," she said. "I wanted to be brave like Adam." She sought advice from her close friends. She prayed about it. Soon, the Gouchers called back and told the journalists that if they would return to Boulder, Kara would sit down for an interview.

Afterward, the couple reached out to Ritzenhein to see if he would also be willing to share his experience with the BBC. Adam and Kara had no knowledge of the infusions at this point, and didn't know what Ritzenhein might say, but figured he'd probably been through something similar to their experience. Ritzenhein said, "I spoke to David Epstein for a while and ultimately decided not to participate since I was still under contract with Nike."

"He knew shit," Kara told me, "and he just hoped it would go away."

Kara made plans to meet with Jordan Hasay before the June USATF National Championships in Iowa. They weren't close but had become casual friends through the running community. Kara had heard that Hasay was considering Nike coaches and she worried that they would funnel her to Salazar. Kara was going to give it to her straight and warn the promising young runner she should stay as far away from Salazar as she could. An hour before they were scheduled to meet at a Panera, Hasay texted Kara that she couldn't make it, then announced on Twitter that she had signed with the Oregon Project.

"And then she raced for them and beat me the next day," Kara told me. "Then I found out she had already been training with them for over a year. She had spent the summer with them and was using an altitude tent. She didn't want to hear it."

Despite the renewed journalistic interest in the team, they continued to perform well. In July, Farah started to show new range by going down to the 1,500 meters and breaking a twenty-eight-year-old British record and capturing the European record as well. His time of 3 minutes and 28.81 seconds made him the sixth-fastest man ever at the distance.

In August, at the 2013 World Championships, Hasay finished twelfth in a 10,000-meter race that saw the top four places go to Ethiopia, Kenya, Ethiopia, Kenya. And in the men's 10,000-meter event, Ritzenhein made a valiant effort to take control of the race with a kilometer remaining, but it would be Farah, again, who took the gold, with Rupp coming in fourth. When Farah won the 5,000-meter event as well, he captured another global competition double-gold.

Kara was working through injury issues, including a—not yet diagnosed—stress fracture in her foot. For her missing Worlds, Nike could exercise a contract clause that allowed them to reduce her salary by 30 percent for the life of the contract. If they took that route, she could only get her full salary back if she set an American record, won a medal at a global competition, or won one of the prestigious marathon "majors." The other option for Nike was to punish Kara for not racing for the past 120 days by suspending her salary for that amount of time. Taking this disciplinary route meant that Nike could reduce her salary by 60 percent. The brand couldn't exercise both, so the Gouchers guessed that they would go for the option that saved them the most money.

In anticipation of this, and as the Gouchers were planning their move back to Boulder, Kara was staging yet another comeback. She was injured and the races were small, but they were all USATF-sanctioned events and fulfilled her contractual obligations with the brand. For whatever reason, Nike didn't notice. They told her they weren't going to take the option. "At that point if they had gone with the failure to make World Champs clause and then taken the two option years, for two years they could have paid me at half salary," said Kara. "But they went with the non-racing thing because they wanted to get more money, and I got off."

Kara, who had been a Nike athlete for more than twelve years at this point, was free to sign with whomever she wanted, and the Gouchers quickly began talking to other sponsors about her next move.

Shortly after Worlds, Salazar, fed up with the rumors about his program and his athletes, did an interview with the *Telegraph*, in which he said too many people are using drugs in sport as an excuse when they don't perform well. "Everyone is so into this drug mania," said Salazar. "Everybody believes that if anybody runs well, they have to be on drugs. That's their excuse for why they don't run well." He claimed his athletes were actually barred from taking sports supplements because of the off chances they might contain banned substances.

"We don't take that much stuff and everything that Mo takes is from UK Athletics," he continued. "None of our athletes are on any sports-specific supplement other than beta-alanine, which is an amino acid. Other than that, it's iron, vitamin D, and that's it. You don't really need anything else."

In November, Salazar received the prestigious Coaching Achievement Award at the IAAF World Athletics Gala in Monaco for his coaching of Farah, Rupp, and Mary Cain, a new addition to the team. The summer of her sophomore year in high school, Cain ran the 1,500 in 4 minutes and 11.1 seconds to win the Junior World Championships in Barcelona and set a new American high school record. In October, she received a call from Salazar. He offered to coach her remotely. The Cain family were also practicing Catholics, and they wondered if God had sent Salazar to them. After getting over the shock and disbelief that she was talking to *the* Alberto Salazar, she accepted.

She was just the type of athlete Salazar looked for. Young enough to mold. Catholic enough to guilt. And naïve enough to convince that his esoteric ways were brilliant.

The Oregon Project @OregonPJT account tweeted, "Congrats to Alberto Salazar. IAAF Coach of the Year." Hasay retweeted it, adding, "So honored to say Alberto is my coach!"

In February 2014, the freshly decorated coach and his Oregon Project headed to the US Indoor National Championship in Albuquerque, New Mexico, with high hopes for Rupp and his two new prodigies, Jordan Hasay and Mary Cain. The stands were filled with signs featuring blown-up photos of Bernard Lagat's head when Rupp lined up against him for the men's 3,000-meter event. The thirty-nine-year-old Lagat won the race, leaving Rupp more than two seconds behind by the finish line.

After the event, Salazar was apoplectic and publicly accused Schumacher's runners, Andrew Bumbalough (eleventh place) and Lopez Lomong (fourth place), of conspiring against his young protégé. He met Lomong in the mixed zone, where the USATF sells merchandise, and got in the athlete's face while he was signing autographs. Salazar claimed that Lomong and Schumacher's athletes were always racing in a way that poorly affected Rupp's performance.

"You're always doing this, and it's bullshit," Salazar screamed.

"I did not touch your athlete," Lomong replied.

"Of course you did!" said Salazar. "You and Bumbalough."

Later, in the athlete area, Salazar charged at Schumacher and was physically restrained by others. Lomong, who was visibly shaken after the incident, told journalist Jon Gugala, "To see the outrage in [Salazar's] eyes, he was very, very mad." Everyone knew that Salazar hated Schumacher and his athletes—the two Nike coaches didn't hang out or consult with each other and they actively avoided having to spend time together at events.

"Lopez was verbally assaulted by Rupp's coach after the three thousand," said Tom Ratcliffe, a track agent who also witnessed the incident. "Anyone else would have had their credential pulled and would have been escorted from the area."

Not only had Salazar not been ejected from the meet, but he successfully petitioned the USATF to have Andrew Bumbalough disqualified from the event. Unfortunately, in his hot-tempered haste the Nike coach had confused Bumbalough for the third-place athlete, Ryan Hill, who was the one who actually made contact with Rupp during the race. A review of the footage showed that neither athlete warranted disqualification, but it didn't matter: Bumbalough—who had done nothing wrong—was out.

The following day, Hasay lined up for the 3,000-meter event in hopes of making her first US national team. She had already run the IAAF qualifying time necessary to go to Worlds, as did Brooks athlete Gabriele "Gabe" Grunewald and fellow Nike athlete Shannon Rowbury. Of the three finishers with the qualifying standard, only two would make the team and represent America at the World Indoor Championships in Sopot, Poland—one of them would leave today terribly disappointed.

Because of the altitude, track meets held in Albuquerque tend to stay reasonably paced until the final stages of the distance. This played to Grunewald's strength—her devastating finishing kick.

Just as she was making her move, however, Grunewald and Hasay made contact with each other. An on-field official raised a yellow flag, indicating that contact had been made between the athletes. It's nearly impossible to tell, exactly, if Hasay slowed suddenly or Grunewald began accelerating before she was clear, but Hasay stumbled a little bit and seemed to deflate. Grunewald, however, blew around her and achieved escape velocity with her impressive kick to win the 2014 USA Indoor Women's 3,000-meter Championship.

Her 20-meter margin of victory left no doubt who the best 3,000-meter female runner in the nation was. Grunewald would be headed to Sopot, Poland, in March to represent her country for the first time at a global championship. Hasay finished in fourth place, missing the team.

Race officials reviewed the tape and quickly determined that no foul had occurred, but Salazar petitioned the judges to have Grunewald disqualified for the inadvertent contact she had made with Hasay 180 meters from the finish. USATF's Jury of Appeals denied the protest and posted the results as official. Grunewald was interviewed on the broadcast as the new national champion.

In the ultimate act of track-and-field skullduggery, Salazar appealed the USATF decision. Gabe's husband, Justin, was told that Salazar was continuing to argue with officials. He walked over to see what was going on and lost his cool with the powerful Nike coach. "She's had cancer twice," he yelled at Salazar. "This is bullshit. How do you sleep at night? You've got a defibrillator in your chest."

Salazar's appeal was denied, at which point, according to USATF's own rules, the process should have ended, unless "new conclusive evidence is presented." But there was no new evidence—just the official race video footage that they'd already reviewed.

Grunewald moved on to Team USA processing. She was fitted for the Nike US National Team's uniform and booked her flight to Poland. Shortly thereafter, the USATF announced it had reviewed "enhanced video evidence" and released a statement that said: "After two reviews, including enhanced video evidence, Gabe Grunewald was disqualified by the Jury of Appeal for clipping and impeding the stride of Jordan Hasay." They stripped Grunewald of the victory and the national championship and disqualified her from the race. This elevated Rowbury to national champion and moved Hasay into third—earning her a slot on the world championship team.

"I was just shocked and definitely confused," Gabe said. "Knowing we had already gone through the protest and appeals process and everything had been ruled in my favor, I just didn't understand."

The next morning, Gabe's husband, Justin, got into an elevator at the Marriott, the event's host hotel, to find Salazar already inside.

"You can only push me so far before I break," said Justin.

"Do I know you?" Salazar replied.

"You're gonna know who I am."

"You're a nobody."

"I'm going to be a doctor. I'm Justin Grunewald."

Accounts differ on what was said before the elevator doors re-opened and the two men went their separate ways, but animosity festered.

Later that same day, Cain and the previous year's outdoor champion and fellow Oregon Project athlete, Treniere Moser, lined up for the women's 1,500-meter event. Cain destroyed the field. The seventeen-year-old closed out the race with a 61-second last lap and finished just 4 seconds shy of Mary Decker Slaney's meet record time.

After Cain's race, seven of the non-Nike-sponsored athletes held hands in solidarity and walked off the track in protest of Grunewald's disqualification. The outcry from the track community was immediate as social media became a buzzing hive of outrage, collectively saying: *Alberto Salazar and the Nike kleptocracy have too much power and have gone too far this time.*

All told, at this one track meet, Salazar verbally assaulted an athlete, physically threatened and intimidated a fellow coach, and had two athletes unjustly disqualified from their national championship races. Two days of discontent later an USATF Appeals Jury, again operating outside of their own rules at this point, reversed the Grunewald decision. Salazar had withdrawn his protest, USATF spokesperson Jill Geer said, and Grunewald was reinstated as the national indoor champion in the 3,000 meters.

Two months later, Nike signed one of the most lucrative and controversial contracts in sporting history, extending their sponsorship of the USATF national team for an additional twenty-three years. The deal, reportedly worth between $450 million and $500 million in cash and goods, assured that American athletes competing in the biggest international sporting events would be required to wear Nike apparel. Non-Nike Olympians could wear whatever shoe they desired, from whatever sponsor they might have, but they would be in Nike kits with Nike swooshes on them until the year 2040, regardless of what sponsor pays their individual salaries.

Through previous agreements, Nike had been providing much

of the operating budget for the USATF since 1991. Nike's sales were up 11 percent in 2013, with the run category contributing $4.2 billion of the $25.3 billion in total sales.

"Nike was founded as a running company," Nike president and CEO Mark Parker said in a statement, "and our passion for track and field is at the core of our DNA."

IN FEBRUARY, SEVERAL REPRESENTATIVES FROM THE SAUCONY SHOE COMPANY FLEW out to Denver to meet with Kara, and Adam, who was now acting as his wife's business manager. The presentation was flashy and impressive. "Would you be comfortable being the face of a brand?" the all-male staff asked Kara, before offering her a six-figure contract.

"It was an awesome pitch," Kara told me, "but they were like another little Nike. They were pitching me sports bras and there's not a single woman there."

The Gouchers moved on and took calls with many major athletic shoe and apparel players, including ASICS, Reebok, and Skechers. No matter who Kara decided to sign her next endorsement deal with, her last contract with Nike stipulated that they had the right to match any other sponsor's offer. This posed quite a problem, since Nike, as the sport's dominant brand, had almost limitless resources. Once they made their decision, Adam got on the phone with Capriotti to let them know.

"Kara's signing with Oiselle," Adam said.

"Oh, Oiselle?" said Capriotti, oozing sarcasm.

"Yep. Look, at the end of the day Kara will not run for Nike again."

The offer from Oiselle (pronounced: wa-zell) was one of the only deals that the couple knew Nike wouldn't match because it gave her part ownership of the company. The brand, which focused exclusively on women's running apparel, was founded in 2007 by Sally Bergesen. According to the company, Oiselle is a French word for *bird* that "alludes to that feeling of freedom and flight that most runners know; when the legs go fast and the heart goes free."

For a time, signing with Oiselle shut down Kara's options with other shoe companies, who had grown accustomed to signing an athlete in ways that would restrict them from working with other clothing brands. Adam reached back out to Rick Higgins, vice president of merchandising and marketing for Skechers Performance Division, to see if he would consider reopening negotiations. Meanwhile, his boss, Skechers founder and CEO Robert Greenberg, sent Higgins a *Wall Street Journal* article about Kara with a note, "You guys got to get her on board."

Two weeks after thirty-eight-year-old Skechers athlete Meb Keflezighi won the Boston Marathon, the brand announced that they had signed Kara Goucher to an undisclosed multiyear deal.

"I started meeting with Skechers back in December. They were actually the first company I met with when I realized I was a free agent," said Kara. "I really liked the people in the performance division a lot. They are really dedicated to the athlete."

Not only had Kara signed a new shoe endorsement contract, but she'd done something unprecedented thus far in elite running: she'd signed a separate apparel deal, by agreeing to represent Oiselle. Even though Keflezighi was head-to-toe in Skechers apparel line, their offerings were limited. Normally, signing with a shoe brand, like Nike, meant there was no opportunity for the athlete to sign a separate apparel contract with another brand, but since Skechers wasn't yet established in apparel, it left the door open for the Gouchers to negotiate.

After moving back to Boulder, Kara rejoined forces with her former coach, Mark Wetmore, and Heather Burroughs, who added the all-important female perspective, which Kara esteemed. They set their next goal on the 2014 New York City Marathon, with the overarching objective of getting Kara on her third Olympic squad at the upcoming 2016 US Olympic Trials Marathon in February. For this, the two coaches added slightly more speed work to Kara's regimen with hopes of improving her "marathon shuffle," as Wetmore described it. Training was fun again, and life was good.

Now, thirty-six-year-old Kara sought out sponsors who saw her value as a role model and ambassador and didn't calibrate her worth by each and every performance. In addition to Skechers and Oiselle, Kara added Nuun, Zensah, and Soleus to the roster of endorsements that collectively allowed her to make a comfortable living. While this side of business was booming, she realized that she hadn't raced a marathon in eighteen months, which raised numerous questions about her abilities by critics and fans, but they did have one recent data point to refer to: six weeks earlier Kara had run a solid 1 hour, 11 minutes, and 39 seconds half marathon time at the Rock 'n' Roll Philadelphia.

When it came to the race in the Big Apple, the lead pack went out harder than expected. Kara ran well early on, if a bit faster than was prudent, but by midway through the race she began to realize that her fitness wasn't quite where it needed to be. She soldiered on but slowed considerably as the finish line drew near. In the final 100 meters, Kara blew kisses to the crowd, but her grimace belied the emotion welling up inside of her. She crossed the finish line crying, in a time of 2 hours, 37 minutes, and 4 seconds, easily the slowest marathon she'd ever run. (Kenyan athletes went one and two.) Kara was nearly ten minutes behind her coaches' estimate.

Back at the hotel with Wetmore, she wanted it straight. They called Burroughs on speakerphone in Boulder, at which point an emotional Kara asked her coaches, "I just want you to tell me if I'm fooling myself," she said. "Am I done? I've had a great career, and I can walk away from it. I don't want to be that person who thinks they have something more to give and they don't. I just want you guys to tell me . . ."

There was a momentary silence, that she said seemed to last an eternity, before both coaches agreed at the same time, "We believe in you. We believe that a year from now you can be a totally different athlete."

"I really needed to hear that at that moment," she later said.

19

OFF TRACK

DUE TO MOUNTING CHATTER WITHIN THE TRACK-AND-FIELD WORLD, BY MARCH 2015 there were now multiple journalistic organizations looking into the Oregon Project. George Arbuthnott, reporting a story for the *Sunday Times* in London, reached out to Nike, Salazar, and Rupp. Their responses were telling.

> . . . with respect to L-Carnitine, I confirmed the use and method with USADA in advance[.] As I've said, L-Carnitine is a widely available, legal nutritional supplement that is not banned by WADA. A few of my athletes tried it but found no benefit so they no longer use it. . . . ~ Alberto Salazar.

Two days later, Rupp responded:

> I have worked and trained hard for over a decade to get to where I am today. I am completely against the use of performance enhancing drugs. L-Carnitine is a widely available, legal nutritional supplement that is not banned by WADA. The first time I tried L-Carnitine was in mid 2011 as a drink and I stopped taking it in 2012. I did not get any infusions or injections in 2011. I found no benefit so I stopped using it.
>
> - Galen Rupp

On Wednesday, June 3, 2015, at 1:03 p.m. Eastern Standard Time, ProPublica, a news website that focuses on investigative journalism, published David Epstein's piece under the headline "Off Track: Former Team Members Accuse Famed Coach Alberto Salazar of Breaking Drug Rules." In a joint effort, that very same Wednesday the British broadcaster, the BBC network, released Mark Daly's hour-long documentary, *Catch Me if You Can*, on the investigation into doping allegations at the Nike Oregon Project.

Innuendo was now replaced with hard facts and first-person accounts in two respectable journalistic outlets, and track and field was forced to deal with their ever-shifting moral landscape.

A few days later, Farah gave an impassioned news conference in which he said he's not leaving his coach. He too was looking for answers from Salazar, and though he'd spoken with him on the phone, he'd be flying back to Portland after his next race to see the proof the coach said he could provide of his innocence. If it was unconvincing, then he'd disconnect from the powerful Nike man. "If Alberto has crossed the line, I'm the first person to leave him," said Farah, adding, "I'm really angry at this situation. It's not fair, it's not right. I haven't done anything but my name's getting dragged through the mud. It's something not in my control but I want to know answers."

For speaking out about her experience with her former coach and athletic sponsor, Kara was bashed online and on social media with comments such as: "She's desperate for attention, she needs her fifteen minutes." "Why are you trying to take down the best group?" and "Why are you trying to take down the people who made you who you are?"

The IAAF's president at the time was a paid Nike ambassador and former English gold medalist named Sebastian Coe. He admitted to not having watched the documentary, but clearly stood by Salazar, his "good friend" of thirty-five years. "Alberto . . . is a first-class coach," he said. "Don't run away with the idea that this [the Nike Oregon Project] is a hole-in-the-wall, circa 1970s Eastern Bloc operation. It's not."

Rupp's parents sat down with the *Oregonian*'s track-and-field reporter, Ken Goe, who has historically contorted his objectivity in an effort to stay in the Nike men's good graces. He was a safe space for them, and in the interview Rupp's parents called the allegations "outrageous and baseless."

Salazar didn't speak to the media, but the longer he remained silent, the more viciously the tongues wagged. He took several weeks to craft a more than 11,000-word rebuttal, which he called an "open letter" and posted to the Nike Oregon Project's official website. The fifty-six-year-old coach denied several of the allegations while ignoring or seeming to misunderstand many others, writing in depth about things that were not claimed at all. To bolster his argument, he linked to copies of email messages, doctors' notes, and medical records. "Former athletes, contractors and journalists make accusations in these stories, harming my athletes," he wrote. "At best they are misinformed. At worst, they are lying. I believe in a clean sport and hard work and so do my athletes. What follows below are the actual facts." He went on to highlight Rupp's somewhat shocking medical maladies: "Galen has suffered severe allergies and breathing issues almost his entire life. He also suffers from Hashimotos disease, a thyroid disease. He has a significant history of hypothyroidism on both sides of his family. Allegations that Galen takes asthma and thyroid medicine for competitive purposes are inaccurate and hurtful . . . Galen takes asthma medication so he can breathe normally—not so he can run better . . . Galen takes a number of prescription medicines to treat his asthma and has received immunotherapy to treat his allergies since 2004. Galen must receive weekly injections to help control his allergies and his asthma." The astute reader is left wondering how Rupp could possibly compete against the best athletes on earth with all that afflicts him.

Salazar's defensiveness peaked when he attacked allegations that were never actually made about his protégé ("The claims that Galen has been on prednisone continuously since he was fifteen are absolutely false"). The BBC and ProPublica covered an instance of Rupp using prednisone as a sixteen-year-old but make no mention of continuous use or any claims of use as a fifteen-year-old, though there is other evidence of this. A former friend and training partner from Portland, named Stuart Eagon, told reporters of the incident. "Galen, myself, and Alberto were outside a hotel room in North Carolina for the national two-mile high school championships," he said. "We were going on our morning run and Alberto said to Galen: 'Have you taken your prednisone yet this morning?'"

Sixteen-year-old Rupp—who Eagon said did not appear to be ill—ran back inside and presumably took the powerful glucocorticoid, which was prohibited during competition at the time. When he returned, ten minutes later, they continued on their run.

When he brought it up two years later, Eagon says that Rupp denied ever taking it. "After that," said Eagon, "I felt as though the dynamic of our relationship changed."

When asked by the BBC for an explanation, Rupp told them, "Earlier in my career, Wada required TUEs for my asthma medication. . . . The few other TUEs I have applied for and received related to the treatment of severe asthma flare-ups."

"Galen only took prednisone when needed to treat an asthma flare-up," said Salazar, "under the direction of his doctor for a limited period of time."

In his rebuttal, Salazar admitted that he damaged his health through overtraining, but saw no apparent parallel from his hypogonadism to the Oregon Project's plague of hypothyroidism, which he denied in any case, by claiming that a statistically insignificant number of his athletes have been diagnosed with hypothyroidism after he started coaching them. He also admitted that he did, in fact, hollow out a book to send drugs through the international post to Rupp in Germany, but said Magness was aware of the type of medications being sent to Rupp and has misled the press and others about those events.

Salazar tackled Kara's claim that he gave her Cytomel to make her lose weight by blaming Brown, who the coach said directed him to give Cytomel to Kara at the World Championships in Daegu, South Korea. But that was August 2011, and Kara's allegations were from March, before the Boston Marathon and her return to racing after childbirth. "His story is a different timeline," Kara told ProPublica. "He's trying to use Daegu to cover up Boston."

Though two other independent sources corroborated Kara's original accusation that he coached Rupp to get a TUE, Salazar wrote, "Kara Goucher's claim that someone can make a couple of statements to a doctor and get a TUE is absurd."

Salazar did confess to testing testosterone on his sons to determine how much would trigger a positive anti-doping test, but did not clear up where the drug used in the tests came from, or how the information would be helpful. It is illegal to share prescription drugs in the United States, and Brown, for his part, denied prescribing the drug for this purpose. "I would never do that," he said.

Salazar blamed the eventual falling out with Kara on Adam, saying that after their argument at the 2008 Olympic Trials, "My relationship with Adam never recovered." He wrote that he gave Kara an ultimatum: if Adam remained involved in her training, then Salazar would not coach Kara.

But Salazar saved his harshest critique for Magness, who he said was incapable of working with elite athletes. He also claimed that Magness's attention to "one female runner" was "to the detriment of the others." Salazar said he challenged Magness about his relationship with Areson (both Magness and Areson told me their relationship didn't become romantic until they had left the Project), writing, "I confronted Magness about it and he denied it. I don't know whether he was telling the truth or not. Nevertheless, I told Magness that whatever was happening had to stop; he had to make it clear to everyone by his actions that he was not having a physical relationship with an Oregon Project athlete. In my opinion, he didn't. Things only got worse.

"He doted on her, and some of his other athletes complained to me about his conduct. In my view, Steve's behavior became more and more unprofessional and counterproductive."

Ultimately, Salazar claimed that Magness did not leave the Oregon Project but was fired by Salazar in 2012. In response,

Magness publicly provided his termination letter, which read, "As discussed, both you and NIKE have decided to terminate your NIKE Contract, entered into on May 4, 2012. As such, this letter shall acknowledge that mutual decision to terminate the Contract, effective immediately."

The embattled coach denounced the journalists involved as "irresponsible," and demanded retractions from the BBC and ProPublica. The track community was incredulous. Many had seen Salazar's antics firsthand, and those that hadn't had been hearing about them for years.

Following Salazar's post, Nike said in a statement: "Both Alberto and Galen have made their position clear and refute the allegations made against them, as shown in Alberto's open letter. Furthermore we have conducted our own internal review and have found no evidence to support the allegations of doping."

After the allegations, former NOP coach John Cook told *Runner's World*, "My general thoughts were that all things eventually come to fruition. I was, frankly, not surprised. Take, for instance, the best female marathoner we have in this country [Shalane Flanagan]. I coached her for two years and she set five US records and medaled in the 10,000 against the East Africans. I'll guarantee to this day that she's as clean as a whistle. Otherwise, why would she have gone with a certain coach and not with a certain other coach?"

The day after Salazar's open letter was posted, the 2015 USA Outdoor Track and Field Championships event began at Hayward Field. By now, the track community had had twenty-two days to digest the shocking BBC and ProPublica revelations. Down on the track, the tension was palpable.

Kara performed poorly, finishing her 5,000-meter final in eighteenth place out of twenty athletes. After the race she entered the media tent behind the bleachers, with her emotions bubbling below the surface, to answer questions. Five minutes in, a reporter asked Kara, "Steve Magness said that Alberto has threatened him in the past. Has he or anyone else threatened you since this has all happened?"

"No," Kara said. "But people have been threatened at this meet."

The details of what she was referencing weren't yet fully known, but the world would come to find out later that Nike's pugnacious global director of athletics, John Capriotti, grabbed Brooks team coach Danny Mackey and threatened to kill him.

With the exception of an incident after the 2014 IAAF World Relay Championships in the Bahamas, Mackey hadn't had much contact with Capriotti since leaving Nike back in 2010. At the relay event, Mackey's wife, Katie, and her team of Brooks professional runners took silver in the 4x1500-meter women's relay. Afterward, she had inadvertently covered up the Nike logo on her World Champ's uniform. The Nike men, taking the least charitable interpretation, assumed she was covering the logo on purpose.

Mackey wasn't at the event, but as the Brooks team coach, he received an angry voice message from Nike's North American Athlete Manager, Paul Moser, after the event.

Then Capriotti called. "Do you know who you're fucking with, motherfucker?" he screamed at Mackey. "You don't fuck with me! Your fucking wife is done with this sport."

To add to Nike's discountenance with Brooks, a couple weeks later Nick Symmonds snuck over the berm and surreptitiously covered the bronze Michael Johnson statue in Brooks clothing.

Now, at the 2015 Track and Field Championships in June, after the 1,500-meter event, Capriotti was seen stalking across the infield with a phalanx of fellow Nike employees including Llewellyn Starks, Ben Cesar, Paul Moser, and Robert Lotwis trailing behind. Mackey was in the medical tent west of Hayward Field talking with his athlete, a former Nike runner named Dorian Ulrey, when Capriotti burst into the tent and accosted the Brooks coach.

Mackey told me that he was sitting on a cooler, talking to Ulrey, when he suddenly felt someone grab his right arm and pull him, almost picking him up out of his seat. It was Capriotti, who said, "We gotta talk right now."

"I'm talking with Dorian," said Mackey. "Gimme a minute."

Capriotti took a step closer and poked Mackey in the chest with two fingers hard enough to push him back. Llewellyn Starks stood over him, "just staring at me," said Mackey.

"Well then we'll go outside, motherfucker," said Capriotti.

"John. Relax. I'm gonna finish talking with Dorian."

Capriotti bent down and, with his nose touching Mackey's right ear, told him, "I know you were the blacked-out-face guy. You're fuckin' dead."

When Mackey stood up, he was chest-to-chest with Starks. "I don't know what you're talkin' about, John. You are going to kill me? For what?"

"Don't you fuckin' lie to me," Capriotti said.

"I don't know what you are talking about. You need to relax."

"You're a pussy, you fuckin' liar. Let's go outside and fuckin' deal with this."

Now Mackey was upset, and his inner kid-from-Chicago came streaming out. He started repeating what Capriotti was saying, really loud, so everyone could hear.

"Outside? Oh, you wanna fight me? For what? You need to relax. Leave."

Capriotti stepped even closer to Mackey and poked him in the chest again, harder this time. "Let's go outside," Capriotti kept saying. "I'm going to fuckin' kill you."

Minutes had passed, and a group began to gather around the two men. Eventually, Starks stepped between them and began to pull Capriotti away, telling him, "We gotta get outta here."

"I know what you fuckin' did. This isn't over," said Capriotti. "You are fuckin' dead." Starks then led Capriotti out of the tent and back to the main area of the track meet. Later that day, Mackey reported the death threat to the police but didn't want to press criminal charges. Capriotti didn't respond to requests for comment, but Mackey told me he assumes he either heard or somehow figured out on his own exactly who had given the anonymous testimony in the BBC documentary. "Only my wife and my brother knew that I spoke with the BBC," said Mackey. "I don't know how Capriotti knew, but he was right." (Through his lawyer, Capriotti unequivocally denied making any of the statements attributed to him by Mackey.)

Some in the media chalked the incident up to "institutional Nike aggression." An unnamed source told reporters after the incident, "I've heard so many accounts of athletes, agents, and meet directors getting treated like absolute dogshit. I kind of wish all the agents would band together and say something . . . but because of the stranglehold he [Capriotti] has on people's paychecks, people won't talk."

Mackey, knowing the Chris Whetstine story all too well, went and bought a gun for protection. "I don't walk with my headphones on anymore," he told me.

Victor Conte, the mastermind behind BALCO, who after being released from prison had rebranded himself as an anti-doping advocate, told the *Japan Times* in August of 2015 that he thought Salazar [had] "studied how to circumvent the system . . . enrolling doctors to assist you to do whatever tests they do to these various exercise-induced asthma medications.

"The fact that somebody is giving testosterone to their own kids and sending in samples," he said, "that's what I did with BALCO. It's exactly what I did. It was diagnostic. It was pre-testing. Why did I do that? Because I was trying to circumvent the testing? Why else would you do it?"

Travis Tygart, the CEO of USADA, told me, "Because the experiments happened on the Nike campus, they realized they had a lot to lose. They lifted the drawbridge and put alligators in the moat." The corporation paid a coterie of lawyers to defend Salazar against the allegations. USADA was in for an exhaustive fight. (Nike has denied obstructing the agency's investigation.)

When USADA asked for email evidence, Nike refused, saying that to share email would expose too much of their intellectual property. Salazar and Rupp wouldn't make themselves available for interviews. And when USADA requested medical documents, Nike cited medical privacy laws.

When his hand was forced, and the famed coach was finally scheduled to sit down with anti-doping officials, Nike and Salazar sent USADA five thousand pages of documents three days beforehand, leaving the anti-doping agency insufficient time to review and prepare. Rupp, who claimed to be an advocate for clean sport, publicly stated that he was "cooperating with whatever officials I need to cooperate with," but privately refused them access to his medical records.

Salazar told the team they should use his Nike lawyer, John Collins, when talking to anyone, especially USADA. Collins came to represent at least seven current and former Oregon Project athletes, including Alvina Begay, Mary Cain, Matt Centrowitz Jr., Dawn Grunnagle, Jordan Hasay, Galen Rupp, and Shannon Rowbury. "As a result of the involvement of Salazar's lawyer, and with limited exceptions as described herein, the foregoing athletes have largely refused to permit USADA to review their medical records," the anti-doping agency wrote in its "Interim Report."

Brown's lawyers were also provided by Nike, and the doctor took an equally antagonistic stance. Even after ten athletes agreed to allow USADA access to their medical records, Brown refused to turn them over, dragging the process out even longer. When he was finally compelled by law, the anti-doping agency noticed that the medical documents were different from some of the same records provided by the athletes. Brown had added annotations to some of the records, most visibly adding infusion amounts where he had neglected to on the originals. All read less than the fifty-milliliter limit set by the WADA code.

"They did everything they could do and obstruct the truth in our opinion," Tygart told me. "It's unprecedented."

The report eventually made its way to the internet for the world to see. When Magness reviewed it, he noticed that his medical records had been altered. Former NOP athlete Dathan Ritzenhein also realized that new notes had been added to his records to make it look like the quantity infused fell below the WADA limit of 50 mL. It seemed as though Dr. Brown was now revising history to better protect himself.

Whatever else might be said or written about Salazar, nobody has ever accused him of modesty; his confidence in himself is legendary. It was as evident when he was a leather-jacket-wearing, shot-calling twentysomething beating the best marathoners in New York as it is today. This was never more apparent than in the days after the BBC and ProPublica investigation into his coaching methods.

In the midst of the firestorm of accusations, Salazar ran into Sean Ingle, a reporter for the *Guardian*, in a hotel lobby. "You should put your money on me being cleared," Salazar said with a grin. "It's a winning bet."

USADA officials would eventually interview nearly everyone involved. Journalists working for the BBC's Panorama program would later uncover documents showing that when asked about his doses of L-carnitine, British athlete Mo Farah denied ever taking it.

UKA's Dr. Rob Chakraverty injected Farah two days before the 2014 London Marathon in his room at The Tower, the official London Marathon hotel, on April 11, 2014. In the hotel room, looking on, were UKA's head of distance running, Barry Fudge, their performance director Neil Black, and Alberto Salazar. Dr. Chakraverty recorded that he gave the runner four injections, spaced out over two hours through a butterfly needle, but failed to record the ever-important quantity.

Days later, Farah finished the race in eighth place. Six days after the hotel room infusion, Farah was drug tested. On his testing form he listed a number of products and medicines but failed to record L-carnitine.

A year later during USADA's investigation, anti-doping officials interviewed Farah for nearly five hours straight. The agent asked him: "If someone said that you were taking L-carnitine injections, are they not telling the truth?"

"Definitely not telling the truth, one hundred percent," said Farah. "I've never taken L-carnitine injections at all."

"Are you sure that Alberto Salazar hasn't recommended that you take L-carnitine injections?"

"No, I've never taken L-carnitine injections."

"You're absolutely sure that you didn't have a doctor put a butterfly needle . . . into your arm . . . and inject L-carnitine a few days before the London marathon?"

"No. No chance."

When Farah left the room he ran into Fudge, though it's not known what was said. Within moments Farah rushed back into the room. "So I just wanted to come clear, sorry, guys, and I did take it at the time and I thought I didn't . . ."

"So you received L-carnitine . . . before the London marathon?" the agent asked.

"Yeah. There was a lot of talk before . . . and Alberto's always thinking about 'What's the best thing?' 'What's the best thing?'"

". . . A few days before the race . . . with . . . Alberto present and your doctor and Barry Fudge and you're telling us all about that now but you didn't remember any of that when I . . . kept asking you about this?"

"It all comes back for me, but at the time I didn't remember."

AS PLANNED, WETMORE AND BURROUGHS HAD REBUILT KARA INTO A WHOLLY NEW athlete a year after her disastrous 2014 New York City Marathon. Fifteen months on, she lined up at the 2016 US Olympic Trials Marathon for a chance to put an exclamation point on an already impressive running career—a third Olympic Team.

In that time, her coaches had decreased her overall mileage, but increased the quality of her workouts, and added renewed emphasis on recovery. Kara was confident and felt as fit as she had ever felt since becoming a mother. And she had recently raced well too, winning a pair of half-marathon events, one in November in a time of 1 hour, 11 minutes, and 13 seconds, and the next a month later in 1 hour, 11 minutes, and 10 seconds. This was even more promising considering that she hadn't tapered into the races and had trained through them as though they were simply hard tempo runs.

"Certainly after the 2014 New York City Marathon," said Wetmore, "nobody would have given Kara much of a chance."

On February 13, Kara lined up with 205 elite marathoners, six of whom had run sub-2:30 personal best times. "This is literally a last chance for me," Kara wrote on her blog about the Olympic Marathon Trials in Los Angeles, a race in which only the top three finishers would make the team. The Southern California weather was the major concern on the day, with pundits assuming that it would slow the athletes by a whopping 2 minutes.

Amy Cragg and Shalane Flanagan, apparently, did not get the memo, and the best friends and training partners ran with what seemed like reckless abandon given the temperature, which was climbing to the seventies as the race progressed. They took to the lead early on in their identical Bowerman Track Club kits: white Nike visors, black bikini bottoms, white half-top shirts emblazoned with the swoosh and the Bowerman Team logo, a capital *B* with a lightning bolt through it. On their feet were prototypes of what would become the Nike Zoom Vaporfly 4% disguised to look like Nike's existing racing flats. At the time, the IAAF's rule on shoes stated: "Shoes must not be constructed so as to give athletes any unfair assistance or advantage—and any type of shoe used must be reasonably available to all in the spirit of the universality of athletics." What we'd find out years later was that the shoes both Cragg and Flanagan were wearing them gave the athletes somewhere between 2 and 5 percent increase in efficiency.

By the eleventh mile Cragg and Flanagan had pulled away completely from the rest of the women in the race, running alone until Cragg noticed Desiree Linden closing on them around mile twenty-five. Flanagan, defending Trials champion, was suffering. After a brief discussion, Cragg took off to secure her victory. Linden, who raced in a Brooks shoe, passed Flanagan at 2 hours, 24 minutes into the race, as Cragg sped away for the win in 2 hours, 28 minutes, and 20 seconds. With Flanagan fading, that third spot was up for grabs.

Flanagan broke the threshold with pure anguish on her face and collapsed into Cragg's arms, unable to stand under her own power. Her husband, Steve, picked her up like a child, carried her away from the finish line, and put her in a wheelchair. Kara emerged 65 seconds later, heartbroken. She would not have her fairy tale Olympic send-off at the Games in Rio.

In her post-race interviews, Kara tipped her hat to the Olympic marathon team. "You know Amy Cragg hasn't had a great year," said Kara. "I kept hoping she was in over her head, but I pretty much knew at twenty-two that it wasn't going to happen. It's a hard pill to swallow, but they were better."

Flanagan's Instagram post following the race read: "I'm not getting any younger but I can honestly say that these flats are helping me run faster. . . . This shoe is a game changer."

Before the men's race, Rupp wore a cooling vest and large elbow-length cooling gloves that looked like robot arms. Though he looked ridiculous, the logic was sound: keep the athlete's core temperature as low as possible for as long as possible before the race, which would turn out to be the hottest US Marathon Trials ever run. Salazar had cut holes into Rupp's lightweight tank top to make it even lighter and more breathable, a tactic the coach favored during his racing days.

Rupp also laced up a pair of the secret Vaporfly prototype shoes for the Trials.

Ritzenhein, now thirty-three years old, emerged in front around mile ten, about fifty minutes into the race. Tyler Pennel made the first consequential surge, at mile sixteen, which shattered the group. Keflezighi and Rupp were the only athletes that went with him, leaving Ritzenhein in the second pack.

Near mile twenty, while cruising along at sub-five-minute miles, Keflezighi offered Rupp the lead by gesturing with his left hand. The two runners exchanged words, which Meb described as "not a very friendly conversation." He later said that Rupp was running too close to him, and he worried he was going to be tripped and aggravate an already tight hamstring. "It's not a track, the road is open," he told the press after the race. The crowd laughed. Rupp declined to take the lead, and despite having the entire road, remained tight on Keflezighi's heels. Ritzenhein dropped out just after the twentieth mile.

Three runners became two, as Keflezighi and Rupp dropped Pennel. They were clipping off mile times of 4 minutes and 59 seconds, 54 seconds, and 52 seconds. Rupp took the lead as Keflezighi grabbed his bottle at an aid station 1 hour and 40 minutes into the race and would not relinquish his place. (Salazar had Rupp practice water pickups at sub-five-minute-mile paces during training.)

Then two finally became one at the twenty-two-mile mark as Rupp gradually began distancing himself from the defending US Marathon Trials champion. He ran the twenty-third mile of the marathon at an astonishing 4-minute-and-47-second pace and finished in 2 hours, 11 minutes, and 12 seconds, punching the air as he broke the tape. "Winning his first marathon in resounding fashion!" the announcers declared. Keflezighi crossed the finish line 68 seconds back in second place while Jared Ward rounded out the Olympic team in third.

AFTER THE NEWS THAT SURROUNDED THE INITIAL BBC AND PROPUBLICA STORIES DIED down, the investigation seemed to languish. There wasn't much new reporting and USADA, like the FBI, doesn't comment on ongoing investigations. To the general public, the case had all but vanished.

In April of 2015, Knight appeared on ABC's *Good Morning America* to promote his new biography, *Shoe Dog*, which he wrote with journalist and ghostwriter John Joseph "J. R." Moehringer. It was a final sign-off for Knight, who at seventy-eight years old would step down from his role as the chairman of Nike's board in June. As planned, Nike president and CEO Mark Parker added the role of chairman to his responsibilities.

During the interview, Knight looked every bit his advanced years, with watery, red, swollen eyes. *Good Morning America* host Robin Roberts fawned over Knight. "Always great to be in your presence, Phil," she said, before telling him how great the book was. "There are times I'm reading it and going, 'these guys aren't going to make it.' "

"We were right on the edge a couple times," Knight said. "The most scary was when we got kicked out of the bank for the second time and they told me they would turn me in to the FBI." Roberts didn't ask him anything about the elite running team that Knight could look out his office window and see, the one that was now under siege.

Back on its campus, Nike continued to funnel the best runners in the world to their famed coach who, it appeared, no longer had contact with Brown and, at least according to his assistant coach, Pete Julian, had stopped doing anything that could be misconstrued as even gray-area tactics. Julian never saw a tube of testosterone cream lying around any of the Oregon Project training camps. He never saw syringes or vials in the refrigerator—"just butter," he told me. And although he admitted he's seen Salazar drunk on more than one occasion, he didn't think the coach had a drinking problem. "I saw nothing in my time that gave me any concern whatsoever," he said.

In June, less than two months before the 2016 Olympics in Rio de Janeiro, Farah again had trouble with coaches making him look bad, when Jama Aden was arrested in a dawn raid of a hotel in Sabadell, Catalonia. There, Spanish authorities found performance-enhancing drugs, including erythropoietin (EPO), in the training groups' rooms. Aden was the Somali-born coach of Ethiopia's Genzebe Dibaba, who broke the 1,500-meter world record in 2015. Pictures of Aden and Farah from a training camp the two had attended the year prior to the arrest soon appeared online. But Farah was able to put that aside and once again performed brilliantly at the Games, winning the Olympic golden-double for a second time, and becoming only the second man to ever achieve the feat twice. NOP athlete Matt Centrowitz Jr. became the first US runner to win the 1,500-meter event since 1908 (though it was the slowest winning time since 1932).

In the last event of the Rio Games, Rupp, again wearing Vaporfly prototype shoes that were still not available to the public, placed third in the marathon. Kenyan athlete Eliud Kipchoge, who won the gold medal, was also in the secret Nike shoe, as was the women's winner, Kenyan Jemima Sumgong.

Collectively, Nike athletes the world over won more than two hundred medals at these Games. And all thirty-two of the Americans who won gold in track and field were Nike athletes.

MY FRONT-PAGE STORY IN THE *NEW YORK TIMES* ON MAY 17, 2017, CONVEYED THE narrative of the leaked USADA document and how many of the athletes seemed coerced by Salazar to do his bidding by holding their financial well-being and professional status over their heads.

Shortly thereafter, I received a tip that fifteen people had been served notification of rules violations by USADA, though every call I made to verify effectively shrunk the number until just two people remained: Alberto Salazar and Dr. Jeffrey Brown.

Working in tandem with Rebecca R. Ruiz at the *New York Times*, we managed to verify that Brown had been notified of anti-doping violations. On June 8, 2017, we published a piece titled "Doctor for Nike Oregon Project Runners Is Notified of Doping Allegations." Verification on Salazar's charges was elusive, however, as he was not speaking to the media and did not respond to calls or emails.

Though we weren't able to verify it at the time, we found out later he was in fact notified of the violations through his lawyer on June 30, 2017, and given a "Charging Letter" that read, in part, "that the USADA Anti-Doping Review Board had met and determined that there was sufficient evidence of anti-doping rule violations."

> (1) Possession of prohibited substances and/or methods including testosterone and prohibited IV infusions and related equipment (such as needles, IV bags and/or syringes, storage containers and other infusion equipment and devices).

(2) Trafficking of testosterone and prohibited IV infusions.
(3) Administration and/or attempted administration of testosterone and prohibited IV infusions.
(4) Assisting, encouraging, aiding, abetting, covering up, and other complicity involving one or more anti-doping rule violations and/or attempted anti-doping rule violations.
(5) Tampering and/or attempted tampering, and
(6) Aggravating circumstances justifying a period of ineligibility greater than the standard sanction.

IN APRIL 2017, KARA WAS NAMED TO THE COLORADO RUNNING HALL OF FAME. LATER that summer, she flew to London to take part in a medal upgrading ceremony from the 10,000-meter race at the 2007 World Championships in Osaka a decade earlier. Years after the event, a reanalysis of the competitors' samples snared Ethiopian-born Turkish athlete Elvan Abeylegesse in a failed test. Abeylegesse was given a two-year competition ban by the IAAF and stripped of her results from August 2007 to 2009. With second place vacated, Kara was upgraded from third to second, bronze to silver—vindication of a kind.

Salazar's athletes continued to perform well while he was being investigated and a new female marathon contender, Jordan Hasay, emerged from his group. Salazar had wooed Hasay while giving a keynote speech at a Christian leadership conference that she attended while still in high school. And after a bumpy start, where the two discussed parting ways, the coach transformed the former middle-distance Oregon Duck, known for her long, braided hair, into a force to be reckoned with in the marathon.

With Salazar as her coach, in January 2017 Hasay ran the second-fastest half marathon debut at the Aramco Houston Half Marathon in a time of 1 hour, 8 minutes, and 40 seconds—the fastest time ever run on the course by an American woman. In April, Hasay broke Kara's debut marathon record by running 2 hours and 33 minutes at the Boston Marathon. Hasay finished third, 10 seconds behind the second-place finisher (Kenyans went one and two).

"I thought that when someone broke it, that it would bother me or I'd get a pang of sadness," said Kara, "but I felt nothing."

In July, the secret prototype Nike shoe that was breaking records was finally released to the public through Nike's retail outlets. The secret of the Nike Zoom Vaporfly was in the proprietary, high-tech foam in the shoe's midsole, making each stride less of a burden on the body by returning more energy. In an age where the average price of a running shoe was closer to $120, Nike was selling the Vaporfly for $250. They flew off the shelves so fast they had trouble keeping them in stock, which spawned a secondary market where they were often selling for twice the retail price. Runners felt faster when wearing the shoe and their times were improving. For once, it seemed, the industry standard hyperbolic marketing hype lived up to its claims.

Hasay had been wearing the Vaporfly 4% during all of her recent running successes. She followed her impressive debut with another third place at a major marathon on October 8 in Chicago. On her way to her Chicago Marathon finishing time of 2 hours, 20 minutes, and 57 seconds she set a new US 25K women's record, which eclipsed Shalane Flanagan's previous record time by 17 seconds.

Whether deserved or not, Salazar's athletes were now seen in a different light, one that questioned everything they did. When asked what she thought about Hasay's emerging speed, Flanagan told *Sports Illustrated*, "There's still an investigation going on, so it's hard to truly and genuinely get excited about the performances that I'm watching. We don't get to choose our parents, but we certainly get to choose our friends and our coaches and who we want to include in our circle and put our faith and our trust in." As if to put a fine, albeit anecdotal, point on it, Shalane Flanagan won the 2017 New York City Marathon by outrunning the field—in a pair of Vaporfly 4% shoes—and ending a four-decade winless streak for American women at the prestigious event.

Eleven days later, researchers from the University of Colorado Boulder published a study in the journal *Sports Medicine* on the shoe, which read in part: "Every single day at every single speed, every runner used less energy with the prototype shoe."

The research showed the sneaker made everyone faster, regardless of their pace or biomechanics. The implications for elite runners were huge. As the researchers concluded—the elusive sub-two-hour marathon was now within reach. Records would begin to fall because of this shoe.

Healthy doubt was hard to kill, however, owing to the fact that Nike had funded the study. Many in the media were still skeptical that a shoe could bestow such a huge advantage.

A spokesperson for the IAAF told the *New York Times* that while it's accurate to say that the Vaporflys are legal, it's actually more accurate to say there is no evidence they shouldn't be. "We need evidence to say that something is wrong with a shoe," he said. "We've never had anyone to bring some evidence to convince us."

Now, people started looking back at races Nike athletes had won wearing the shoes before they were a known entity. When the math was run on Kara's Olympic Marathon Trials time, she had been beaten that day by two athletes wearing a Vaporfly 4% prototype shoe. She missed third place and that final Olympic spot by just 0.7 percent to Flanagan.

"It is highly likely that Goucher is the first known athlete to miss the Olympics due to shoe technology," University of Colorado's professor of sports governance, Roger Pielke, told the *Telegraph*.

20
BANNED IN DOHA

ON MAY 3, 2018, AMID ONE OF THE MOST TURBULENT PERIODS IN NIKE'S HISTORY, shares of their stock hit a record $86.30. Nothing, it seemed, could diminish investors' belief in the sportswear giant.

The same day, in the Tiger Woods Conference Center at Nike's headquarters, CEO Mark Parker apologized to company employees for the corporate culture which excluded some staff, then assured them that departures related to the unfolding scandal would be completed soon. Earlier in the year, a group of female employees, fed up with the status quo at the male-dominated company, covertly surveyed their colleagues to see who had fallen victim to sexual harassment or gender discrimination within the company. The results were shocking and detailed years of ignored sexual harassment and unfair promotional practices that left many women watching their male counterparts climb the corporate ladder in their place.

The packet of complete questionnaires was presented to Parker on March 5, ultimately leading him to usher at least six high-ranking men unceremoniously out of the company. Among them was Trevor Edwards, the man who was widely seen as the likely successor to the CEO.

The May employee meeting came five days after the *New York*

Times published reporting with more than fifty current and former employees that described widespread discrimination against women, including sexual harassment and stymied careers based on gender, and a dysfunctional workplace that was demeaning to female employees, and one in which claims of impropriety were simply ignored by Nike's human resources department. Nike explained the culprits to the *New York Times* as "an insular group of high-level managers" who "protected each other and looked the other way."

"Why did it take an anonymous survey to make change?" former employee Amanda Shebiel asked in the *New York Times*. "Many of my peers and I reported incidences and a culture that were uncomfortable, disturbing, threatening, unfair, gender-biased and sexist—hoping that something would change that would make us believe in Nike again. No one went just to complain. We went to make it better."

Speaking with me under the condition of anonymity due to ongoing litigation, a former employee told me she heard gay coworkers called "dykes" by high-ranking men. "It's crazy the amount of money these guys make. In Portland they start to feel like they are God's gift," she said. "That played out in who they hired. If they didn't like you, you didn't stand a chance."

Parker took an apologetic tone and promised a systematic review of company processes in an effort to "restore trust in places where it has eroded."

NEITHER SALAZAR NOR BROWN WAS GOING TO TAKE A BAN FROM SPORT WITHOUT A fight and simply go away quietly. With Nike's wallet, they both contested the USADA findings, and through the spring and summer of 2018 underwent the arbitration process. This was their last opportunity to beat the USADA case. If they failed to convince two of the three arbitrators, sanctions would be imposed and announced to the world. If they succeeded, the general public would never hear a word about it. (This is so innocent parties aren't wrongly assumed guilty in the court of public opinion without due process.)

During the third week of May 2018, Danny Mackey opened Skype on his computer and connected to the arbitration hearing of Alberto Salazar. He would be asked to call in twice, once for ninety minutes and a second time for forty-five, during which he revealed the Nike lawyers tried to paint him as a coaching competitor who was jealous of Salazar's success. "They presented emails out of sequence to make it look like I was lying," Mackey told me. "As though I was trying to coerce Alberto into doing something wrong."

Mackey calmly explained the timeline discrepancy. Also, he pointed out he wasn't a coach when he left that message on the USADA hotline; he was a fellow Nike employee who loved his job. As a matter of fact, he continued to work for the shoe brand for two more years after he made the call, and he wouldn't start coaching for another four years.

Kara and Adam flew to Los Angeles so that Kara could testify in person before the panel in the case. Adam was not subpoenaed and was therefore not allowed inside during Kara's testimony. She walked into the room alone, and was immediately confronted with all six of Salazar's lawyers and his wife, Molly, who the couple assumed was only there to intimidate Kara, and to possibly soften her testimony. Seated on the same side of the table as Salazar, Kara was questioned for four hours straight, with the coach often leaning over the table to stare at his former athlete and show his displeasure with what she was saying.

While trying to avoid eye contact with Molly for fear that she'd break down and cry, Kara detailed her time inside the Nike Oregon Project.

During cross-examination, Salazar's lawyers tried to make her look like a liar by calling attention to even the slightest perceived inconsistency. Additionally, they tried to smear her as the unstable spouse of Adam Goucher, a former elite runner whose career had imploded under Salazar. For that, they hypothesized, Adam carried an unfair hatred for his former coach.

Kara admitted that although it was accurate to say Adam and

Alberto grew to hate each other, that didn't change the facts or make her testimony any less true.

Salazar's lawyers played a video of an interview Kara gave after the 2016 Olympic Marathon Trials, in which she used foul language. "People ask, 'How did you come back?' Letting go of that shit is how I came back," she said in the video. "I lost two hundred pounds of fucking baggage I've been carrying around." They then distributed an email Kara had written, in which the subject line read, "Fuck."

She had a sudden realization that this might be all they had, the fact that she swears on occasion. "I may swear when I'm upset," Kara said, "but I'm not a liar."

A few months after the arbitration hearing, the Gouchers were told that the results of the case were imminent. Then, *By the end of the year* became *A few more months* and *Definitely, no later than March*.

On September 30, 2019, fourteen years and four months after the first call to USADA's hotline, more than twelve years after Mackey left his voicemail, six years and nine months since Magness's first email, and four years and three months since the joint BBC and ProPublica investigation was published, I received a tip that the anti-doping agency was finally going to announce that Alberto Salazar and Dr. Jeffrey Brown were being banned from sports for doping violations. I began calling sources to verify and worked with the *New York Times* to break the news. When our article titled "Alberto Salazar, Coach of the Nike Oregon Project, Gets a 4-Year Doping Ban" went live, it was just past two in the morning in Qatar where Salazar was with his current Oregon Project athletes in Doha, including Ethiopian Yomif Kejelcha, Ethiopian-born Dutch Sifan Hassan, and American Clayton Murphy.

Oregon Project assistant coach Pete Julian got up to use the bathroom at four in the morning. The room was dark so he turned his phone on to use it to light his way, then noticed it was flooded with messages.

"It was so shocking because it was one hundred and eighty degrees different from what I thought was going to happen," Julian told me. "Alberto had told us the night before, he had a premonition it was going to come out and he said, 'We're all good. I've been assured everything's fine. Pete, this is going to be great because we can put this behind us.' It was absolutely shocking." Julian is certain Salazar also had no idea he would be banned, saying he doubts he would have come all the way to Doha: "Why would you come somewhere that you are going to be thrown out?"

That morning, Salazar's athlete, American 800-meter runner Clayton Murphy, woke up to the news that his coach was banned from sports for the next four years. This was not the way he wanted to start the day, to say the least. Later, he would have to race the fastest men on earth in the championship's 800-meter final.

Julian called a team meeting for the Oregon Project staff and athletes to talk through the shocking news before they had to face the public. He told them what he knew about the situation, which wasn't much more than what the news was reporting, owing to the fact that he was barred from communicating with Salazar. Trying to process what this meant for their careers and the day's races meant there was apprehension, and tears. Julian reassured the athletes that although they were losing Salazar they still had each other. That they'd be able to stay together as a group, and that he'd always be there for them.

Officials from the IAAF—which had rebranded itself World Athletics in June—emailed an advisory to the athletes, then canvassed the team hotels to deliver notifications, by hand, to ensure that everyone was aware of Salazar's ban. From the moment it was posted on USADA's website, his athletes were told to sever ties with Salazar. The watchdog organization, the Athletics Integrity Unit, contacted all of the other Nike Oregon Project athletes and forbade them from contact with the former team leader, then made each of Salazar's athletes sign an agreement to that effect.

No last-minute advice. No pep talks. No pre-race pills. Instantly, Salazar was radioactive. He was forced to leave the host hotel and was barred from all training venues, including the Khalifa International Stadium where the track events were taking place.

Murphy packed his bags and moved out of the room he was sharing with American middle-distance runner and 2019 NCAA Champion Bryce Hoppel, to avoid having to see the other athletes and answer questions about his wayward coach.

Salazar's name was in everyone's mouth when they lined up to run the event, passing through the start line like a tarnished secret. Still, the NOP athletes had to compete.

An American man had not won an 800-meter race at a global championship since 1985, but the anticipation was high for another NOP athlete, Donavan Brazier, who was on the Oregon Project team, but coached by Julian. He won the event in 1 minute and 42.34 seconds, and in doing so broke a thirty-four-year-old American record for the distance.

After the event, Brazier, who said he was unaware of the investigation, was asked about Salazar. "The ban on Alberto is, I think, very disappointing, because at the end of the day, he's the founder of the Oregon Project," he said. When asked if he would stay with the disgraced program, he told them, "Yeah, if Pete [Julian] takes control."

Murphy finished dead last in 1 minute and 47.84 seconds. When asked about Salazar after the event he said, "We developed a great relationship, but I was never asked to do anything, was never told to do anything, and everything that I was a part of was up to code, standard, and clean," before he walked away mid–follow-up question. Distraught, he flew home to the States later that day.

THE USADA ANNOUNCEMENT WAS SEIZED UPON BY EVERY MAJOR NEWS ORGANIZAtion. The American Arbitration Association panel found that the most famous coach in track and field committed the following violations of the WADA Code:

1. Administration of a Prohibited Method (with respect to an infusion in excess of the applicable limit),

2. Tampering and/or attempted tampering with the doping control process, and
3. Trafficking of testosterone through involvement in a testosterone testing program in violation of the rules.

The Panel also found Dr. Brown guilty on the charges of administration of a prohibited method, tampering of records with respect to the L-carnitine infusions, and complicity in Salazar's trafficking of testosterone.

Travis Tygart, USADA Chief Executive Officer, said in the statement, "The athletes in these cases found the courage to speak out and ultimately exposed the truth. While acting in connection with the Nike Oregon Project, Mr. Salazar and Dr. Brown demonstrated that winning was more important than the health and well-being of the athletes they were sworn to protect."

Later, Salazar put out a three-paragraph statement on the Nike Oregon Project website saying that he was "shocked by the outcome," and taking a similar tack to his 2015 rebuttal to the news by singling out a section of the report that he could spin in his favor.

Alberto Statement
September 30, 2019

I am shocked by the outcome today. Throughout this six-year investigation my athletes and I have endured unjust, unethical and highly damaging treatment from USADA. This is demonstrated by the misleading statement released by Travis Tygart stating that we put winning ahead of athlete safety. This is completely false and contrary to the findings of the arbitrators, who even wrote about the care I took in complying with the World Anti-Doping code:

"The Panel notes that the Respondent does not appear to have been motivated by any bad intention to commit the violations the Panel found. In fact, the Panel was struck by the amount of care generally taken by Respondent to ensure that whatever new technique or method or substance he was going to try was lawful under the World Anti-Doping Code, with USADA's witness characterizing him as the coach they heard from the most with respect to trying to ensure that he was complying with his obligations."

I have always ensured the WADA code is strictly followed. The Oregon Project has never and will never permit doping. I will appeal and look forward to this unfair and protracted process reaching the conclusion I know to be true. I will not be commenting further at this time.

Nike, firmly standing by their man, put out a statement that echoed Salazar's comments: "Today's decision had nothing to do with administering banned substances to any Oregon Project athlete. As the panel noted, they were struck by the amount of care Alberto took to ensure he was complying with the World Anti-Doping Code," without acknowledging what the report also said, that unfortunately, Salazar's "desire to provide the very best results and training for athletes . . . clouded his judgment in some instances, when his usual focus on the rules appears to have lapsed."

Salazar packed up his bags, left the official event hotel, and disappeared. A wanted poster with a picture of his event badge was hung near the entrance to Khalifa International Stadium that read, in red text and all capitals:

"THIS BADGE MUST BE WITHDRAWN. THIS RECORD IS CANCELLED FROM IAAF."

EPILOGUE

The anti-doping fight has moved on, but it hasn't moved far enough because what you see now is riders and teams operating in what we call the gray areas, using a legal medication for the wrong reason.

—Investigative journalist David Walsh, 2018

DESPITE THE FACT THAT HER COACH HAD BEEN BANNED FROM THE EVENT IN DOHA, Salazar's athlete, Dutch runner Sifan Hassan, achieved an unprecedented double victory at the championship event, winning gold in the 1,500-meter and 10,000-meter races, becoming the first person in history to win both at a single World Championships or Olympic Games. Acknowledging the improbability of an athlete having success at events that require such disparate abilities, LetsRun wrote that her performance would elevate Hassan to "all-time legend status in the sport of track & field."

Suddenly mired in the Salazar scandal, Hassan professed her innocence and frustration after the event. "It has been a hard week for me," she told reporters. "I was so angry. I couldn't talk to anyone. I ran all-out. I wanted to show that hard work can beat everything."

Nike runners were in a tough position. The brand was standing behind their coach, which left athletes who were happy to see Salazar banned forced to awkwardly avoid saying as much in interviews.

Days later, in an email to employees, Nike's chief executive, Mark Parker, wrote that the company would continue to support the embattled coach. However, Parker announced the following week that Nike was shuttering the program entirely: "This situation, along with ongoing unsubstantiated assertions, is a distraction for many of the athletes and is compromising their ability to focus on their training and competition needs. I have therefore made the decision to wind down the Oregon Project. We will help all of our athletes in this transition as they choose the coaching set up that is right for them.

"And, as we have said, we will continue to support Alberto in his appeal as a four-year suspension for someone who acted in good faith is wrong."

Three weeks later, on October 22, 2019, Parker then announced he would be stepping down as chief executive of Nike in January, a coincidence he said had nothing to do with the recent doping ban handed to Nike's most prominent coach and brand stalwart.

USADA's chief executive, Travis Tygart, recently told me he believes Salazar's final appeal with the Court of Arbitration for Sport (CAS) won't be heard until November of 2020, delaying what has already been a difficult, expensive, and protracted battle. CAS has the power to exonerate Salazar, but the appeal also leaves him open to new charges. "We're seeking a lifetime ban," said Tygart. "We think that's appropriate for the culture that was created as well as the specific anti-doping rule violations."

During my reporting, more recent Oregon Project athletes, like Mary Cain, Jordan Hasay, and Matthew Centrowitz Jr., did not respond to interview requests. I had chased unsubstantiated reports about an altercation in 2015, where Salazar publicly berated Cain for weighing too much to run fast, but could find no one to go on the record. After Jeré Longman and I reported Salazar's ban in the *New York Times*, on September 30, 2019, Cain had a change of heart about her Nike experience. She said that it wasn't until she saw the news of the ban that she realized what had happened to her in a new context. She now believes she had been abused by Salazar and the powerful swoosh-emblazoned men around him. She reached out to an editor named Lindsay Crouse at the *New York Times* to tell her story. Crouse, who works on opinion documentaries for the newspaper, published a dramatic Op-Ed video, titled "I Was the Fastest Girl in America, Until I Joined Nike," on November 7, 2019.

In the video, Cain describes how Salazar courted her as a high school senior, telling her she was "the most talented athlete he'd ever seen." She moved to Oregon her freshman year of college to join the Nike Oregon Project. Throughout her time there, Cain says Salazar became convinced that for her to improve she needed to lose weight. He focused her attention on an arbitrary number—114 pounds. Anything over this weight was unacceptable. Salazar would weigh Cain in front of her teammates and would publicly shame her if she came in too heavy. When she did, she said he pushed birth control pills and diuretics on her to assist in weight loss.

This obsessive focus on her weight drove Cain to levels of stress and malnourishment that caused her to lose her period for three years, a condition called amenorrhea, that resulted in a syndrome called Relative Energy Deficiency in Sports (RED-S). Without normal hormone fluctuation, a female body can develop issues with estrogen levels that can then negatively affect bone health, which Cain says happened to her, telling reporters that she broke five different bones in her time on the Nike Oregon Project.

She felt trapped. She started to cut herself, and considered suicide.

At a meet in 2015, Cain said Salazar did in fact have an altercation with her. She said he screamed at her publicly and told her she had clearly gained five pounds. That night, she admitted to Salazar and Treasure that she was cutting herself. She told the *Times* that the two men basically told her they just wanted to go to bed, leaving her to her psychological struggles alone.

"I got caught in a system designed by and for men," said Cain in the video, "which destroys the bodies of young girls."

As I write this epilogue, the seeds of change continue to grow at Nike, where, in December of 2019, several hundred employees picketed at the rededication of the Alberto Salazar building to protest the company's treatment of women. The office building, which sits across from the John McEnroe building and next to the Steve Prefontaine building, on the shores of Lake Nike, had been under renovation and was being reopened in Salazar's honor when it's estimated that more than four hundred employees walked the campus. Signs in the crowd of both men and women read "We Believe Mary" and "Do the Right Thing." A flyer circulated around that read "No employee is permitted to speak with the news media on any Nike-related matter, either on or off the record, without prior approval from Nike Global Communications" (though a company spokesperson said that the flyer was "not officially distributed by Nike").

In January of 2020, Salazar was placed on a temporary banned list by the US Center for SafeSport, an organization that investigates emotional, physical, and sexual misconduct in sports.

Salazar responded to Cain's allegations in a statement released to the *Oregonian* and *Sports Illustrated*, which said: "My foremost goal as a coach was to promote athletic performance in a manner that supported the good health and well-being of all my athletes. On occasion, I may have made comments that were callous or insensitive over the course of years of helping my athletes through hard training. If any athlete was hurt by any comments that I have made, such an effect was entirely unintended, and I am sorry. I do dispute, however, the notion that any athlete suffered any abuse or gender discrimination while running for the Oregon Project."

Some of Nike's highest-profile athletes, like middle-distance runner Colleen Quigley, who is coached by Jerry Schumacher, have begun to speak their minds more freely. "I think most of us are glad that it happened," she said during a podcast interview on the *Clean Sport Collective*. "We all knew that it should happen. We have always felt like we were the good guys and they were the bad guys, you know? And so it makes us feel better about what we do, like yes, we are doing this right and people who aren't doing it right are gonna have consequences."

There is also a groundswell of athletes from the United Kingdom now voicing their concerns over the UK Athletics' use of Salazar as a coaching consultant. Emma Jackson, a former 800-meter runner, said she believes high doses of the thyroid hormone thyroxine prescribed while under the care of a medical official linked to UKA, cut her running career short. In addition, a former Nike-sponsored 800-meter runner, British athlete Jo Pavey, the 2014 European 10,000-meter champion, is calling for a full investigation of UKA's use of Alberto Salazar as a coaching consultant.

Before finishing my book, I spoke to Salazar's biographer, contributing *Runner's World* writer John Brant. He has a complicated relationship with the Cuban coach and wrote about it for *Runner's World* shortly after the ban was announced. He told me Salazar was always a gentleman around him and that he still has respect for the man. "I think his achievements need to stand next to his shortcomings and his wrongdoings," he said.

Though the popularity of running as a pastime and a mode of simple exercise continues to grow, for many it's becoming harder and harder to appreciate the spectacle of professional running. Will fans stick around through all this turmoil? At Salazar's last stand, the 2019 World Championships, Khalifa International Stadium had swaths of empty seats. And although America is a celebrity worshipping culture, the dangers of this idolatry are many, and now, manifest.

As I'm closing the writing of this book, on March 20, 2020, Kara finally received her upgraded Osaka World Championship silver medal in the mail—twelve and a half years late. She once told me she used to see running as a meritocracy, where commitment and hard work mattered above all else—but cheating and doping had muddied the waters. "I don't even know what place I got anymore," she said. "At the end of the day I love that I've discovered so much about myself and I discovered that I'm stronger than I thought, but it sure would be nice to know how good I was. Was I fifth in the world? Was I tenth? I can live with either one, but I just want to know."

ACKNOWLEDGMENTS

TO ALLISON DEVEREUX, MY AGENT AND THE FIRST PERSON TO SEE AN INKLING OF promise in a proposal by an unknown, first-time author. Thankfully, the resulting book looks nothing like that initial proposal, and I owe that to you. Thanks for taking a chance on me.

I'm deeply indebted to my editor, Matthew Daddona at Dey Street Books, who believed in this project, treated me like I was already good at this, and pushed me toward the finish line. Your skill in both shaping narratives and dealing with neurotic writers became more apparent each week that passed. I could not have done this without your patience and hard work.

HarperCollins and Dey Street supported this book through a reporting and fact-checking process that took longer than envisioned. Consequential books don't get published without courageous publishers, so thank you for helping me usher this to completion.

This book would not have been possible without the help of many sources close to the story at hand. I was struck by both the courage and, far less often, the cowardice of some involved in the story, but came to terms with the latter in light of my growing understanding of the immense power of the world's most dominant sports brand. I am especially grateful to those who agreed to talk to me for this book at tremendous personal and professional peril. You have risked much, and for that I applaud your courage.

I worked hard to include most of the newsworthy and revelatory information, but I must crib a sentiment from award-winning foreign correspondent and author Stephen Kinzer: "Everything in this book is true, but not everything that is true is in this book."

I deployed three wonderful fact-checkers to help me assure the first clause of that last sentence. Thank you, Amy Blumuth and Stephanie Hayes, for all your help. Special thanks to Parker Henry, who worked closest to me in the effort toward factual verification. You were an essential component. I can honestly say that having this much talent helping me fact-check was the only thing that allowed me to sleep at night.

To Tim Catalano, whom I deeply admire, you consistently gave me that extra push I needed throughout the process of writing this narrative. Now we can finally get out and ski together, no?

Special thanks have to go to my fellow comrades-in-arms in the profession of written words. I owe a debt of gratitude to the many reporters and publications that helped report this story over the years. Two journalists in particular were especially kind, Brian Metzler, who has founded magazines and top-edited many others, and investigative reporter for the *Oregonian*, Jeff Manning. Both of you gave me time, ideas, and hard-earned contacts in my efforts to unveil the truth—thank you.

Of course, David Epstein and Mark Daly were the two journalists who really put this story on the map. It was their professional standard that I attempted to live up to every day that I worked on this book. Thank you for your persistent and dogged reportage.

Thank you, Kenny Moore, John Brant, J. B. Strasser, and Laurie Becklund, who disproportionately wrote many of the source materials that I used for the early Nike days. And Jon Gugala, who did some of the precise reporting around events in this book, which I was then able to lean on. Additionally, thank you for commiserating with me and being unfailingly supportive.

Class acts, all.

Christopher Hitchens, whose courage, intellect, and bravado are as inspiring as they are unachievable for us mere mortals, but whose high-water mark gives us all something to strive for. No single intellectual has had more of an influence on my development than you have, and I'm saddened that I never had a chance to tell you that in person.

Sam Harris, who's shown that it's not only possible, but necessary, to have good faith debates with people you don't agree with, while leaving aside all the theatrical hysterics and lingering hurt feelings. Your work has fundamentally changed my life.

Cal Newport, whom I've never met but whose books have given me the space to do the work it took for me to write my own. Thank you.

And finally, special thank-you to my family, Lana and the kids, Josh, Alesha, Sam, and Jackson. To my poor, long-suffering mother, who raised us without much help or money, and who probably thought her son was finally set when he got a high-paying job at Microsoft in the late '90s, only to watch him throw it all away to move into a van to become an athlete and a writer: I'm convinced that your half of my genes contains every bit of redeemable DNA that I possess. You continually inspire me to be a better person (and a faster reader). It is your joie de vivre that I hope to emulate as I age. It was the examples you set in my life—a dedication to both fitness and literature—that have left me with the skills that now allow me to make a living. I love you.

Last, but most important . . . to my wife, Tessa, the sweetest and most empathetic person I've ever met. I could not have done this without your love and support as I sacrificed our financial (and mental) well-being for a project that, at times, didn't look as though it would ever get made. On top of being my sounding board for almost every decision during the writing and reporting of this book, your honesty, integrity, and marketing savvy are without peer. And my mother-in-law, Leanne DuBois Harrow (I told you I'd get you in here somehow), honestly, I owe you everything for bringing Tessa into this world and for raising her to personify many of the traits that I lack and admire.

Much love, all.

A NOTE ON SOURCES

***WIN AT ALL COSTS* IS BASED ON MORE THAN THREE YEARS OF REPORTING, DRAWING** on interviews and conversations with close to one hundred sources. It was subjected to a rigorous fact-checking process with the current deputy research chief at the *Atlantic*, and the former deputy fact-checker at the *New Yorker*.

Some of the included dialogue has been drawn from historical documents, like emails and transcripts of speeches, as well as contemporaneous newspapers, magazines, television reports, and books, while the rest was written with at least one person who was present.

I reached out to all the key figures in *Win at All Costs* prior to publication to offer them an opportunity to respond to any allegations being made about them. For sources who replied, the narrative reflects their responses. For those that did not, a good faith effort was made to include existing public statements or the like. For written material quoted throughout the book, an attempt was made to include the original language, including spelling and copy errors. No sources were given fake names or pseudonyms in the text, though a few were spoken to for background and given anonymity.

It should be noted that certain facts presented in this book will inevitably be disputed by some of the parties concerned. In each case, I've done my best to objectively weigh the totality of evidence and my team's ability to corroborate, or not, what was being asserted to decide which sources to privilege. Some facts have been obscured by time, clouded by mythmaking, or simply misremembered by key figures. In those instances, I used my best judgment to discern what should be included in the narrative or cut entirely, and I'm comfortable with the choices I've made.

NOTES

EPIGRAPH

ix **"Sport is no longer a release":** Roger Angel, "Down the Drain," *New Yorker*, June 23, 1975.

PROLOGUE

1 **"Someone's coming to the door":** Kara Goucher. Interview by author. Tape recording. Boulder, Colorado, October 2, 2017.

4 **Yuliya asked her coach Vladimir Mokhnev:** John Brant, "The Marriage that Led to the Russian Track Team's Olympic Ban," *New York Times Magazine*, June 22, 2016.

4 **"I train like they say, I take drugs like they say":** John Brant, "The Marriage that Led to the Russian Track Team's Olympic Ban," *New York Times Magazine*, June 22, 2016.

5 **the sentiment in her home country is one of contempt:** Lucy Ash, "Yuliya Stepanova: What do Russians think of doping whistleblower?" BBC, December 30, 2016, https://www.bbc.com/news/magazine-38406627.

8 **"abusing prescription medications":** "INTERIM REPORT OF THE U.S. ANTI-DOPING AGENCY," USADA, March 17, 2016, Page 22 of 269.

8 **give her prescription Celebrex:** "INTERIM REPORT OF THE U.S. ANTI-DOPING AGENCY," USADA, March 17, 2016, Page 23 of 269.

8 **"I don't know if Alberto did something to me":** "INTERIM REPORT OF THE U.S. ANTI-DOPING AGENCY," USADA, March 17, 2016, Page 268 of 269.

11 **more than 2,300 feet of uphill:** GPS file, https://connect.garmin.com/modern/activity/2833549578.

01: THE BEST FEELING I'VE EVER HAD IN MY LIFE

13 **twenty-nine men from eighteen different countries:** Olympic, "Mo Farah Wins 10,000m Gold—London 2012 Olympics," August 4, 2012, YouTube video, https://www.youtube.com/watch?v=9-gOCOu_KGU.

13 **"the world's greatest distance runner":** Olympic, "Mo Farah Wins 10,000m Gold—London 2012 Olympics," August 4, 2012, YouTube video, https://www.youtube.com/watch?v=9-gOCOu_KGU.

14 **110 gold medals:** "London Olympics by the numbers," CNN, July 27, 2012, https://www.cnn.com/2012/07/27/world/olympics-numbers/index.html.

14 **American record holder in the 10,000 meters:** http://www.usatf.org/halloffame/TF/showBio.asp?HOFIDs=143.

14 **an Englishman never had:** Rob Draper, "Golden Mo-ment! Farah wins 10,000m to complete stunning night for Britain," *The Mail*, August 4, 2012.

15 **"wouldn't be trying to win it":** Ken Goe, "London Olympics: Silver for Galen Rupp; gold for Mo Farah; 'overwhelming' feeling for their coach, Alberto Salazar," *Oregonian*, August 4, 2012.

15 **considers himself British:** When asked he's quoted as saying, "Nah, mate - this is my country," Judy Richard, "The Games Made Us Great Again," *The Express*, August 18, 2012, Accessed January 23, 2019, www.LexisNexis.com.

15 **$37,500 for a gold medal:** Victor Mather, "Who Won the Most Medals and More Olympics Questions, Answered," *New York Times*, https://www.nytimes.com/guides/sports/2018-olympics-questions.

15 **$1 million for a gold medal, $500,000 for silver, and $250,000 for bronze:** Kathleen Elkins, "Here's how much Olympic athletes earn in 12 different countries," CNBC.com, February 25, 2018, https://www.cnbc.com/2018/02/23/heres-how-much-olympic-athletes-earn-in-12-different-countries.html.

16 **take the lead and set the pace:** Olympic, "Mo Farah Wins 10,000m Gold - London 2012 Olympics," August 4, 2012, YouTube video, https://www.youtube.com/watch?v=9-gOCOu_KGU.

17 **he ran the final 400 meters in about 53 seconds:** Rob Draper, "Golden Moment! Farah wins 10,000m to complete stunning night for Britain," *The Mail*, August 4, 2012.

17 **Farah, who is a practicing Muslim:** Jack De Menezes, "Mo Farah targeted with racist abuse after posting a Merry Christmas greeting," *The Independent*, December 27, 2017.

17 **"Look at the scalps of Africa taken by Mo Farah, and of course, Galen Rupp":** Olympic, "Mo Farah Wins 10,000m Gold—London 2012 Olympics," August 4, 2012, YouTube video, https://www.youtube.com/watch?v=9-gOCOu_KGU.

17 **"It was the greatest feeling, perhaps, I've ever had":** Ken Goe, "London Olympics: Silver for Galen Rupp; gold for Mo Farah; 'overwhelming' feeling for their coach, Alberto Salazar," *Oregonian*, August 4, 2012.

18 **"I'm so proud. Congrats. TC":** No attribution, "Nike's Oregon Project," *1859 Oregon* magazine, March 1, 2013, https://1859oregonmagazine.com/explore-oregon/outdoors/2013-march-april-1859-magazine-portland-oregon-alberto-salazar-nike-oregon-project/.

18 **lit up with texts:** Steve Magness. Interview by author. Tape recording. Houston, Texas, November 11, 2018.

18 **"the most disheartening moments of my life":** David Epstein, "Off Track: Former Team Members Accuse Famed Coach Alberto Salazar of Breaking Drug Rules," Propublica.org, June 3, 2015.

19 **"stayed up past ten p.m.":** Steve Magness, ScienceOfRunning.com, June 2019, https://www.scienceofrunning.com/2019/06/the-real-struggle-for-athletes-moving-on-from-their-sport.html.

19 **fastest mile run by any high schooler that year:** Peter Gambaccini, "Chat: Alan Webb's Amazing 3:53.43 High School Mile," *Runner's World*, October 17, 2006, https://www.runnersworld.com/races-places/a20794145/chat-alan-webbs-amazing-3-53-43-high-school-mile.

21 **Salazar was obsessed with beating Jerry Schumacher:** Steve Magness. Interview by author. Tape recording. Houston, Texas, November 11, 2018.

21 **undergraduate degree was in business and marketing:** Business and marketing degree: Alberto Salazar and John Brant, *14 Minutes: A Running Legend's Life and Death and Life* (New York: Rodale Books, 2013), 104–105.

22 **a fact the world wouldn't find out about until the report went public in 2017:** "Interim Report of the US Anti-Doping Agency to the Texas Medical Board Concerning the U.S. Anti-Doping Agency's Investigation Regarding Dr. Jeffrey Stuart Brown, Alberto Salazar and the Nike Oregon Project," United States Anti-Doping Agency, March 17, 2016.

22 **graduated summa cum laude:** Steve Magness, "About," ScienceOfRunning .com, https://www.scienceofrunning.com/about?v=7516fd43adaa.

22 **Clive Cussler novel:** Steve Magness. Interview by author. Tape recording. Houston, Texas, November 11, 2018.

23 **"not even on the same sphere of talent level of Alan Webb":** Steve Magness. Interview by author. Tape recording. Houston, Texas, November 11, 2018.

24 **"for the greater good, there is some shit that needs to be fixed here":** Steve Magness. Interview by author. Tape recording. Houston, Texas, November 11, 2018.

02: FORT KNOX WEST

25 **four-hundred-acre corporate headquarters:** No attribution, Nike.com, https://jobs.nike.com/nike-world-headquarters-oregon.

25 **"a topographical map of Nike's history and growth":** Phil Knight, *Shoe Dog* (New York: Simon & Schuster, 2016), 365.

25 **$1 billion expansion project:** Tim Newcomb, "An In-Depth Look Inside Nike's Sprawling Oregon Headquarters," Complex.com, August 2, 2018, https://www.complex.com/sneakers/2018/08/an-in-depth-look-inside-nike-sprawling-oregon-headquarters.

26 **"in utero":** Kenny Moore, "The Story of Oregon's Legendary Coach and Nike's Co-Founder," University of Oregon, March 5, 2008 (1:09), https://www.youtube.com/watch?v=sah--lyZR1k.

26 **a green Plymouth Valiant:** Phil Knight, *Shoe Dog* (New York: Simon & Schuster, 2016), 55. And Kenny Moore, *Bowerman and the Men of Oregon* (New York: Rodale Books, 2006), 160.

26 **$39 billion:** "NIKE, Inc. Reports Fiscal 2019 Fourth Quarter and Full Year Results," Nike.com, https://news.nike.com/news/nike-inc-reports-fiscal-2019-fourth-quarter-and-full-year-results.

26 **$35 billion:** No attribution, "#25 Phil Knight & family," Forbes.com, https://www.forbes.com/profile/phil-knight/#260c76d61dcb.

26 **48 percent of Americans:** J. B. Strasser and Laurie Becklund, *Swoosh: The Unauthorized Story of Nike and the Men Who Played There* (New York: HarperCollins, 1993), 229.

27 **"sheep's piss":** Kenny Moore, *Bowerman and the Men of Oregon* (New York: Rodale Books, 2006), 184.

27 **biblical verse:** J. B. Strasser and Laurie Becklund, *Swoosh: The Unauthorized Story of Nike and the Men Who Played There* (New York: HarperCollins, 1993), 24.

27 **urinate on his young athletes:** J. B. Strasser and Laurie Becklund, *Swoosh:*

The Unauthorized Story of Nike and the Men Who Played There (New York: HarperCollins, 1993), 23.

27 **brand them with a hot key in the sauna:** "PBS Oregon Experience: Bill Bowerman," PBS Oregon Broadcast season 1, episode 104, February 12, 2007, https://www.pbs.org/video/oregon-experience-bill-bowerman.

27 **shoes were central to athletic performance:** Phil Knight, *Shoe Dog* (New York: Simon & Schuster, 2016), 44.

27 **"Crap. Crap. Crap":** Kenny Moore, *Bowerman and the Men of Oregon* (New York: Rodale Books, 2006), 121.

27 **fought the Germans:** Pat Putnam, "The Freshman and the Great Guru," *Sports Illustrated*, June 15, 1970.

27 **resented paying Adidas:** J. B. Strasser and Laurie Becklund, *Swoosh: The Unauthorized Story of Nike and the Men Who Played There* (New York: HarperCollins, 1993), 27–28.

27 **one hundred:** J. B. Strasser and Laurie Becklund, *Swoosh: The Unauthorized Story of Nike and the Men Who Played There* (New York: HarperCollins, 1993), 41.

28 **without concern for political affiliation, religious faith, or ethnicity:** Adi & Käthe Dassler Memorial Foundation, Adidassler.org, https://www.adidassler.org/en/life-and-work/chronicle.

28 **make marching boots for the Nazi troops:** J. B. Strasser and Laurie Becklund, *Swoosh: The Unauthorized Story of Nike and the Men Who Played There* (New York: HarperCollins, 1993), 25.

28 **systematic disenfranchisement of the Jews:** Oliver Hilmes, *Berlin 1936: Sixteen Days in August* (United Kingdom: Vintage, 2018), 140.

28 **suspending distribution of *Der Stürmer*:** Oliver Hilmes, *Berlin 1936: Sixteen Days in August* (United Kingdom: Vintage, 2018), 214.

29 **eighteen hundred journalists from fifty-nine countries:** Oliver Hilmes, *Berlin 1936: Sixteen Days in August* (United Kingdom: Vintage, 2018), 47.

29 **and sales took off:** Barbara Smit, *Sneaker Wars* (New York: Harper Perennial, 2009), 29.

29 **shocked the Allied Forces:** Amalie Kvame Holm, "Amphetamine gets the job done," University of Oslo, January 9, 2015, https://www.sv.uio.no/iss/english/research/news-and-events/news/2015/amphetamine-gets-the-job-done.html.

30 **employed to build *panzers* (tank armor) and bazookas:** Barbara Smit, *Sneaker Wars* (New York: Harper Perennial, 2009), 20.

30 **new names such as "Blitz" and "Kampf":** Barbara Smit, *Sneaker Wars* (New York: Harper Perennial, 2009), 16.

30 **would later rename his company Puma:** Nick Carbone, "Adidas vs. Puma," *Time* magazine, August 23, 2011, http://content.time.com/time/specials/packages/article/0,28804,2089859_2089888_2089889,00.html.

30 **kangaroo skin:** Kenny Moore, *Bowerman and the Men of Oregon* (New York: Rodale Books, 2006), 125.

30 **like Spalding and Rawlings:** Kenny Moore, *Bowerman and the Men of Oregon* (New York: Rodale Books, 2006), 159.

30 **"nobody would have anything to do with it":** "PBS Oregon Experience: Bill

Bowerman." PBS Oregon Broadcast, season 1, episode 104, February 12, 2007, https://www.pbs.org/video/oregon-experience-bill-bowerman.

31 **three separate times:** Kenny Moore, *Bowerman and the Men of Oregon* (New York: Rodale Books, 2006), 160.

31 **"I ran a personal record":** J. B. Strasser and Laurie Becklund, *Swoosh: The Unauthorized Story of Nike and the Men Who Played There* (New York: HarperCollins, 1993), 127.

31 **"he saw character":** J. B. Strasser and Laurie Becklund, *Swoosh: The Unauthorized Story of Nike and the Men Who Played There* (New York: HarperCollins, 1993), 127.

31 **personal best of 4 minutes and 13 seconds:** "The David Rubenstein Show: Phil Knight," *Bloomberg*, June 28, 2017, https://www.bloomberg.com/news/videos/2017-06-28/the-david-rubenstein-show-phil-knight-video.

31 **seven-tenths of a second:** Jesse Squire, "Notable Celebrities that Have Competed at Penn Relays or Drake Relays," *Citius Mag*, April 24, 2018, http://citiusmag.com/penn-relays-drake-relays-celebrities-alumni/.

31 **"a good squad man":** J. B. Strasser and Laurie Becklund, *Swoosh: The Unauthorized Story of Nike and the Men Who Played There* (New York: HarperCollins, 1993), 10.

31 **on Thanksgiving Day of 1962:** J. B. Strasser and Laurie Becklund, *Swoosh: The Unauthorized Story of Nike and the Men Who Played There* (New York: HarperCollins, 1993), 15.

31 **"90 percent of Americans still had never been on an airplane":** Phil Knight, *Shoe Dog* (New York: Simon & Schuster, 2016), 11–12.

32 **more than $8 million:** Joshua Hunt, *University of Nike: How Corporate Cash Bought American Higher Education* (New York: Melville House Publishing, 2018), 18.

32 **poor Adidas knockoff:** Joshua Hunt, *University of Nike: How Corporate Cash Bought American Higher Education* (New York: Melville House Publishing, 2018), 17.

32 **candles from a Buddhist shrine:** J. B. Strasser and Laurie Becklund, *Swoosh: The Unauthorized Story of Nike and the Men Who Played There* (New York: HarperCollins, 1993), 21.

32 **Pabst Blue Ribbon:** J. B. Strasser and Laurie Becklund, *Swoosh: The Unauthorized Story of Nike and the Men Who Played There* (New York: HarperCollins, 1993), 17.

32 **"faked out Tiger Shoe Co.":** J. B. Strasser and Laurie Becklund, *Swoosh: The Unauthorized Story of Nike and the Men Who Played There* (New York: HarperCollins, 1993), 17.

32 **four factories:** J. B. Strasser and Laurie Becklund, *Swoosh: The Unauthorized Story of Nike and the Men Who Played There* (New York: HarperCollins, 1993), 40.

33 **$500 a month:** Kenny Moore, *Bowerman and the Men of Oregon* (New York: Rodale Books, 2006), 158.

33 **he wanted to be cut in on the deal:** Kenny Moore, *Bowerman and the Men of Oregon* (New York: Rodale Books, 2006), 158–159.

33 **both men investing $500 into the venture:** Kenny Moore, *Bowerman and the Men of Oregon* (New York: Rodale Books, 2006), 159.

33 **"no pissing on partners in the shower":** Kenny Moore, *Bowerman and the Men of Oregon* (New York: Rodale Books, 2006), 159–160.

33 **opened in October 1990:** No attribution, "Nike and Oregon city hope to get along again," *Herald Tribune*, October 2, 2006, https://www.nytimes.com/2006/10/02/business/worldbusiness/02iht-nike.2998782.html.

34 **nearly $2 billion:** No attribution, "Nike adds nearly $2 billion to Oregon economy," Nike.com, February 14, 2005, https://news.nike.com/news/nike-adds-nearly-2-billion-to-oregon-economy.

34 **eight hundred athletic and outdoor industry companies:** Kati Chitrakorn, "Why You Need to Move to Footwear's 'Silicon Valley'," *The Business of Fashion*, March 3, 2016, https://www.businessoffashion.com/articles/careers/why-you-need-to-move-to-footwears-silicon-valley.

34 **only *Fortune* 500 company:** Scott Bernard Nelson, "Oregon-based Lithia Motors joins Nike, Precision Castparts on Fortune 500," *Oregonian*, June 4, 2015, https://www.oregonlive.com/business/2015/06/fortune_500_oregon_lithia_moto.html#incart_river.

34 **added an athletic and outdoor certificate:** Matthew Kish, "Portland State launches online Athletic & Outdoor certificate," *Portland Business Journal*, September 19, 2018.

34 **sensible diners to flee restaurants:** J. B. Strasser and Laurie Becklund, *Swoosh: The Unauthorized Story of Nike and the Men Who Played There* (New York: HarperCollins, 1993), 1.

35 **"*Saturday Night Live* of the *Fortune* 500":** J. B. Strasser and Laurie Becklund, *Swoosh: The Unauthorized Story of Nike and the Men Who Played There* (New York: HarperCollins, 1993), 1.

35 **"the company's first significant NCAA rule violation":** J. B. Strasser and Laurie Becklund, *Swoosh: The Unauthorized Story of Nike and the Men Who Played There* (New York: HarperCollins, 1993), 231–232.

35 **Match Point shoes:** Phil Knight, *Shoe Dog* (New York: Simon & Schuster, 2016), 215.

35 **on the heel:** Barbara Smit, *Sneaker Wars* (New York: HarperCollins Perennial Edition, 2009), 212.

36 **"we'd need top athletes wearing and talking about our brand":** Phil Knight, *Shoe Dog* (New York: Simon & Schuster, 2016), 215.

36 **"David versus Goliath mentality":** Bill Saporito, "How Phil Knight Built Nike into a $100 Billion Global Empire," *Maxim*, September 29, 2016, https://www.maxim.com/entertainment/how-phil-knight-became-sultan-of-swoosh-2016-9.

36 **wearing Converse shoes:** Barbara Smit, *Sneaker Wars* (New York: HarperCollins Perennial Edition, 2009), 197.

36 **the running guys:** Josh Peter, "Error Jordan: Key figures still argue over who was responsible for Nike deal," *USA Today*, September 30, 2015, https://www.usatoday.com/story/sports/nba/2015/09/30/error-jordan-key-figures-still-argue-over-who-responsible-nike-deal/72884830/.

36 **urging of his mother:** Barbara Smit, *Sneaker Wars* (New York: HarperCollins Perennial Edition, 2009), 197.

36 **"give the kid everything you got":** Joshua Hunt, *University of Nike: How Corporate Cash Bought American Higher Education* (New York: Melville House Publishing, 2018), 61.

37 **$2.5 million:** J. B. Strasser and Laurie Becklund, *Swoosh: The Unauthorized Story of Nike and the Men Who Played There* (New York: HarperCollins, 1993), 431.

37 **$7 million:** J. B. Strasser and Laurie Becklund, *Swoosh: The Unauthorized Story of Nike and the Men Who Played There* (New York: HarperCollins, 1993), 431.

37 **on each Nike Air basketball shoe sold:** J. B. Strasser and Laurie Becklund, *Swoosh: The Unauthorized Story of Nike and the Men Who Played There* (New York: HarperCollins, 1993), 432.

37 **reshape the American endorsement business:** Barbara Smit, *Sneaker Wars* (New York: HarperCollins Perennial Edition, 2009), 309.

37 **2.3 million pairs:** Charles P. Pierce, "The Next Superstar," *New York Times*, November 15, 1992.

37 **generated $100 million in sales:** Joshua Hunt, *University of Nike: How Corporate Cash Bought American Higher Education* (New York: Melville House Publishing, 2018), 61.

38 **Stanford thesis into a company:** J. B. Strasser and Laurie Becklund, *Swoosh: The Unauthorized Story of Nike and the Men Who Played There* (New York: HarperCollins, 1993), 384.

38 **"an imaginative guy":** Jeff Johnson. Telephone interview by author. April 26, 2020.

38 **"the first corporate entity to be involved":** Joshua Hunt, *University of Nike: How Corporate Cash Bought American Higher Education* (New York: Melville House Publishing, 2018), 63.

38 **Nike's first all-school deal:** Joshua Hunt, *University of Nike: How Corporate Cash Bought American Higher Education* (New York: Melville House Publishing, 2018), 62.

38 **revenue doubled, profits quadrupled:** Joshua Hunt, *University of Nike: How Corporate Cash Bought American Higher Education* (New York: Melville House Publishing, 2018), 62.

38 **migrated from the bottom of the sleeve:** Michael MacCambridge, *The Franchise: A History of Sports Illustrated Magazine* (New York: Hyperion, 1997), 354–355.

39 **czar of athletic footwear:** Richard Sandomir, "Scrappy Reebok Aims at Aloof Nike," *New York Times*, February 9, 1993.

39 **"They're insane, sick, disgusting, I think":** Barbara Smit, *Sneaker Wars* (New York: HarperCollins Perennial Edition, 2009), 334.

39 **it looked as if Nike had bent time and space:** Donald Katz, *Just Do It: The Nike Spirit in the Corporate World* (Newark, NY: Audible, 1994).

39 **"Athletes are everything at Nike":** No attribution, "Inside the Swoosh," Nike.com, Accessed May 23, 2019, https://jobs.nike.com/nike-world-headquarters-oregon.

40 **a key tactical decision:** Barbara Smit, *Sneaker Wars* (New York: HarperCollins Perennial Edition, 2009), 188.

41 **"we have to walk that line":** Bill Saporito, "How Phil Knight Built Nike into a $100 Billion Global Empire," *Maxim*, September 29, 2016, https://www.maxim.com/entertainment/how-phil-knight-became-sultan-of-swoosh-2016-9.

41 **"an unclimbable pyramid":** Tiffany Hsu, "Ex-Employees Sue Nike, Alleging Gender Discrimination," *New York Times*, August 10, 2018.

42 **Ferrari parked in Ray Allen's reserved spot:** Person reporting on Nike campus. September 2019.

42 **twelve thousand employees:** Aaron Mesh and Sophie Peel, "Don't Worry, Be Nike: Oregon's Flagship Company Has Done Wrong. Why Aren't Its Employees Resisting?" wweek.com, November 20, 2019, https://www.wweek.com/news/2019/11/20/oregons-flagship-company-has-done-wrong-why-arent-its-employees-resisting.

03: WHAT ARE YOU ON?

44 **only 4 percent:** Andrew Tilin, "The Ultimate Running Machine," *Wired*, October 2008.

44 **"sixth place is considered an accomplishment":** Jennifer Kahn, "The Perfect Stride," *New Yorker*, November 8, 2010.

45 **"Tom got out a pen":** No attribution, "Nike's Oregon Project: Putting US Runners Back on the Podium," *1859 Oregon Magazine*, March–April 2013.

45 **doctorate in biomechanics:** Andrew Tilin, "The Ultimate Running Machine," *Wired*, October 2008.

46 **into superstar athletes:** Andrew Tilin, "The Ultimate Running Machine," *Wired*, October 2008.

47 **Decker Slaney "and her coach":** Dave Kuehls, "Mary Slaney Tries Again," *Runner's World*, June 1996.

47 **impermissible ratio of testosterone to epitestosterone:** Juliet Macur, "Testosterone Seems to Be Enhancer of Choice," *New York Times*, July 31, 2006.

47 **two-year retroactive ban:** Danielle Eurich. Email with author. Monday, June 5, 2017.

47 **since she was twelve years old:** *Encyclopedia Britannica*, https://www.britannica.com/biography/Marion-Jones.

49 **"sweatshop":** Guy Raz, "How I Built This with Guy Raz: Live Episode! BuzzFeed: Jonah Peretti," NPR, July 27, 2017, https://www.npr.org/2017/10/03/539523369/live-episode-buzzfeed-jonah-peretti.

49 **Personal iD:** Guy Raz, "How I Built This with Guy Raz: Live Episode! BuzzFeed: Jonah Peretti," NPR, July 27, 2017, https://www.npr.org/2017/10/03/539523369/live-episode-buzzfeed-jonah-peretti.

52 **he forwarded the email string to a dozen friends:** Jonah Peretti, "My Nike Media Adventure," *Nation*, March 22, 2001.

52 **millions of inboxes:** Guy Raz, "How I Built This with Guy Raz: Live Episode! BuzzFeed: Jonah Peretti," NPR, July 27, 2017, https://www.npr.org/2017/10/03/539523369/live-episode-buzzfeed-jonah-peretti.

53 **the message was sent from the Nike campus:** Tom Farrey, "Just Don't Do It," ESPN.com, https://www.espn.com/page2/s/farrey/010227.html.

53 **"Sales are up":** Tom Farrey, "Just Don't Do It," ESPN.com, https://www.espn.com/page2/s/farrey/010227.html.

53 **a month before:** No attribution, "Nike Names Building After Lance Armstrong," TotalBike.com, February 28, 2001, http://www.totalbike.com/news/article/28/.

54 **emailed Howard Slusher:** Bob Babbitt and Paul Huddle, "Alberto Salazar Interview," *Competitor Radio*, March 15, 2011, https://web.archive.org/web/20121001002052/http://content.blubrry.com/competitor_radio/AlbertoSalazar-3-15-11.mp3.

54 **"as I climb the mountains ahead in my life":** No attribution, "Nike Names Building After Lance Armstrong," TotalBike.com, February 28, 2001, http://www.totalbike.com/news/article/28/.

54 **"no more inspiring stories than Lance's":** No attribution, "Nike Names Building After Lance Armstrong," TotalBike.com, February 28, 2001, http://www.totalbike.com/news/article/28/.

55 **his wife, Kristin; mother, Linda; son Luke:** No attribution, "Nike Names Building After Lance Armstrong," TotalBike.com, February 28, 2001, http://www.totalbike.com/news/article/28/.

55 **paid cyclists on other teams to let him win:** Juliet Macur, *Cycle of Lies: The Fall of Lance Armstrong* (New York: HarperCollins, 2014), 43.

55 **mega-corp paid the then-head of the Union Cycliste Internationale:** No attribution, "Verbruggen denies Kathy LeMond's story of cover-up payment, *Cycling News*, October 18, 2012, https://www.cyclingnews.com/news/verbruggen-denies-kathy-lemonds-story-of-cover-up-payment/.

55 **28 minutes and 30 seconds, the accepted benchmark:** Andrew Tilin, "The Ultimate Running Machine," *Wired*, October 2008.

56 **Davis said he was completely broke:** Andrew Tilin, "The Ultimate Running Machine," *Wired*, Octobter 2008.

57 **"the greatest concentration of elite athletic talent":** Robert Siegel, "How One Kenyan Tribe Produces the World's Best Runners," NPR, November 1, 2013, https://www.npr.org/templates/transcript/transcript.php?storyId=241895965.

57 **altitudes of 7,000 to 8,000 feet above sea level:** Jill Barker, "Are Kenyans Born to Run?" *Gazette*, May 11, 2004.

58 **could damage them:** Rob Beamish and Ian Ritchie, *Fastest, Highest, Strongest: A Critique of High-Performance Sport* (New York: Routledge, 2006), 110.

58 **twelve hours each day:** Andrew Tilin, "The Ultimate Running Machine," *Wired*, October 2008.

59 **canceled their Pilates sessions:** Andrew Tilin, "The Ultimate Running Machine," *Wired*, October 2008.

59 **It violates the spirit of sport:** Maggie Durand. E-mail communication with author. Sent Tuesday, July 2, 2019, at 12:12 p.m.

60 **"B-plus" competitors:** Jennifer Kahn, "The Perfect Stride," *New Yorker*, November 8, 2010.

04: TAKING RUNNING OFF THE BACK PAGE

61 **Hollywood, Florida:** Adam Goucher. Interview by author. Tape recording. Boulder, Colorado, September 26, 2017.

61 **advantageous to curing tuberculosis:** Scott Rappol, "America's Greatest Sanitarium," *The Gazette*, June 17, 2007.

61 **dog and horse racing tracks:** Adam Goucher. Telephone interview by author. March 11, 2020.

62 **"That was hard":** Chris Lear, "Learning to Fly," *Runner's World*, May 1, 2001.

62 **cleared six feet:** John Brant, "For Better or For Worse," *Runner's World*, June 2008.

62 **again in ninth grade:** Adam Goucher. Telephone interview by author. March 11, 2020.

62 **spent most of her time at her boyfriend's house:** Adam Goucher and Tim Catalano, *Running the Edge* (Boulder, CO: Maven Publishing, 2011), 3.

63 **working with the local:** Adam Goucher. Telephone interview by author. March 11, 2020.

63 **he hung a handmade poster:** Chris Lear, "Learning to Fly," *Runner's World*, May 1, 2001.

64 **future running star:** Keflezighi is a three-time national champion in cross-country running, having won the USA Cross Country Championships in 2001, 2002, and 2009. He's also won Boston and New York City Marathons.

64 **he would have likely just flipped again:** Tim Catalano. Interview by author. Boulder, Colorado, May 8, 2019.

64 **second-best freshman finish:** Marc Bloom, "The Young Coloradan With Endless Oxygen," *New York Times*, November 18, 1995.

65 **"best young runner":** Marc Bloom, "The Young Coloradan With Endless Oxygen," *New York Times*, November 18, 1995.

65 **Donald Trump's eponymous buildings:** Kara Goucher. Interview by author. Tape recording. Boulder, Colorado, October 5, 2017.

66 **"We believe in education":** Jim Ferstle, "Girl from the North Country," *Runner's World*, March 31, 2008.

67 **Tollefson added to her long list of championships:** Tollefson became the first (and only) high school runner in the nation to win five consecutive state cross-country titles.

69 **20 percent scholarship:** Kara Goucher. Telephone interview by author. March 11, 2020.

70 **than anyone else in the history of running:** Jeff Manning, "Capriotti: Nike's hot-tempered track chief makes waves," *The Oregonian/OregonLive*, July 6, 2016, https://www.oregonlive.com/business/2016/07/post_242.html.

70 **"I knew what I was doing":** Jeff Manning, "Capriotti: Nike's hot-tempered track chief makes waves," *The Oregonian/OregonLive*, July 6, 2016, https://www.oregonlive.com/business/2016/07/post_242.html.

70 **reached out to Capriotti to join him:** No attribution, "Cal Poly Dedicates Steve Miller and John Capriotti Athletics Complex," March 28, 2018, https://www.gopoly.com/sports/track/2017-18/releases/20180328m5rt27.

71 **Nike offered Adam a $90,000:** Adam Goucher. Interview by author. Tape recording. Boulder, Colorado, September 14, 2017.

72 **in six years:** YouTube. "Adam Goucher outkicks Bob Kennedy - Men's 5,000m

(finish) - 1999 USA Championships," Jim Muchmore, Published on November 15, 2018, https://www.youtube.com/watch?v=acZUJkgd_Qc.

76 **in front of twenty thousand:** Weldon Johnson, "Q&A With Weldon Johnson on the Trials," LetsRun.com, https://www.letsrun.com/weldontrialsqa.html.

05: JUST A COACH DOING THE RIGHT THING

78 **tenth place:** WorldAthletics.org, https://www.worldathletics.org/results/world-athletics-championships/2001/8th-iaaf-world-championships-2639/men/5000-metres/final/result.

78 **arguably the most decorated:** CUBuffs.com, https://cubuffs.com/sports/2008/7/7/1506580.aspx.

79 **cracked her kneecap:** Bruce Barcott, "Mind Gains," *Runner's World*, February 11, 2010.

79 **"either going to be a train wreck or something incredibly strong":** John Brant, "For Better or For Worse," *Runner's World*, June 2008.

79 **Nike signed Kara for $35,000:** Adam Goucher. Interview by author. Phone Call. Boulder, Colorado, June 11, 2019.

79 **"ended up dying":** Bruce Barcott, "Mind Gains," *Runner's World*, February 11, 2010.

80 **used to call the Trials "ordeals":** Kenny Moore, *Track Town, USA: Hayward Field: America's Crown Jewel of Track and Field* (Rich Clarkson, Denver, 2010), 83.

80 **first American athlete to ever test positive:** M. Nicole Nazzaro, "Out of a Doping Scandal, Redemption and Progress," *New York Times*, August 17, 2013.

81 **the prescription drug Modafinil in her system:** John Crumpacker, "Kelli White suspended," SFgate.com, Published 4:00 a.m., May 20, 2004, https://www.sfgate.com/sports/article/SPORTS-AND-DRUGS-Kelli-White-suspended-The-2757572.php.

81 **hyper-focused, energetic, and confident:** Mark Fainaru-Wada and Lance Williams, *Game of Shadows* (New York: Penguin Group, 2006), 139.

81 **15 minutes and 28 seconds:** USATF.org, http://oldserver.usatf.org/events/2004/OlympicTrials-TF/entry/qualifyingStandards.asp.

81 **"I got dead last":** Adam and Kara Goucher. Interview by author. Tape recording. Boulder, Colorado, September 18, 2017.

82 **between 0.45 to 4.12:** Jeffrey Garber. Telephone interview by author. February 27, 2020.

83 **a time of 13 minutes and 27.36 seconds:** USA Track and Field, "Men 5000 Meter Run Finals," usatf.org, http://www.usatf.org/events/2004/OlympicTrials-TF/results/F11.asp.

83 **bodyguard escorted her out:** Mark Fainaru-Wada and Lance Williams, *Game of Shadows* (New York: Penguin Group, 2006), 234.

83 **"dark and deep hole":** Mark Fainaru-Wada and Lance Williams, *Game of Shadows* (New York: Penguin Group, 2006), 236.

84 **"an athlete capable of beating Marion":** Mark Fainaru-Wada and Lance Williams, *Game of Shadows* (New York: Penguin Group, 2006), 237.

84 **wandered through Borders Bookstore:** Brian Meltzer, originally published as "Back on Track" in *Running Times*, July/August 2006; available at https://www.runnersworld.com/advanced/a20832485/back-on-track-1/.

85 **"'What a great guy'":** Chris McClung, Kara and Adam Goucher, "Episode #14: Kara and Adam Goucher on the 4-Year Bans for Salazar/Brown for Doping Violations," *Clean Sport Collective* podcast, October 7, 2019, https://podcasts.apple.com/us/podcast/episode-14-kara-adam-goucher-on-4-year-bans-for-salazar/id1466187704?i=1000452633967.

86 **"you aren't as good as you hoped you were":** Kara and Adam Goucher. Interview by author. Tape recording. Boulder, Colorado, September 28, 2017.

87 **the 1976 Olympics were overrun:** Kenny Moore, *Bowerman and the Men of Oregon* (New York: Rodale Books, 2006), 345.

87 **deep voices, broad shoulders:** Frank Shorter. Interview by author. Tape recording. Boulder, Colorado, September 11, 2018.

87 **sore losers:** Jeré Longman, "Just Following Orders, Doctors' Orders," *New York Times*, April 22, 2001.

87 **"beaten by men":** Jeré Longman, "Just Following Orders, Doctors' Orders," *New York Times*, April 22, 2001.

88 **Adidas on his feet:** Frank Shorter. Interview by author. Tape recording. Boulder, Colorado, September 11, 2018.

88 **"card-carrying running-shoe company":** Phil Knight, *Shoe Dog* (New York: Simon & Schuster, 2016), 294.

88 **father of the running boom:** Frank Shorter and John Brant, *My Marathon: Reflections on a Gold Medal Life* (New York: Rodale Books, 2016), Kindle ed., Location 1,510.

89 **in tandem with the shadow of performance-enhancing drugs:** Frank Shorter and John Brant, *My Marathon: Reflections on a Gold Medal Life* (New York: Rodale Books, 2016), Kindle ed., Location 2,205.

89 **"spiritual and moral decline":** Rob Beamish and Ian Ritchie, *Fastest, Highest, Strongest: A Critique of High-Performance Sport* (New York: Routledge, 2006), 7.

90 **Montreal was well suited to impress:** David Halberstam, *The Best American Sports Writing of the Century* (New York: Houghton Mifflin, 1999), 458.

90 **"a giant trading floor":** J. B. Strasser and Laurie Becklund, *Swoosh: The Unauthorized Story of Nike and the Men Who Played There* (New York: HarperCollins, 1993), 204.

90 **"We felt like pawns":** Frank Shorter and John Brant, *My Marathon: Reflections on a Gold Medal Life* (New York: Rodale Books, 2016), Kindle ed., Location 2,001.

91 **the brand didn't make a track spike:** J. B. Strasser and Laurie Becklund, *Swoosh: The Unauthorized Story of Nike and the Men Who Played There* (New York: HarperCollins, 1993), 204.

91 **behind three runners:** Barbara Smit, *Sneaker Wars* (New York: HarperCollins Perennial Edition, 2009), 103.

91 **$15,000 for the year:** J. B. Strasser and Laurie Becklund, *Swoosh: The Unauthorized Story of Nike and the Men Who Played There* (New York: HarperCollins, 1993), 205.

91 **seventeen miles a day:** David Epstein, *Sports Illustrated*, August 05, 2008, https://www.si.com/olympics/2008/08/05/shorter-cuw.

92 **orange Onitsuka Tiger Obori marathon flats:** Frank Shorter. Interview by author. Tape recording. Boulder, Colorado, September 11, 2018.

92 **"No, NO!" he cried:** Phil Knight, *Shoe Dog* (New York: Simon & Schuster, 2016), 295.

92 **switched from soda to vodka:** Phil Knight, *Shoe Dog* (New York: Simon & Schuster, 2016), 295.

93 **before every kid had a backpack:** John Brant, "Frank's Story," *Runner's World*, June 21, 2016.

93 **ran the oval once:** Frank Shorter and John Brant, *My Marathon: Reflections on a Gold Medal Life* (New York: Rodale Books, 2016), Kindle ed., Location 2,095.

93 **a recent hamstring injury:** Frank Shorter and John Brant, *My Marathon: Reflections on a Gold Medal Life* (New York: Rodale Books, 2016), Kindle ed., Location 1,987.

93 **"strongly suspected he was blood doping":** Frank Shorter and John Brant, *My Marathon: Reflections on a Gold Medal Life* (New York: Rodale Books, 2016), Kindle ed., Location 1,866.

94 **world-record pace:** Frank Shorter and John Brant, *My Marathon: Reflections on a Gold Medal Life* (New York: Rodale Books, 2016), Kindle ed., Location 2,106.

94 **"parachuted in":** Frank Shorter and John Brant, *My Marathon: Reflections on a Gold Medal Life* (New York: Rodale Books, 2016), Kindle ed., Location 2,193.

95 **50 seconds later:** International Olympic Committee, "Montreal Marathon Men results," Accessed October 17, 2018, https://www.olympic.org/montreal-1976/athletics/marathon-men.

95 **before being ushered to drug testing:** Frank Shorter and John Brant, *My Marathon: Reflections on a Gold Medal Life* (New York: Rodale Books, 2016), Kindle ed., Location 2,157.

95 **"Shogun of shoes":** Phil Knight, *Shoe Dog* (New York: Simon & Schuster, 2016), 63.

95 **everyone in the world wearing athletic shoes:** Phil Knight, *Shoe Dog* (New York: Simon & Schuster, 2016), 64.

06: IT WON'T BE PRETTY

96 **reputed budget of $7 million:** J. B. Strasser and Laurie Becklund, *Swoosh: The Unauthorized Story of Nike and the Men Who Played There* (New York: HarperCollins, 1993), 204.

96 **Geoff Hollister, suggested:** Kenny Moore, *Bowerman and the Men of Oregon* (New York: Rodale Books, 2006), 260.

96 **"the price of an ad in *Sports Illustrated*":** J. B. Strasser and Laurie Becklund, *Swoosh: The Unauthorized Story of Nike and the Men Who Played There* (New York: HarperCollins, 1993), 229–230.

97 **pursue suspensions:** J. B. Strasser and Laurie Becklund, *Swoosh: The Unau-*

thorized Story of Nike and the Men Who Played There (New York: HarperCollins, 1993), 230.

97 **Harry Johnson, was picked:** Kenny Moore, *Bowerman and the Men of Oregon* (New York: Rodale Books, 2006), 349.

97 **travel expenses:** George Malley, "RE: Athletics West," LetsRun.com, May 10, 2009, Accessed March 30, 2019, https://www.letsrun.com/forum/flat_read.php?thread=3007030&page=2.

97 **Nike's first poster boy:** Craig Virgin. Interview by author. Phone call recording. September 5, 2019.

98 **nine-time All-American:** Craig Virgin. Interview by author. Phone call recording. September 5, 2019.

99 **"very poor implementation":** George Malley, "RE: Athletics West," LetsRun.com, May 10, 2009, Accessed March 30, 2019, https://www.letsrun.com/forum/flat_read.php?thread=3007030&page=2.

99 **"It tasted like shit":** George Malley, "RE: Athletics West," LetsRun.com, May 10, 2009, Accessed March 30, 2019, https://www.letsrun.com/forum/flat_read.php?thread=3007030&page=2.

99 **"ignoring this undercurrent":** Craig Virgin. Interview by author. Phone call recording. September 5, 2019.

99 **It won't be pretty:** Photo taken of the Principles. #3 was affirmed by Liz Dolan interview. (Ep. 342). *Freakonomics*, freakonomics.com/podcast/lance-armstrong.

100 **"'I have to tolerate you'":** Kenny Moore, *Bowerman and the Men of Oregon* (New York: Rodale Books, 2006), 349.

100 **"distributing a lot of crap":** Kenny Moore, *Bowerman and the Men of Oregon* (New York: Rodale Books, 2006), 349.

100 **eased its pursuit of those in violation:** Taylor Branch, "The Shame of College Sports," *Atlantic,* October 2011.

101 **investing in more athletes directly:** J. B. Strasser and Laurie Becklund, *Swoosh: The Unauthorized Story of Nike and the Men Who Played There* (New York: HarperCollins, 1993), 229.

102 **wives passionately protested:** J. B. Strasser and Laurie Becklund, *Swoosh: The Unauthorized Story of Nike and the Men Who Played There* (New York: HarperCollins, 1993), 230.

103 **"I'd bet my life":** Jeff Johnson. Telephone interview by author. April 26, 2020.

103 **"here is a doctor that is going to help you":** Craig Virgin. Interview by author. Phone call recording. September 5, 2019.

104 **"I wouldn't be surprised if he weren't clean":** Ron Tabb. Interview by author. Phone call. January 5, 2018.

104 **$12,000 a year:** Craig Virgin. Interview by author. Phone call recording. September 5, 2019.

104 **Nike shuttered the Athletics West:** Walter Drenth. Interview by author. Phone call. November 1, 2019.

105 **"I'm never doing that":** Walter Drenth. Interview by author. Phone call. November 1, 2019.

105 **"loss of a wonderful person":** Scott Pengelly. Telephone interview by author. May 1, 2019.

07: NOTHING TO LOSE

106 **"If we hated it we'd just move back":** Adam and Kara Goucher. Interview by author. Tape recording. Boulder, Colorado, September 18, 2017.

107 **sleep in sleeping bags:** Adam and Kara Goucher. Interview by author. Tape recording. Boulder, Colorado, September 18, 2017.

107 **popular among NBA players:** Juliet Macur, *Cycle of Lies: The Fall of Lance Armstrong* (New York: HarperCollins, 2014), 210.

108 **Sportswriter of the Year:** Claire Cozens, "Top cyclist to sue Sunday Times over doping claims," *Guardian*, June 15, 2004.

108 **lie, steal, and threaten:** Juliet Macur, *Cycle of Lies: The Fall of Lance Armstrong* (New York: HarperCollins, 2014), 211.

108 **eighty million:** Phil Wahba, "Nike drops partnership with Lance Armstrong-founded charity," Reuters, May 28, 2013, https://www.reuters.com/article/us-cycling-armstrong-livestrong/nike-drops-partnership-with-lance-armstrong-founded-charity-idUSBRE94R0PR20130529.

108 **as did Armstrong's cycling rival:** Anushka Asthana, "How a yellow wristband became a fashion must," *Guardian*, August 7, 2004.

108 **helped sell them:** Juliet Macur, *Cycle of Lies: The Fall of Lance Armstrong* (New York: HarperCollins, 2014), 387.

108 **"all sports at the elite level are a fraud":** Mark Fainaru-Wada and Lance Williams, *Game of Shadows* (New York: Penguin Group, 2006), 239.

109 **Jones filed a $25 million lawsuit:** Mark Kreidler, "Hunter was right—Jones got big boost from steroids," ESPN, October 5, 2007.

109 **"by the book":** Scott Douglas, "Former Nike Oregon Project Coach 'Not Surprised' by Doping Allegations," RunnersWorld.com, June 8, 2015, https://www.runnersworld.com/news/a20806570/former-nike-oregon-project-coach-not-surprised-by-doping-allegations/.

109 **three men were tasked by Salazar:** Scott Douglas, "Former Nike Oregon Project Coach 'Not Surprised' by Doping Allegations," RunnersWorld.com, June 8, 2015, https://www.runnersworld.com/news/a20806570/former-nike-oregon-project-coach-not-surprised-by-doping-allegations/.

110 **"in from Japan":** Vern Gambetta. Interview by author by telephone. March 21, 2019.

110 **"Salazar's the greatest":** Vern Gambetta. Interview by author by telephone. March 21, 2019.

110 **"The rest of the world is doing that":** Dick Patrick, "Choosing running over college," *USA Today*, Posted February 11, 2005, 12:43 a.m, https://usatoday30.usatoday.com/sports/olympics/summer/2005-02-11-distance-running-cover_x.htm.

111 **sub-five-minute mile as a freshman:** Alberto Salazar. Interview by Bob Babbitt and Paul Huddle. Original interview on *Competitor Radio*, March 15, 2011, available at the Internet Archive.

111 **4 minutes and 32 seconds:** Alberto Salazar and John Brant, *14 Minutes:*

A Running Legend's Life and Death and Life (New York: Rodale Books, 2013), 206.

111 **"he's so far ahead":** Alberto Salazar and John Brant, *14 Minutes: A Running Legend's Life and Death and Life* (New York: Rodale Books, 2013), 206.

111 **"lucky for me":** John Brant, "Galen Rupp Feels Lucky," *Runner's World*, June 2012.

112 **"he just kept improving and improving":** Bob Babbitt and Paul Huddle, "Alberto Salazar Interview," *Competitor Radio*, March 15, 2011.

112 **until Salazar called:** Dick Patrick, "Choosing running over college," *USA Today*, Posted February 11, 2005, 12:43 a.m., https://usatoday30.usatoday.com/sports/olympics/summer/2005-02-11-distance-running-cover_x.htm.

112 **"a level playing field for all":** NCAA.org, Accessed August 13, 2019, https://www.ncaa.org/student-athletes/play-division-i-sports.

113 **José worked as a civil engineer:** Neil Amdur, "Salazar, Mrs. Roe Set Records," *New York Times,* October 26, 1981.

113 **religion ran contrary to the Marxist philosophy:** "1961 Castro declares himself a Marxist-Leninist," *History Channel*, November 13, 2009, https://www.history.com/this-day-in-history/castro-declares-himself-a-marxist-leninist.

113 **"no room for God in the revolution":** Alberto Salazar and John Brant, *14 Minutes: A Running Legend's Life and Death and Life* (New York: Rodale Books, 2013), 10.

113 **"death, honor, and blood loyalty":** Alberto Salazar and John Brant, *14 Minutes: A Running Legend's Life and Death and Life* (New York: Rodale Books, 2013), 11.

114 **in 1960:** Kenny Moore, "There Are Only 26 Miles to Go," *Sports Illustrated*, November 3, 1980.

114 **Wayland, Massachusetts:** Alberto Salazar and John Brant, *14 Minutes: A Running Legend's Life and Death and Life* (New York: Rodale Books, 2013), 17.

114 **"strife, conflict, and intense emotion":** Alberto Salazar and John Brant, *14 Minutes: A Running Legend's Life and Death and Life* (New York: Rodale Books, 2013), 3.

114 **"unlike that of other American families":** Alberto Salazar and John Brant, *14 Minutes: A Running Legend's Life and Death and Life* (New York: Rodale Books, 2013), 4.

114 **"forces beyond my understanding":** Alberto Salazar and John Brant, *14 Minutes: A Running Legend's Life and Death and Life* (New York: Rodale Books, 2013), 3.

114 **"the one most in need of grace":** John Brant, "Duel in the Sun," *Runner's World*, April 4, 2004.

116 **John Denver tape:** Gerald Scott, "THE LEGEND LIVES ON: Even though Steve Prefontaine died almost 10 years ago, the memory of his life and controversy surrounding his death are as alive as ever," *Los Angeles Times*, May 6, 1985.

120 **"man of honor":** Alberto Salazar and John Brant, *14 Minutes: A Running Legend's Life and Death and Life* (New York: Rodale Books, 2013), 99.

121 **3 minutes and 47.33 seconds:** Neil Amdur, "Coe Over Ovett. Ovett Over Coe," *New York Times*, August 28, 1981, http://archive.nytimes.com/www.nytimes

.com/packages/html/sports/year_in_sports/08.28.html?scp=14&sq=field%252
0day&st=cse.

126 **considered pseudoscience nonsense:** Mark Fainaru-Wada and Lance Williams, *Game of Shadows* (New York: Penguin Group, 2006), 13.

126 **"have better sex":** Mark Fainaru-Wada and Lance Williams, *Game of Shadows* (New York: Penguin Group, 2006), 1.

128 **often cruel to the staff:** John Brant, "Why I Still Have Faith in Alberto Salazar," RunnersWorld.com, October 24, 2019.

128 **considered suicide:** John Brant, "The Marathoner Speaks to His God," *New York Times Play*, October 28, 2007.

09: LOYALTY OVER COMPETENCY

140 **Rupp too was diagnosed with a thyroid condition:** "INTERIM REPORT OF THE U.S. ANTI-DOPING AGENCY," USADA, March 17, 2016, page 34.

141 **working on a graduate school engineering project:** Brian Gillis, "A Brief History of AlterG's NASA Technology—The Early Years," AlterG.com, April 22, 2015, Accessed July 9, 2019, https://www.alterg.com/treadmill-training-rehab/athletics/a-brief-history-of-altergs-nasa-technology-the-early-years.

141 **15 to 25 percent more training volume:** AlterG, "Alberto Salazar from Nike Oregon Project uses the AlterG," March 17, 2010, YouTube video, https://www.youtube.com/watch?v=40Vh7HdMrFI.

142 **"he had a falling out":** Adam and Kara Goucher. Interview by author. Tape recording. Boulder, Colorado, September 18, 2017.

142 **"I wish I would have backtracked":** Vern Gambetta. Interview by author. Recorded phone call. Boulder, Colorado, September 11, 2018.

143 **an altitude tent at 14,000 feet:** Vern Gambetta. Interview by author. Phone call. March 21, 2019.

143 **both coach and director were gone:** Joshua Hunt, *University of Nike: How Corporate Cash Bought American Higher Education* (New York: Melville House Publishing, 2018), 171.

143 **Smith was replaced:** *Seattle Times* staff, "Former Stanford coach Vin Lananna takes over Oregon track program," *Seattle Times*, July 13, 2005.

144 **Moos would retire a year later:** Joshua Hunt, *University of Nike: How Corporate Cash Bought American Higher Education* (New York: Melville House Publishing, 2018), 171.

144 **"the monster that ate me":** Jim Moore, "How a Monster Ate Former UO Athletic Director Moos," *Seattle Post-Intelligencer*, October 18, 2007.

144 **"making the arena a reality":** Associated Press, "Knight's $100 million gift to bankroll Oregon athletics fund," ESPN.com, August 21, 2007, https://www.espn.com/college-sports/news/story?id=2984161.

146 **a new junior standard:** USATF.org, Page last updated: 12/26/2018, Accessed October 16, 2019, http://www.usatf.org/statistics/records/view.asp?division=american&location=outdoor%20track%20%26%20field&age=junior&sport=TF.

147 **he was anemic:** Adam and Kara Goucher. Interview by author. Tape recording. Boulder, Colorado, September 18, 2017.

149 **"it felt dangerous":** David Epstein, "Elite Runner Had Qualms When Alberto

Salazar Told Her to Use Asthma Drug for Performance," ProPublica.org, June 17, 2015.

149 **"drug testing can be circumvented":** Scott Douglas, "Former Nike Oregon Project Coach 'Not Surprised' by Doping Allegations," *Runner's World*, June 8, 2015, https://www.runnersworld.com/news/a20806570/former-nike-oregon-project-coach-not-surprised-by-doping-allegations/.

150 **"I don't think they'd be happy with what's going on":** Scott Douglas, "Former Nike Oregon Project Coach 'Not Surprised' by Doping Allegations," *Runner's World*, June 8, 2015, https://www.runnersworld.com/news/a20806570/former-nike-oregon-project-coach-not-surprised-by-doping-allegations/.

150 **"my amigos":** Scott Douglas, "Former Nike Oregon Project Coach 'Not Surprised' by Doping Allegations," *Runner's World*, June 8, 2015, https://www.runnersworld.com/news/a20806570/former-nike-oregon-project-coach-not-surprised-by-doping-allegations/.

150 **"I have a [sic] NDA":** Dan Pfaff in email with author. March 20, 2019, 8:40 p.m. MST.

150 **"experience tells me to stay away":** Dan Pfaff in email with author. March 21, 2019, 6:30 p.m. MST.

10: YOU HAVE NO IDEA

151 **"group training":** John Galvin, "The Hansons-Brooks Distance Project," *Runner's World*, June 27, 2007, https://www.runnersworld.com/races-places/a20834663/the-hansons-brooks-distance-project/.

152 **between $200,000 and $250,000:** Chris Chavez, "Keith and Kevin Hanson on 20 Years of Advancing American Distance Running," *CITIUS MAG Podcast with Chris Chavez*, April 6, 2020, https://podcasts.apple.com/us/podcast/keith-kevin-hanson-on-20-years-advancing-american-distance/id1204506559?i=1000470689600.

152 **it was fine to drink:** Tim Catalano. Interview by author. Tape recording. Erie, Colorado. November 5, 2019.

153 **"the most fulfilled I've ever felt as a coach":** Kara Goucher. Interview by author. Tape recording. Boulder, Colorado, September 28, 2017.

153 **found a 10,000-meter event in Helsinki:** Adam and Kara Goucher. Interview by author. Tape recording. Boulder, Colorado, September 18, 2017.

153 **"You have no idea":** Bruce Barcott, "Mind Gains," *Runner's World*, February 11, 2010.

154 **"to legitimize the program":** Adam and Kara Goucher. Interview by author. Tape recording. Boulder, Colorado, September 18, 2017.

155 **Dathan Ritzenhein signed:** Dennis Young, "Dathan Ritzenhein Is No Longer with Nike," FloTrack.org, May 13, 2017, https://www.flotrack.org/articles/5065285-dathan-ritzenhein-is-no-longer-with-nike.

155 **"'If you don't have anything nice to say'":** Ryan Hall, email with author, March 16, 2018.

155 **a figure his agent has disputed:** Jennifer Kahn, "The Perfect Stride," *New Yorker*, November 8, 2010.

155 **Hollywood actors as distractions:** Juliet Macur, "Exhausted and Nearly Walking, Armstrong Reaches His Goal," *New York Times*, November 5, 2006.

156 **"his carpe diem day":** Juliet Macur, *Cycle of Lies: The Fall of Lance Armstrong* (New York: HarperCollins, 2014), Kindle ed., Location 257.

156 **"CHEAT TO WIN":** Tom Hoffarth, "What to do with our Livestrong wristbands now?," *Los Angeles Daily News*, October 21, 2012, https://www.dailynews.com/2012/10/21/tom-hoffarth-what-to-do-with-our-livestrong-wristbands-now-2/.

156 **"cardiovascularly, it was very easy":** Juliet Macur, "Exhausted and Nearly Walking, Armstrong Reaches His Goal," *New York Times*, November 5, 2006.

157 **"the way I feel now":** Juliet Macur, "Exhausted and Nearly Walking, Armstrong Reaches His Goal," *New York Times*, November 5, 2006.

157 **"why he's so good":** Jennifer Kahn, "The Perfect Stride," *New Yorker*, November 8, 2010.

158 **"Sprint 101 biomechanics":** Jennifer Kahn, "The Perfect Stride," *New Yorker*, November 8, 2010.

159 **testosterone to epitestosterone was eleven to one:** Juliet Macur, "Backup Sample on Landis Is Positive," *New York Times*, August 5, 2006.

159 **wearing a yellow Livestrong bracelet:** Image accompanying article. Lynn Zinser, "Record Holder in 100 Meters Failed a Drug Test," *New York Times*, July 30, 2006.

159 **"I am the best of the best":** No attribution, "Gatlin sets new 100m world record," CNN, May 12, 2006, https://edition.cnn.com/2006/SPORT/05/12/athletics.gatlin/.

159 **genuinely believable:** Duncan Mackay, "Gatlin turns into the fastest falling hero in the world," *Guardian*, July 30, 2006, https://www.theguardian.com/sport/2006/jul/31/athletics.sport.

160 **"change track for the better":** Eric Adelson, "Gatlin: I knew I had won," ESPN, August 22, 2004, https://tv5.espn.com/olympics/summer04/trackandfield/columns/story?id=1865367.

160 **new, labor-intensive, and costly testing method:** Philip Hersh, "Gatlin was caught by advanced test," *Chicago Tribune*, August 1, 2006.

160 **"willingness to take responsibility for his actions":** Image accompanying article. Lynn Zinser, "Record Holder in 100 Meters Failed a Drug Test," *New York Times*, July 30, 2006.

160 **syringe containing "The Clear":** Mark Fainaru-Wada and Lance Williams, *Game of Shadows* (New York: Penguin Group, 2006), 282.

161 **"Nike guy":** Wayne Drehs, "Inside the masseur-with-a-grudge conspiracy theory," ESPN.com, August 4, 2006, https://www.espn.com/olympics/trackandfield/news/story?id=2539489.

161 **Gatlin and his lawyers cringed:** Mark Fainaru-Wada and Lance Williams, *Game of Shadows* (New York: Penguin Group, 2006), 282.

161 **Gatlin would eventually use this defense:** UNITED STATES ANTI-DOPING AGENCY v. JUSTIN GATLIN, American Arbitration Association No. 30 190 00170 07 North American Court of Arbitration for Sport Panel. Page 1.

161 **"find out who our friends are":** No attribution, "Chris Whetstine Issues Statement to LetsRun.com," LetsRun.com, August 2, 2006, https://www.letsrun.com/2006/whets.php.

162 **"There's no way in the world":** Wayne Drehs, "Inside the masseur-with-a-

grudge conspiracy theory," ESPN.com, August 4, 2006, https://www.espn.com/olympics/trackandfield/news/story?id=2539489.

162 **hired by Nike in 2004:** Llewellyn Starks. Interview by author. Phone call. September 20, 2019.

162 **hung up the phone:** Llewellyn Starks. Interview by author. Phone call. September 20, 2019.

162 **Nike finally suspended Graham's contract:** Lynn Zinser, "Nike Drops Track Coach and Star Amid Scandal," *New York Times*, August 26, 2006.

11: EVEN DYING WON'T KEEP HIM

164 **"sit down and meet with him":** Bruce Barcott, "Mind Gains," *Runner's World*, February 11, 2010.

165 **"had to establish belief":** Bruce Barcott, "Mind Gains," *Runner's World*, February 11, 2010.

166 **"now they're hungrier than ever":** Bluehilltec, "Keeping up with the Gouchers—Episode I," May 29, 2007, YouTube video, https://www.youtube.com/watch?v=G7-EtNXs6jc&t=26s.

167 **"she's there to win":** Bruce Barcott, "Mind Gains," *Runner's World*, February 11, 2010.

168 **four charges from the resuscitation paddles:** Amby Burfoot, "The Day Alberto Salazar Died," *Runner's World*, September 7, 2007.

168 **"thirteen or fourteen minutes":** Amby Burfoot, "The Day Alberto Salazar Died," *Runner's World*, September 7, 2007.

168 **He was in the hospital:** Amby Burfoot, "The Day Alberto Salazar Died," *Runner's World*, September 7, 2007.

168 **"Of course. Galen":** Phil Knight, *Shoe Dog* (New York: Simon & Schuster, 2016), 370.

168 **"He bristled at prying questions":** Amby Burfoot, "The Day Alberto Salazar Died," *Runner's World*, September 7, 2007.

168 **confirmed his belief:** Amby Burfoot, "The Day Alberto Salazar Died," *Runner's World*, September 7, 2007.

169 **"I wanted to be worthy of it":** John Brant, "The Marathoner Speaks to His God," *New York Times Play*, October 28, 2007.

169 **dropped from the 1980 high of 208:** J. B. Strasser and Laurie Becklund, *Swoosh: The Unauthorized Story of Nike and the Men Who Played There* (New York: HarperCollins, 1993), 437.

169 **alarming increase of men using testosterone:** Institute of Medicine (US) Committee on Assessing the Need for Clinical Trials of Testosterone Replacement Therapy; Liverman, CT, Blazer DG, editors. Testosterone and Aging: Clinical Research Directions. Washington (DC): National Academies Press (US); 2004. 1, Introduction, https://www.ncbi.nlm.nih.gov/books/NBK216164/.

169 **increased their chances of a heart attack:** Katrina Karkazis, "The Testosterone Myth," *Wired*, April 2018.

170 **yellow Livestrong bracelet:** Amby Burfoot, "The Day Alberto Salazar Died," *Runner's World*, September 7, 2007.

170 **over his left pectoral:** Tim Catalano. Interview by author. Tape recording. Erie, Colorado. November 5, 2019.

170 **swoosh tattoo on his shoulder:** YouTube, *Guardian Sport*, "Nike Oregon Project closed down in wake of Salazar scandal," October 11, 2019, https://www.youtube.com/watch?v=Nrqpv5kZPKE.

170 **one thousand samples:** No attribution, "Drug testing to increase at Osaka," BBC, Friday, August 3, 2007, http://news.bbc.co.uk/sport2/hi/athletics/6929518.stm.

170 **August 25, 2007:** Nike Corporation, "Nike Assembles All-Star Cast Calling for Equality in Women's Sports," October 1, 2019, https://news.nike.com/news/nike-assembles-all-star-cast-calling-for-equality-in-womens-sports.

170 **young female athletes across America still felt "unequal":** Nike Corporation, "Nike Assembles All-Star Cast Calling for Equality in Women's Sports," October 1, 2019, https://news.nike.com/news/nike-assembles-all-star-cast-calling-for-equality-in-womens-sports.

170 **high school coach Bill Ressler:** Richard Sandomir, "Nike Puts Women Back on the Pedestal," *New York Times*, August 24, 2007.

170 **three-foot-wide mouthpiece:** Richard Sandomir, "Nike Puts Women Back on the Pedestal," *New York Times*, August 24, 2007.

171 **"Nike Puts Women Back on the Pedestal":** Richard Sandomir, "Nike Puts Women Back on the Pedestal," *New York Times*, August 24, 2007.

171 **humidity around 80 percent:** IAAF.org, https://web.archive.org/web/20070926025449/http://www.iaaf.org/newsfiles/40110.pdf.

172 **"It was unbelievable":** "11th IAAF World Championships—Osaka 2007—Day 1," FloTrack.org, https://www.flotrack.org/events/5000066-11th-iaaf-world-championships-osaka-2007-day-1/videos?playing=5086116.

172 **"even dying won't keep him":** "11th IAAF World Championships—Osaka 2007—Day 1," FloTrack.org, https://www.flotrack.org/events/5000066-11th-iaaf-world-championships-osaka-2007-day-1/videos?playing=5086115.

173 **"it couldn't have helped":** John Brant, "The Marathoner Speaks to His God," *New York Times Play*, October 28, 2007.

173 **Farah took the lead:** Wayne Middlesteadt, "2007 IAAF World Championships Men's 5,000m," October 25, 2014, YouTube video, https://www.youtube.com/watch?v=EGnyDZVaWR0.

174 **"I'd eventually work there":** Adam and Kara Goucher. Interview by author. Tape recording. Boulder, Colorado, October 3, 2017.

174 **"You have nothing to lose":** Bruce Barcott, "Mind Gains," *Runner's World*, February 11, 2010.

174 **nearly two-year hiatus:** No attribution, "Athletics: Kara Goucher first, Paula Radcliffe second in Great North Run," *New York Times*, September 30, 2007.

175 **a couple of sub-five-minute miles:** Matthew Brown, "Radcliffe's return derailed by debutante Goucher—Great North Run report," IAAF.org, September 30, 2007, https://www.iaaf.org/news/report/radcliffes-return-derailed-by-debutante-gouch-1.

175 **"good to be back":** No attribution, "Athletics: Kara Goucher first, Paula Radcliffe second in Great North Run," *New York Times*, September 30, 2007.

176 ***I'm coming back next year*:** Bruce Barcott, "Mind Gains," *Runner's World*, February 11, 2010.

176 **dysfunctional joint:** "Adam Goucher Bio," USATF.org, http://www.usatf.org/athletes/bios/TrackAndFieldArchive/2008/Goucher_Adam.asp.

176 **"I was trapped":** Adam and Kara Goucher. Interview by author. Tape recording. Boulder, Colorado, September 18, 2017.

177 **1 minute and 58 seconds for the 800 meters:** Danny Mackey, "5 Questions with Danny Mackey," RunnerSpace.com, https://www.runnerspace.com/gprofile.php?mgroup_id=44907&do=news&news_id=195465.

177 **six years of collegiate eligibility:** Mario Fraioli, "Going Long: An Interview with Danny Mackey," Medium.com, May 29, 2017, https://medium.com/the-morning-shakeout/going-long-an-interview-with-danny-mackey-b9efb5d0ebde.

177 **published biomechanics research:** "Between the Beginning and End of a Repetition: How Intrinsic and Extrinsic Factors Influence the Intensity of a Biceps Curl," National Strength and Conditioning Association, vol. 2, no. 5.

178 **applied online:** Danny Mackey. Interview by author. Phone call. September 6, 2019.

178 **psychological side of shoe innovation:** Mario Fraioli, "Going Long: An Interview with Danny Mackey," Medium.com, May 30, 2017, https://medium.com/the-morning-shakeout/going-long-an-interview-with-danny-mackey-b9efb5d0ebde.

178 **$45,000 a year:** Danny Mackey. Interview by author. Phone call. September 6, 2019.

178 **"get in trouble with Michael Jordan":** "Physical Preparation Podcast: Danny Mackey on Trial, Effect, and Coaching Runners," Robertson Training Systems, January 11, 2019, http://robertsontrainingsystems.com/blog/danny-mackey-podcast/.

179 **"dark side to that too":** Danny Mackey. Interview by author. Phone call. September 6, 2019.

179 **resurrect the Bowerman Athletic Club:** Mario Fraioli, "Going Long: An Interview with Danny Mackey," Medium.com, May 30, 2017, https://medium.com/the-morning-shakeout/going-long-an-interview-with-danny-mackey-b9efb5d0ebde.

179 **Nike is known for patenting everything:** Amby Burfoot, "Those Superfast Nike Shoes Are Creating a Problem," *New York Times*, October 18, 2019, https://www.nytimes.com/2019/10/18/sports/marathon-running-nike-vaporfly-shoes.html.

180 **"easy it is for smart chemists":** Sean Ingle, "How cheats cheat: why dopers have the edge in athletics' war on drugs," *Guardian*, August 20, 2015, https://www.theguardian.com/sport/2015/aug/20/doping-world-athletics-championships-cheats.

181 **"My voice is shaking":** Danny Mackey. Interview by author. Phone call. September 6, 2019.

181 **Marion Jones's involvement in a check-fraud scheme:** Lynn Zinser and Michael S. Schmidt, "Jones Admits to Doping and Enters Guilty Plea," *New York Times*, October 6, 2007.

181 **"biggest frauds in sporting history":** Dave Zirin, "In the Year of #BlackGirlMagic, Marion Jones Is Missing," *Nation*, September 11, 2015, https://

www.thenation.com/article/in-the-year-of-blackgirlmagic-the-missing-marion-jones/.

181 **distributing steroids and money laundering:** Robert Siegel, "BALCO's Conte Pleads Guilty to Steroid Charges," *NPR*, July 15, 2005, https://www.npr.org/templates/story/story.php?storyId=4756365.

182 **"my stomach had a different agenda":** Danny Mackey, "Renegades of Science," FloTrack.org, September 18, 2008, https://www.flotrack.org/articles/5016883-renegades-of-science.

183 **behind Ritzenhein:** USA Track & Field, https://www.usatf.org/events/2008/OlympicTrials-Marathon-Men/results.asp.

12: AM I WORKING FOR THE NIKE MAFIA?

186 **separation between coach and protégé:** Sean Ingle, "John Rohatinsky backs drug allegations made against Alberto Salazar," *Guardian*, June 14, 2015.

186 **used the word "weird":** Tim Catalano. Interview by author. Tape recording. Erie, Colorado. November 5, 2019.

186 **"paced by Galen Rupp":** John Z. Stiner. Interview by author. Phone call. December 21, 2018.

186 **other pacers had been fired:** John Z. Stiner. Interview by author. Phone call. December 21, 2018. Al Kupczak and Amy Begley also attested to this.

187 **"I'm sure that's not a violation":** Adam and Kara Goucher. Interview by author. Tape recording. Boulder, Colorado, September 18, 2017.

187 **gone through about five sports massage therapists:** John Z. Stiner. Interview by author. Phone call. December 21, 2018.

188 **"like working on a woman":** John Z. Stiner. Interview by author. Phone call. December 21, 2018.

189 **"busting Rupp's balls":** John Z. Stiner. Interview by author. Phone call. December 21, 2018.

189 **"business acquaintances" only:** Matthew Futterman, "Another of Alberto Salazar's Runners Says He Ridiculed Her Body for Years," *New York Times*, November 14, 2019.

190 **between 106 and 116 pounds:** Amy Begley. Telephone interview by Parker Henry. March 23, 2020.

190 **"my husband was a lucky guy":** Matthew Futterman, "Another of Alberto Salazar's Runners Says He Ridiculed Her Body for Years," *New York Times*, November 14, 2019, https://www.nytimes.com/2019/11/14/sports/olympics/alberto-salazar-nike.html.

190 **"I was the favorite":** Erin Strout, "Why Women Will Save Running," Womens Running.com, November 14 2019, https://www.womensrunning.com/2019/11/news/why-women-will-save-running_103411.

191 **does not deny using the topical testosterone cream:** Alberto Salazar, "Alberto Open Letter Part 1," NikeOregonProject.com, June 24, 2015, https://web.archive.org/web/20150627071828/https://nikeoregonproject.com/blogs/news/35522561-alberto-open-letter-part-1.

191 **"might be EPO":** John Z. Stiner. Interview by author. Phone call. December 21, 2018.

192 "possible increased risk of heart attack or stroke": Androgel.com, https://www.androgel.com.

192 Jerry Schumacher as his successor: Sara Germano, "Inside a Nike Family Feud," *Wall Street Journal*, March 7, 2014, https://www.wsj.com/articles/inside-a-nike-family-feud-alberto-salazar-and-jerry-schumacher-1394234890.

192 "coach insurance for Galen": No attribution, "Alberto Salazar Brings Jerry Schumacher to Nike Oregon Project," tracktownusa.com, http://www.tracktownusa.com/track.item.55/Alberto-Salazar-Brings-Jerry-Schumacher-to-Nike-Oregon-Project.html.

193 "Kara, this is Ken Goe": Kara Goucher. Interview by author. Tape recording. Boulder, Colorado, September 28, 2017.

193 first test her fitness: USATF.com, Friday, June 27, 2008, 9:20 p.m., https://www.usatf.org/events/2008/OlympicTrials-TF/schedule.asp.

193 Adam advanced to the finals: USATF.com, accessed Monday, June 30, 2008, 9:40 p.m., https://www.usatf.org/events/2008/OlympicTrials-TF/quotes/F14.asp.

194 Carl Lewis and Michael Johnson were there walking around: John Z. Stiner. Interview by author. Phone call. December 21, 2018.

194 high school best time of 4 minutes and 14.50 seconds: SLOBoe, "Jordan Hasay Olympic Trials 1500 m Track & Field Record," January 25, 2099, YouTube video, https://www.youtube.com/watch?v=_aQNvv1PY7A.

194 She finished tenth in the final: USATF.org, accessed Monday, June 20, 2008, https://www.usatf.org/events/2008/OlympicTrials-TF/results/F10.asp.

195 the most promising distance runner since Steve Prefontaine: Tim Breault, "Galen Rupp chases 10,000-meter dreams in U.S. Olympic trials," *Oregonian*, July 4, 2008.

195 nervously trying to cover every move: Jennie McCafferty, "U.S. Olympic Team Trials—Track & Field '08—Day 6 Recap," July 7, 2008, YouTube video, https://www.youtube.com/watch?v=YAzet7DeBlk.

195 "how intense the training is": Gordon Bakoulis, "Racing: Adam Goucher," *Runner's World*, October 24, 2008, https://www.runnersworld.com/advanced/a20792833/racing-adam-goucher/.

196 "I wasn't going to kiss his ass": Adam and Kara Goucher. Interview by author. Tape recording. Boulder, Colorado, September 18, 2017.

196 "to make us feel like we had autonomy": Adam and Kara Goucher. Interview by author. Tape recording. Boulder, Colorado, September 18, 2017.

197 "I'm disappointed": K. Pates, duluthnews.com, August 16, 2008.

197 "should I just go home?": Adam and Kara Goucher. Interview by author. Tape recording. Boulder, Colorado, September 18, 2017.

198 He increased her mileage: Bruce Barcott, "Mind Gains," *Runner's World*, February 11, 2010.

198 Her new keyword: Bruce Barcott, "Mind Gains," *Runner's World*, February 11, 2010.

199 if she could finish the distance: Bruce Barcott, "Mind Gains," *Runner's World*, February 11, 2010.

199 "You're putting her on the bus": Bruce Barcott, "Mind Gains," *Runner's World*, February 11, 2010.

199 She dropped the first one: No attribution, "Goucher 3rd as fastest American woman ever at NYC Marathon," FloTrack.org, November 2, 2008, https://www.flotrack.org/articles/5016947-goucher-3rd-as-fastest-american-woman-ever-at-nyc-marathon-4-americans-in-mens#!/dashboard.

200 **"It's a 10K":** Bruce Barcott, "Mind Gains," *Runner's World*, February 11, 2010.

200 **Adam took a picture of coach and athlete:** Sally Bergesen, "State of the Sport, Kara + Sally, Part 2 podcast," Oiselle.com, December 4, 2019, https://www.oiselle.com/blogs/oiselle-blog/state-of-sport-kara.

200 **"I was devastated":** Adam and Kara Goucher. Interview by author. Tape recording. Boulder, Colorado, September 18, 2017.

200 **Adam keeps taking cold showers:** Adam and Kara Goucher. Interview by author. Tape recording. Boulder, Colorado, September 18, 2017.

13: LET'S RUN

201 **he starts his day reading the site:** Tim Ferriss, "Dissecting the Success of Malcolm Gladwell (#168)," *The Tiim Ferriss Show*, June 21 2016, https://tim.blog/2016/06/21/malcolm-gladwell/, 50:40.

202 **"Why did Hall DNF?":** LetsRun.com, November 3, 2019, https://www.letsrun.com/forum/flat_read.php?thread=9681839.

202 **"Cancelled is REPUGNANT":** LetsRun.com, November 4, 2019, https://www.letsrun.com/forum/flat_read.php?thread=9683045.

202 **"Why do many here hate on Kara Goucher":** LetsRun.com, June 19, 2019, https://www.letsrun.com/forum/flat_read.php?thread=9450092.

203 **one hundred and twelfth running:** Frank Litsky, "Kenyan Wins His 4th Boston Marathon," *New York Times*, April 22, 2008.

203 **American flag tattoos:** No attribution, "Boston Marathon 2009: The Elites," *Runner's World*, October 1, 2012, https://www.runnersworld.com/races-places/g20819544/boston-marathon-2009-the-elites/?slide=5.

203 **distance running's most fabled race:** Frank Litsky, "Kenyan Wins His 4th Boston Marathon," *New York Times*, April 22, 2008.

203 **"I felt so hurt":** Adam and Kara Goucher. Interview by author. Tape recording. Boulder, Colorado, September 18, 2017.

204 **"*Do I really want to be just mediocre*":** Jennifer Kahn, "The Perfect Stride," *New Yorker*, November 8, 2010.

204 **Oregon Twilight Track and Field Meet:** GoDucks.com, https://goducks.com/roster.aspx?rp_id=3614.

204 **felt something wet on his back:** United States Anti-Doping Agency, "AAA Panel Imposes 4-Year Sanctions on Alberto Salazar and Dr. Jeffrey Brown for Multiple Anti-Doping Rule Violations," Page 103, Paragraph 381, Accessed October 20, 2019, https://www.usada.org/sanction/aaa-panel-4-year-sanctions-alberto-salazar-jeffrey-brown.

204 **not only plausible, but probable:** United States Anti-Doping Agency, "AAA Panel Imposes 4-Year Sanctions on Alberto Salazar and Dr. Jeffrey Brown for Multiple Anti-Doping Rule Violations," Page 103, Paragraph 382, Accessed October 20, 2019, https://www.usada.org/sanction/aaa-panel-4-year-sanctions-alberto-salazar-jeffrey-brown.

205 **"really, really trust":** United States Anti-Doping Agency, "AAA Panel Im-

poses 4-Year Sanctions on Alberto Salazar and Dr. Jeffrey Brown for Multiple Anti-Doping Rule Violations," Page 104, Paragraph 384, Accessed October 20, 2019, https://www.usada.org/sanction/aaa-panel-4-year-sanctions-alberto-salazar-jeffrey-brown.

205 **left unattended was dumped out:** Pete Julian. Interview by author. Tape recording. Boulder, Colorado, November 7, 2019.

205 **"possibly rubbed something onto Galen":** United States Anti-Doping Agency, "AAA Panel Imposes 4-Year Sanctions on Alberto Salazar and Dr. Jeffrey Brown for Multiple Anti-Doping Rule Violations," Page 103, Paragraph 381, Accessed October 20, 2019, https://www.usada.org/sanction/aaa-panel-4-year-sanctions-alberto-salazar-jeffrey-brown.

205 **"plant the Nike flag":** Larry Eder, "The RBR Interview: Nike's John Capriotti on the Pre Classic, and Bekele's 10k WR attempt, by Larry Eder," RunBlogRun.com, May 13, 2008, http://www.runblogrun.com/track-field/the-rbr-interview-nikes-john-capriotti-on-the-pre-classic-and-bekeles-10k-wr-attempt-by-larry-eder.html.

206 **basement pharmacy full of supplements:** "INTERIM REPORT OF THE U.S. ANTI-DOPING AGENCY," USADA, March 17, 2016, page 21.

206 **testosterone cream on his son:** United States Anti-Doping Agency, "AAA Panel Imposes 4-Year Sanctions on Alberto Salazar and Dr. Jeffrey Brown for Multiple Anti-Doping Rule Violations," Page 106, Paragraph 397, Accessed October 20, 2019, https://www.usada.org/sanction/aaa-panel-4-year-sanctions-alberto-salazar-jeffrey-brown.

206 **Aegis Labs:** United States Anti-Doping Agency, "AAA Panel Imposes 4-Year Sanctions on Alberto Salazar and Dr. Jeffrey Brown for Multiple Anti-Doping Rule Violations," Page 105, Paragraph 391, Accessed October 20, 2019, https://www.usada.org/sanction/aaa-panel-4-year-sanctions-alberto-salazar-jeffrey-brown.

206 **Dr. Brown was worried:** United States Anti-Doping Agency, "AAA Panel Imposes 4-Year Sanctions on Alberto Salazar and Dr. Jeffrey Brown for Multiple Anti-Doping Rule Violations," Page 105, Paragraph 390, Accessed October 20, 2019, https://www.usada.org/sanction/aaa-panel-4-year-sanctions-alberto-salazar-jeffrey-brown.

207 **"repeat it using three pumps":** United States Anti-Doping Agency, "AAA Panel Imposes 4-Year Sanctions on Alberto Salazar and Dr. Jeffrey Brown for Multiple Anti-Doping Rule Violations," Page 105, Paragraph 391, Accessed October 20, 2019, https://www.usada.org/sanction/aaa-panel-4-year-sanctions-alberto-salazar-jeffrey-brown.

207 **"a positive test":** United States Anti-Doping Agency, "AAA Panel Imposes 4-Year Sanctions on Alberto Salazar and Dr. Jeffrey Brown for Multiple Anti-Doping Rule Violations," Page 105, Paragraph 391, Accessed October 20, 2019, https://www.usada.org/sanction/aaa-panel-4-year-sanctions-alberto-salazar-jeffrey-brown.

207 **an increased T/E ratio to 2.8:** United States Anti-Doping Agency, "AAA Panel Imposes 4-Year Sanctions on Alberto Salazar and Dr. Jeffrey Brown for Multiple Anti-Doping Rule Violations," Page 105, Paragraph 393, Accessed October

20, 2019, https://www.usada.org/sanction/aaa-panel-4-year-sanctions-alberto-salazar-jeffrey-brown.

207 **"I don't think it's worth it":** United States Anti-Doping Agency, "AAA Panel Imposes 4-Year Sanctions on Alberto Salazar and Dr. Jeffrey Brown for Multiple Anti-Doping Rule Violations," Page 106, Paragraph 397, Accessed October 20, 2019, https://www.usada.org/sanction/aaa-panel-4-year-sanctions-alberto-salazar-jeffrey-brown.

207 **"I can't prevent you from doing anything":** United States Anti-Doping Agency, "AAA Panel Imposes 4-Year Sanctions on Alberto Salazar and Dr. Jeffrey Brown for Multiple Anti-Doping Rule Violations," Page 105, Paragraph 394, Accessed October 20, 2019, https://www.usada.org/sanction/aaa-panel-4-year-sanctions-alberto-salazar-jeffrey-brown.

208 **a plain envelope:** United States Anti-Doping Agency, "AAA Panel Imposes 4-Year Sanctions on Alberto Salazar and Dr. Jeffrey Brown for Multiple Anti-Doping Rule Violations," Page 106, Paragraph 398, Accessed October 20, 2019, https://www.usada.org/sanction/aaa-panel-4-year-sanctions-alberto-salazar-jeffrey-brown.

208 **"cream that he used on one of his sons":** United States Anti-Doping Agency, "AAA Panel Imposes 4-Year Sanctions on Alberto Salazar and Dr. Jeffrey Brown for Multiple Anti-Doping Rule Violations," Page 107, Paragraph 398, Accessed October 20, 2019, https://www.usada.org/sanction/aaa-panel-4-year-sanctions-alberto-salazar-jeffrey-brown.

208 **"full fledged research protocol":** United States Anti-Doping Agency, "AAA Panel Imposes 4-Year Sanctions on Alberto Salazar and Dr. Jeffrey Brown for Multiple Anti-Doping Rule Violations," Page 106, Paragraph 395, Accessed October 20, 2019, https://www.usada.org/sanction/aaa-panel-4-year-sanctions-alberto-salazar-jeffrey-brown.

209 **"I'm gonna throw up":** Ryan from FloTrack, "Kara Goucher 10th in marathon at 2009 IAAF World Championships," FloTrack.org, August 23, 2009, https://www.flotrack.org/video/5160791-kara-goucher-10th-in-marathon-at-2009-iaaf-world-championships.

209 **"they were better":** Ryan from FloTrack, "Kara Goucher part 2 after marathon at 2009 IAAF World Championships," FloTrack.org, August 23, 2009, https://www.flotrack.org/video/5160787-kara-goucher-part-2-after-marathon-at-2009-iaaf-world-championships.

210 **"I'm willing to take that risk":** Jennifer Kahn, "The Perfect Stride," *New Yorker*, November 8, 2010.

210 **He took odd supplements:** Kara Goucher. Interview by author. Tape recording. Boulder, Colorado, September 28, 2017.

210 **received regular shots:** Doug Beghtel, "Watching Rupp Go," oregonlive.com, June 9, 2009, https://www.oregonlive.com/trackandfield/2009/06/watching_rupp_go_the_ducks_are.html.

211 **heat turned up to 78 degrees:** Doug Beghtel, "Watching Rupp Go," oregonlive.com, June 9, 2009, https://www.oregonlive.com/trackandfield/2009/06/watching_rupp_go_the_ducks_are.html.

211 **the inaugural Bowerman Award:** Tom Lewis, "Rupp, Barringer Honored as

Inaugural Winners of The Bowerman," ustfccca.org, December 17, 2009, http://www.ustfccca.org/2009/12/featured/rupp-barringer-honored-as-inaugural-winners-of-the-bowerman.

211 **"not far off that":** Doug Beghtel, "Watching Rupp Go," oregonlive.com, June 9, 2009, https://www.oregonlive.com/trackandfield/2009/06/watching_rupp_go_the_ducks_are.html.

212 **beating nine of the ten Kenyan runners:** Matt Fitzgerald, "Dathan Ritzenhein Smashes 5000m American Record!" *Competitor Running* (now PodiumRunner.com), August 28, 2009, https://www.podiumrunner.com/dathan-ritzenhein-smashes-5000m-american-record_5045.

212 **catching up to the Africans:** Mike Wilson, "Dathan Ritzenhein's 5,000-meter record shows U.S. is 'closing the gap,'" OregonLive.com, August 28, 2009, https://www.oregonlive.com/trackandfield/2009/08/dathan_ritzenheins_5000meter_r.html.

212 **"putting the frosting on it":** Mike Wilson, "Dathan Ritzenhein's 5,000-meter record shows U.S. is 'closing the gap'," *Oregonian*, August 28, 2009, https://www.oregonlive.com/trackandfield/2009/08/dathan_ritzenheins_5000meter_r.html.

212 **the school record:** Alan Webb. Interview by author by phone. Boulder, Colorado, March 26, 2019.

213 **"really cared":** Alan Webb. Interview by author by phone. Boulder, Colorado, March 26, 2019.

213 **his Nike contacts:** Alan Webb. Interview by author by phone. Boulder, Colorado, March 26, 2019.

213 **The record changed Webb's life:** Alan Webb. Interview by author by phone. Boulder, Colorado, March 26, 2019.

213 **"I'm on the *Letterman Show*!":** Sam Thompson, "Alan Webb on Letterman," January 7, 2006, YouTube video, https://www.youtube.com/watch?v=ouv3sM60aaE.

213 **yearly salary of $250,000:** "Alan Webb signs rich deal with Nike," *USA Today*, August 15, 2002.

214 **beginning of a new era:** No attribution, "The Alan Webb Story—USATF.tv," USATF.org, https://www.usatf.tv/gprofile.php?mgroup_id=45365&do=videos&video_id=106500.

214 **Nike Air Zoom Milers:** No attribution, "The Visual History of the Nike Track Spike," Nike.com, August 12, 2016, https://news.nike.com/news/nike-track-spike-history.

214 **"totally tapped out":** Alan Webb. Interview by author by phone. Boulder, Colorado, March 26, 2019.

214 **to set a new world record:** Peter Gambaccini, "The Phenom: Alan Webb," *Runner's World*, December 7, 2007.

214 **was the favorite to win:** Dave Ungrady, "For Webb, Road to Recovery Will Head Down Fifth Avenue," *New York Times*, September 26, 2010.

214 **finished a disappointing eighth:** Sportsnetwork, "Lagat ends U.S. drought with 1,500 win," August 1, 2007, YouTube video, https://www.youtube.com/watch?v=P0hkE4qhPrg.

215 **"It fell mighty short":** Dave Ungrady, "For Webb, Road to Recovery Will Head Down Fifth Avenue," *New York Times*, September 26, 2010.

215 **"He has to lean out":** Dick Patrick, "Alan Webb leaves longtime coach to join Alberto Salazar in Oregon," *USA Today*, August 6, 2009.

216 **"we've been very separated":** Matthew Futterman, "Shalane Flanagan Was Not Surprised by Alberto Salazar's Ban," *New York Times*, October 30, 2019, https://www.nytimes.com/2019/10/30/sports/shalane-flanagan-alberto-salazar.html.

216 **Nike Coach of the Year Award:** "Nike Coach of the Year," USATF.org, https://www.usatf.org/statistics/awards/Administrative_kentico/Overall/CoachOfTheYear.asp.

217 **more than one million copies:** Brian Metzler, *Kicksology: The Hype, Science, Culture, and Cool of Running Shoes* (Boulder, CO: VeloPress, 2019), 78.

14: I PAY YOU TO RUN

219 **lifetime deals:** Ahiza Garcia, "Cristiano Ronaldo is the third athlete to sign Nike 'lifetime' deal," CNN.com, November 9, 2016, https://money.cnn.com/2016/11/09/news/companies/cristiano-ronaldo-nike-lifetime-contract/index.html.

219 **$20,000 a year:** Adam and Kara Goucher. Interview by author. Tape recording. Boulder, Colorado, October 3, 2017.

219 **arduous fertility treatment:** Kara Goucher, Instagram post, @karagoucher, January 19, 2020, https://www.instagram.com/p/B7g-_NLpUuO/.

220 **ten months postpartum:** Tara Parker-Pope, "Two Running Stars Train While Pregnant," *New York Times*, August 19, 2010.

220 **"out of the house":** Tara Parker-Pope, "Two Running Stars Train While Pregnant," *New York Times*, August 19, 2010.

220 **60 percent:** No attribution, "Running USA Releases 2019 U.S. Running Trends Report," RunningUSA.org, https://runningusa.org/RUSA/News/2019/Running_USA_Releases_2019_U.S._Running_Trends_Report.aspx.

221 **were big fans:** Sally Bergesen, "State of the Sport, Kara + Sally, Part 1," Oiselle.com, Podcast, December 4, 2019, https://www.oiselle.com/blogs/oiselle-blog/state-of-sport-kara.

222 **"was Alberto lying?":** Adam and Kara Goucher. Interview by author. Tape recording. Boulder, Colorado, October 3, 2017.

222 **between twelve to seventeen months:** John Slusher, "Nike Contract," email message to Kara Goucher, March 15, 2011.

222 **Salazar forwarded a BBC article:** "INTERIM REPORT OF THE U.S. ANTI-DOPING AGENCY," USADA, March 17, 2016, page 46.

223 **"Talk to you later":** "INTERIM REPORT OF THE U.S. ANTI-DOPING AGENCY," USADA, March 17, 2016, page 46.

223 **skip the consultation:** "INTERIM REPORT OF THE U.S. ANTI-DOPING AGENCY," USADA, March 17, 2016, page 47.

223 **165 pounds:** No attribution, "The Week That Was In Running—The 26:59.60 That Shocked The World," LetsRun.com, May 4, 2010, https://www.letsrun.com/2010/weekthatwas0504.php.

223 **Keflezighi had held the American record:** No attribution, LetsRun.com, https://www.letsrun.com/2010/solinsky0502.php.

224 **"Go Galen!":** Sara Germano, "Inside a Nike Family Feud," *Wall Street Journal*, March 7, 2014, https://www.wsj.com/articles/inside-a-nike-family-feud-alberto-salazar-and-jerry-schumacher-1394234890.

224 **passed him with authority:** FloTrack, "Chris Solinsky Breaks 10K American Record," April 28, 2017, YouTube video, https://www.youtube.com/watch?v=5Rkvd5dfwMQ.

224 **first American-born runner:** Sara Germano, "Inside a Nike Family Feud," *Wall Street Journal*, March 7, 2014, https://www.wsj.com/articles/inside-a-nike-family-feud-alberto-salazar-and-jerry-schumacher-1394234890.

224 **27 minutes and 10.74 seconds for 10,000 meters:** No attribution, "Galen Rupp Talks About His 2010 Season," LetsRun.com, August 27, 2010.

225 **Webb would run faster than ever:** Editors, "Alberto Salazar Confident About Alan Webb's Future," LetsRun.com, August 26, 2010, https://www.letsrun.com/2010/salazar-0826.php.

225 **"waddling":** Dave Ungrady, "For Webb, Road to Recovery Will Head Down Fifth Avenue," *New York Times*, September 26, 2010.

225 **"the systematic approach":** Editors, "Alberto Salazar Confident About Alan Webb's Future," LetsRun.com, August 26, 2010, https://www.letsrun.com/2010/salazar-0826.php.

225 **"Webb will run faster":** Editors, "Alberto Salazar Confident About Alan Webb's Future," LetsRun.com, August 26, 2010, https://www.letsrun.com/2010/salazar-0826.php.

227 **Radcliffe's push-off:** Steve Magness. Interview by author. Tape recording. Houston, Texas, November 11, 2018.

228 **at a distinct disadvantage:** Steve Magness. Interview by author. Tape recording. Houston, Texas, November 11, 2018.

228 **very top end of the range:** Steve Magness. Interview by author. Tape recording. Houston, Texas, November 11, 2018.

228 **at that level:** Danny Mackey. Interview by author. Phone Call. September 6, 2019.

229 **last day was October 30, 2010:** United States Anti-Doping Agency, "AAA Panel Imposes 4-Year Sanctions on Alberto Salazar and Dr. Jeffrey Brown for Multiple Anti-Doping Rule Violations," page 105, paragraph 394, Accessed October 20, 2019, https://www.usada.org/sanction/aaa-panel-4-year-sanctions-alberto-salazar-jeffrey-brown.

230 **"He seemed nice to me":** Steve Magness. Interview by author. Tape recording. Houston, Texas, November 11, 2018.

230 **Salazar, whom he revered:** Anna Kessel, "Mo Farah gets new coach Alberto Salazar for London 2012 Olympics," *Guardian*, February 18, 2011, https://www.theguardian.com/sport/2011/feb/18/mo-farah-alberto-salazar-olympics.

230 **"matchstick man from south-west London":** Simon Turnbull, "Meet the holistic guru who is helping Mo fly," *Independent*, July 22, 2011.

230 **"'why are you wasting time'":** "INTERIM REPORT OF THE U.S. ANTI-DOPING AGENCY," USADA, March 17, 2016, page 16. Transcript of Inter-

view (Under Oath) of Alberto Salazar (February 4, 2016), p. 37, line 6 through p. 38, line 19.

231 **failed anti-doping test:** Tim Layden, "Paralysis by Urinalysis Mary Slaney's Disputed Drug Test Proves One Thing: Track and Field Isn't Making the Grade," *Sports Illustrated*, May 26, 1997.

231 **"That's the question":** Matt Slater and Samuel Smith, "Questions mount over Alberto Salazar's links to Mary Slaney," BBC.com, June 12, 2015, https://www.bbc.com/sport/athletics/33096367.

231 **"specifically an endurance place":** Anna Kessel, "Mo Farah gets new coach Alberto Salazar for London 2012 Olympics," *Guardian*, February 18, 2011, https://www.theguardian.com/sport/2011/feb/18/mo-farah-alberto-salazar-olympics.

231 **"back door at Nike":** Ken Goe, "Oregon track & field rundown: Now, at long last, Mo Farah can use the front entrance at Nike," *Oregonian*, January 10, 2019.

231 **he would have argued:** Ken Goe, "Oregon track & field rundown: Now, at long last, Mo Farah can use the front entrance at Nike," *Oregonian*, January 10, 2019.

232 **underwater treadmill every other day:** No attribution, "RW Interviews: Mo Farah," RunnersWorld.com, April 8, 2011, https://www.runnersworld.com/uk/training/motivation/a767081/rw-interviews-mo-farah.

232 **"Big-time results":** Chris McClung, Kara and Adam Goucher, "Episode #14: Kara and Adam Goucher on the 4-Year Bans for Salazar/Brown for Doping Violations," *Clean Sport Collective* podcast, October 7, 2019, https://podcasts.apple.com/us/podcast/episode-14-kara-adam-goucher-on-4-year-bans-for-salazar/id1466187704?i=1000452633967.

232 **"I was still bleeding":** Sally Bergesen, "State of the Sport, Kara + Sally, Part 1," Oiselle.com, December 4, 2019, https://www.oiselle.com/blogs/oiselle-blog/state-of-sport-kara.

233 **with his stopwatch:** Adam and Kara Goucher. Interview by author. Tape recording. Boulder, Colorado, October 3, 2017.

233 **what to expect from her body:** Mario Fraioli, "Kara Goucher Returns to Racing," PodiumRunner.com, January 16, 2011, https://www.podiumrunner.com/photo-gallery-kara-goucher-returns-to-racing_20421.

233 **still the same athlete:** Mario Fraioli, "Podcast: Episode 27 with Kara Goucher," *Morning Shakeout*, August 21, 2018, https://themorningshakeout.com/podcast-episode-27-with-kara-goucher.

234 **"had no fight, nothing":** Liz Robbins, "The Runner Kara Goucher Is Back and Ready to Run," *New York Times*, March 13, 2011.

234 **2 minutes and 13 seconds behind her:** Mario Fraioli, "Kara Goucher Returns to Racing," *Podium Runner*, January 16, 2011, https://www.podiumrunner.com/photo-gallery-kara-goucher-returns-to-racing_20421.

234 **custom LunarGlide+ 2:** *Sneaker Freaker Magazine*, "Oprah Receives Nike LunarGlide 2 Customs from Phil Knight," May 3, 2011, YouTube video, https://www.youtube.com/watch?v=UmQc8Jnd41Q.

235 **lubricated by wine or beer:** John Brant, "Why I Still Have Faith in Alberto Salazar," RunnersWorld.com, October 24, 2019.

235 **Mountain View:** John Brant. Telephone interview by fact-checker Parker Henry. March 20, 2020.

235 ***She has a kid and a husband*:** Steve Magness. Interview by author. Tape recording. Houston, Texas, November 11, 2018.

235 **"obsessed with the size of my breasts":** Mark Daly, "Alberto Salazar's spectacular fall from grace," BBC, February 24, 2020, https://www.bbc.com/sport/athletics/51599747.

236 **Deliverables from Nike's SPARQ:** AMERICAN ARBITRATION ASSOCIATION ("AAA") COMMERCIAL ARBITRATION TRIBUNAL AAA Case No. 01-17-0004-0880, paragraph 79, https://www.usada.org/wp-content/uploads/Salazar-AAA-Decision-1.pdf.

236 ***Fast Company* magazine called it:** Austin Carr, "The Truth Behind "Secret" Innovation At Nike, Apple, Google X," March 5, 2013, https://www.fastcompany.com/3006383/truth-behind-secret-innovation-nike-apple-google-x.

237 **"a benefit":** AMERICAN ARBITRATION ASSOCIATION ("AAA") COMMERCIAL ARBITRATION TRIBUNAL AAA Case No. 01-17-0004-0880, paragraph 79, https://www.usada.org/wp-content/uploads/Salazar-AAA-Decision-1.pdf.

238 **"cutting-edge sports science":** AMERICAN ARBITRATION ASSOCIATION ("AAA") COMMERCIAL ARBITRATION TRIBUNAL AAA Case No. 01-17-0004-0880, paragraph 81, https://www.usada.org/wp-content/uploads/Salazar-AAA-Decision-1.pdf.

238 **"would not gain any performance benefits":** AMERICAN ARBITRATION ASSOCIATION ("AAA") COMMERCIAL ARBITRATION TRIBUNAL AAA Case No. 01-17-0004-0880, paragraph 81, https://www.usada.org/wp-content/uploads/Salazar-AAA-Decision-1.pdf.

15: DO YOU HAVE ANYTHING TO CONFESS?

239 **"cut back on the Cytomel":** "INTERIM REPORT OF THE U.S. ANTI-DOPING AGENCY," USADA, March 17, 2016, page 76 of 269.

240 **if he would fail a drug test:** David Epstein, "Off Track: Former Team Members Accuse Famed Coach Alberto Salazar of Breaking Drug Rules," ProPublica.org, June 3, 2015, https://www.propublica.org/article/former-team-members-accuse-coach-alberto-salazar-of-breaking-drug-rules.

240 **anodyne champion:** Martin Fritz Huber, "Galen Rupp Is Hard to Love," *Outside* magazine, May 17, 2018.

241 **Clive Cussler:** Steve Magness. Interview by author. Tape recording. Houston, Texas, November 11, 2018.

241 **"get it through customs":** Steve Magness. Interview by author. Tape recording. Houston, Texas, November 11, 2018.

241 **"zero sense scientifically":** Steve Magness. Interview by author. Tape recording. Houston, Texas, November 11, 2018.

241 **"not going to fall apart":** Steve Magness. Interview by author. Tape recording. Houston, Texas, November 11, 2018.

241 **six weeks prior:** Anna Kessel, "Mo Farah hits the heights as new methods pay off in Birmingham," *Guardian*, February 19, 2011, https://www.theguardian.com/sport/2011/feb/20/mo-farah-athletic-grand-prix.

241 **covered the three-stripe logo:** Runners Awesome, "Mo Farah vs Galen Rupp at 5000m UK Indoor 2011," April 2, 2018, YouTube video, https://www.youtube.com/watch?v=W61gohcgYzk.

242 **held by Nick Rose:** Anna Kessel, "Mo Farah hits the heights as new methods pay off in Birmingham," *Guardian*, February 19, 2011, https://www.theguardian.com/sport/2011/feb/20/mo-farah-athletic-grand-prix.

242 **13 minutes and 11.44 seconds:** No attribution, "Galen Rupp Athlete Profile," WorldAthletics.org, https://worldathletics.org/athletes/united-states/galen-rupp-196507.

242 **high-tempo training:** Simon Turnbull, "Meet the holistic guru who is helping Mo fly," *Independent*, July 22, 2011.

242 **"it has made a difference":** Anna Kessel, "Mo Farah hits the heights as new methods pay off in Birmingham," *Guardian*, February 19, 2011, https://www.theguardian.com/sport/2011/feb/20/mo-farah-athletic-grand-prix.

242 **$2.5 million per year:** Juliet Macur, *Cycle of Lies: The Fall of Lance Armstrong* (New York: HarperCollins, 2014), Kindle ed., Location 257.

243 **Landis told me:** Floyd Landis. Interview by author. Tape recording. Golden, Colorado. November 17, 2017.

243 **testosterone patches:** Floyd Landis. Telephone interview by author. January 28, 2018.

243 **same reasons for retiring:** Juliet Macur, "Lance Armstrong Retires From Cycling," *New York Times*, February 16, 2011.

243 **thirty-nine-year-old:** Juliet Macur, "Lance Armstrong Retires from Cycling," *New York Times*, February 16, 2011.

243 **one of the toughest athletes:** Bob Babbitt and Paul Huddle, "Alberto Salazar Interview," *Competitor Radio*, March 15, 2011, https://web.archive.org/web/20121001002052/http://content.blubrry.com/competitor_radio/AlbertoSalazar-3-15-11.mp3.

243 **plenty of it:** Steve Magness, "Magness: My Interactions with Lance Armstrong," PodiumRunner.com, January 16, 2013, https://www.podiumrunner.com/magness-my-interactions-with-lance-armstrong_64596.

244 **kept in touch:** Juliet Macur, "Exhausted and Nearly Walking, Armstrong Reaches His Goal," *New York Times*, November 5, 2006.

244 **testing in the Nike lab:** Steve Magness. Interview by author. Tape recording. Houston, Texas, November 12, 2018.

244 **videotaped his running form:** Bob Babbitt and Paul Huddle, "Alberto Salazar Interview," *Competitor Radio*, March 15, 2011, https://web.archive.org/web/20121001002052/http://content.blubrry.com/competitor_radio/AlbertoSalazar-3-15-11.mp3.

244 **"'Ask Galen'":** Kara and Adam Goucher. Interview by author. Tape recording. Boulder, Colorado, September 28, 2017.

245 **"You do not need that":** Kara and Adam Goucher. Interview by author. Tape recording. Boulder, Colorado, September 28, 2017.

245 **"stop trying to be the doctor":** David Epstein, "Alberto Salazar Disputes Allegations—Some of Which Were Never Made," ProPublica.org, June 24, 2015, https://www.propublica.org/article/alberto-salazar-disputes-allegations-some-of-which-were-never-made.

245 **"asking me to lie":** Sally Bergesen, "State of the Sport, Kara + Sally, Part 1," Oiselle.com, podcast, December 4, 2019, https://www.oiselle.com/blogs/oiselle-blog/state-of-sport-kara.

246 **crashing to the pavement:** Doug Binder, "Mo Farah, Galen Rupp, Kara Goucher shine at New York Half Marathon," OregonLive.com, March 21, 2011.

246 **"excited for Kara's race":** Doug Binder, "Mo Farah, Galen Rupp, Kara Goucher shine at New York Half Marathon," OregonLive.com, March 21, 2011.

246 **11 seconds behind:** Doug Binder, "Mo Farah, Galen Rupp, Kara Goucher shine at New York Half Marathon," *Oregonian*, March 21, 2011, https://www.oregonlive.com/trackandfield/2011/03/mo_farah_galen_rupp_kara_gouch.html.

246 **previous course record:** Aimee Berg, "Newcomer Wins New York City Half Marathon," *New York Times*, March 20 2011.

246 **more than $47.6 million:** Tess Stynes, "Nike CEO Mark Parker's Total Pay Soars to $47.6 Million for Fiscal Year 2016," *Wall Street Journal*, July 25, 2016, https://www.wsj.com/articles/nike-ceo-mark-parkers-total-pay-soars-to-47-6-million-for-fiscal-year-2016-1469483249?ns=prod/accounts-wsj.

247 **was in high school:** Steve Magness. Interview by author. Tape recording. Houston, Texas, November 11, 2018.

248 **Myhre received an award:** David Epstein, "Off Track: Former Team Members Accuse Famed Coach Alberto Salazar of Breaking Drug Rules," ProPublica.org, June 3, 2015, https://www.propublica.org/article/former-team-members-accuse-coach-alberto-salazar-of-breaking-drug-rules.

248 **the last he ever heard of the issue:** Brad Wilkins. Telephone interview by fact-checker Parker Henry. April 28, 2020.

248 **"trying to do my best":** John Slusher, "Nike Contract," email message to Kara Goucher, March 15, 2011.

248 **"You win the Boston Marathon":** Adam and Kara Goucher. Interview by author. Tape recording. Boulder, Colorado, October 3, 2017.

248 **"*why are you reading LetsRun*":** Steve Magness. Interview by author. Tape recording. Houston, Texas, November 11, 2018.

249 **"tenfold":** Steve Magness. Interview by author. Tape recording. Houston, Texas, November 11, 2018.

249 **Duniway Park Track:** John Brant, "Why I Still Have Faith in Alberto Salazar," RunnersWorld.com, October 24, 2019.

249 **"trust your speed":** John Brant, "Why I Still Have Faith in Alberto Salazar," RunnersWorld.com, October 24, 2019.

250 **"34.9 would not have been good":** Mario Fraioli, "Podcast: Episode 27 with Kara Goucher," *Morning Shakeout*, August 21, 2018, https://themorningshakeout.com/podcast-episode-27-with-kara-goucher.

251 **"special relationship":** TheBostonMarathon, "Kara Goucher, 2011 Boston Marathon Press Conference," April 15, 2011, YouTube video, https://www.youtube.com/watch?v=ykIn64HIfkg.

251 **another marathon race:** Liz Robbins, "The Runner Kara Goucher Is Back and Ready to Run," *New York Times*, March 13, 2011.

251 **50 percent downhill:** Neil Amdur, "The Heartbreak of Technicalities," *New York Times*, April 18, 2011.

251 **"the cradle":** Neil Amdur, "The Heartbreak of Technicalities," *New York Times*, April 18, 2011.

251 **2 hours, 4 minutes, and 58 seconds and a fourth-place:** Neil Amdur, "The Heartbreak of Technicalities," *New York Times*, April 18, 2011.

251 **breakthrough race:** Mario Fraioli, "Top American Finishers at the 2011 Boston Marathon," PodiumRunner.com, April 26, 2011, https://www.podiumrunner.com/photo-gallery-top-american-finishers-at-the-2011-boston-marathon_26009.

252 **"just judgment on her body":** @Adam_Goucher, Twitter.com, November 7, 2019, 8:58 p.m., https://twitter.com/Adam_Goucher/status/1192652647821504512.

252 **"I wouldn't have trained":** Sally Bergesen, "State of the Sport, Kara + Sally, Part 1 podcast," Oiselle.com, December 4, 2019, https://www.oiselle.com/blogs/oiselle-blog/state-of-sport-kara.

252 **add another year:** Sally Bergesen, "State of the Sport, Kara + Sally, Part 1 podcast," Oiselle.com, December 4, 2019, https://www.oiselle.com/blogs/oiselle-blog/state-of-sport-kara.

252 **"In the end I took":** Sally Bergesen, "State of the Sport, Kara + Sally, Part 1 podcast," Oiselle.com, December 4, 2019, https://www.oiselle.com/blogs/oiselle-blog/state-of-sport-kara.

253 **inject EPO in preparation:** CBS News, "Teammate: Lance Armstrong cheated," May 19, 2011, YouTube video, https://www.youtube.com/watch?v=ejaiWZX8p40.

253 **"I rest my case":** Lance Armstrong, Twitter.com, https://twitter.com/lancearmstrong/status/713587504344023O6.

253 **"Congratulations to @eki_ekimov":** Lance Armstrong, Twitter.com, https://twitter.com/lancearmstrong/status/713876055094722257.

253 **on multiple occasions:** Juliet Macur, "Armstrong Accused of Drug Use by Ex-Teammate," *New York Times*, May 19, 2011.

253 **"just trying to sell books":** Sally Bergesen, "State of the Sport, Kara + Sally, Part 1 podcast," Oiselle.com, December 4, 2019, https://www.oiselle.com/blogs/oiselle-blog/state-of-sport-kara.

254 **much-reduced contract:** No attribution, "Alan Webb Leaves Alberto Salazar," LetsRun.com, March 30, 2011, https://www.letsrun.com/2011/webb-0330.php.

254 **"kind of counter to that":** Alan Webb. Interview by author. Phone call. April 2, 2019.

254 **he became exasperated:** Alan Webb. Interview by author. Phone call. April 2, 2019.

254 **PR of 4 minutes and 18 seconds:** Mario Fraioli, "5 Questions with Jackie Areson," ESPN.com, January 31, 2013, https://www.espn.com/blog/endurance/post/_/id/699/5-questions-with-jackie-areson.

254 **Division I 1,500-meter outdoor championships:** No attribution, "Jackie Areson," UHCougars.com, https://uhcougars.com/sports/track-and-field/roster/coaches/jackie-areson/124.

255 **contingent on her joining the Oregon Project:** Jackie Areson. Interview by author. Phone call. October 24, 2019.

255 **contract for $30,000:** Jackie Areson. Interview by author. Phone call. October 24, 2019.

255 **personal best time of 4 minutes and 12 seconds:** No attribution, "Jackie Areson," UHCougars.com, https://uhcougars.com/sports/track-and-field/roster/coaches/jackie-areson/124.

256 **"a deal breaker":** Jackie Areson. Interview by author. Phone call. October 24, 2019.

256 **master's of science:** Darren Treasure, LinkedIn.com, https://www.linkedin.com/in/dr-darren-treasure-06a80251/.

16: INFUSED

257 **10,000 to 12,000 feet above sea level:** "INTERIM REPORT OF THE U.S. ANTI-DOPING AGENCY," USADA, March 17, 2016, page 38.

258 **Farah on numerous new substances:** "INTERIM REPORT OF THE U.S. ANTI-DOPING AGENCY," USADA, March 17, 2016, page 38.

258 **tested on his own children:** Steve Magness. Interview by author. Tape recording. Houston, Texas, November 12, 2018.

258 **26 minutes and 46 seconds:** Simon Turnbull, "Meet the holistic guru who is helping Mo fly," *Independent*, July 22, 2011.

258 **"Do you think they are cheating?":** Chris McClung, Kara and Adam Goucher, "Episode #14: Kara and Adam Goucher on the 4-Year Bans for Salazar/Brown for Doping Violations," *Clean Sport Collective* podcast, October 7, 2019, https://podcasts.apple.com/us/podcast/episode-14-kara-adam-goucher-on-4-year-bans-for-salazar/id1466187704?i=1000452633967.

259 **black allergy mask:** Matt Scherer, "Mens 10,000m—USA Outdoor Track and Field Championships 2011," RunnerSpace.com, June 24, 2011, https://www.runnerspace.com/video.php?video_id=49837-Mens-10000-USA-Outdoor-Track-and-Field-Championships-2011.

259 **Tegenkamp in second:** Matt Scherer, "Mens 10,000m—USA Outdoor Track and Field Championships 2011," RunnerSpace.com, June 24, 2011, https://www.runnerspace.com/video.php?video_id=49837-Mens-10000-USA-Outdoor-Track-and-Field-Championships-2011.

259 **had to petition her way into:** Matt Scherer, "Womens 10,000m—USA Outdoor Track and Field Championships 2011," ArmoryTrack.com, June 23, 2011, https://www.armorytrack.com/gprofile.php?mgroup_id=31488&do=videos&video_id=49805.

259 **Flanagan took it out hard:** Matt Scherer, "Womens 10,000m—USA Outdoor Track and Field Championships 2011," ArmoryTrack.com, June 23, 2011, https://www.armorytrack.com/gprofile.php?mgroup_id=31488&do=videos&video_id=49805.

259 **lapped many of the other women:** Matt Scherer, "Womens 10,000m—USA Outdoor Track and Field Championships 2011," ArmoryTrack.com, June 23, 2011, https://www.armorytrack.com/gprofile.php?mgroup_id=31488&do=videos&video_id=49805.

259 **"the biggest butt on the start line":** Matthew Futterman, "Another of Alberto Salazar's Runners Says He Ridiculed Her Body for Years," *New York Times*, November 14, 2019.

259 **Bernard Lagat and Chris Solinsky going one-two:** Ian Terpin, "Mens 5000—USA Outdoor Track and Field Championships 2011," RunnerSpace.com, June 24, 2011, https://www.runnerspace.com/video.php?video_id=50176-Mens-5000-USA-Outdoor-Track-and-Field-Championships-2011.

259 **we're going to train smarter:** Simon Turnbull, "Meet the holistic guru who is helping Mo fly," *Independent*, July 22, 2011.

260 **unconventional uses for prescription medications:** "INTERIM REPORT OF THE U.S. ANTI-DOPING AGENCY," USADA, March 17, 2016, page 39.

260 **According to partial records reviewed by USADA:** "INTERIM REPORT OF THE U.S. ANTI-DOPING AGENCY," USADA, March 17, 2016, page 52.

260 **require treatment at much lower levels:** "INTERIM REPORT OF THE U.S. ANTI-DOPING AGENCY," USADA, March 17, 2016, page 84.

260 **two weeks on the drugs:** "INTERIM REPORT OF THE U.S. ANTI-DOPING AGENCY," USADA, March 17, 2016, page 52.

260 **began weaning her:** Arianna Lambie, email with author, February 20, 2020.

261 **"one period every three months":** Mark Daly, "Alberto Salazar's spectacular fall from grace," BBC, February 24, 2020, https://www.bbc.com/sport/athletics/51599747.

261 **"irregular heart rhythms":** Heidi Godman, "For borderline underactive thyroid, drug therapy isn't always necessary," *Harvard Health Publishing*, October 9, 2013, https://www.health.harvard.edu/blog/for-borderline-underactive-thyroid-drug-therapy-isnt-always-necessary-201310096740.

261 **create kidney stones:** "INTERIM REPORT OF THE U.S. ANTI-DOPING AGENCY," USADA, March 17, 2016, page 45.

262 **vitamin D toxicity:** No attribution, "What is vitamin D toxicity, and should I worry about it since I take supplements?" Mayo Clinic, https://www.mayoclinic.org/healthy-lifestyle/nutrition-and-healthy-eating/expert-answers/vitamin-d-toxicity/faq-20058108.

262 **Cortisol is an antagonist to testosterone:** "Relationship Between Circulating Cortisol and Testosterone: Influence of Physical Exercise," Sports Science Medicine, PMCID: PMC3880087. PMID: 24431964, March 2005, 4(1), 76–83, https://www.ncbi.nlm.nih.gov/pmc/articles/PMC3880087/.

263 **"I felt reassured":** "INTERIM REPORT OF THE U.S. ANTI-DOPING AGENCY," USADA, March 17, 2016, page 38.

263 **"I'm not saying that it didn't happen":** Kara Goucher. Telephone interview by author. April 14, 2020.

264 **no better than sixth:** No attribution, "Unbeatable: The Amazing Mo Farah Wins the 10,000 at 2017 Worlds, Earns 10th Global Gold in 'Toughest Race' of His Life," LetsRun.com, August 4, 2017, https://www.letsrun.com/news/2017/08/unbeatable-amazing-mo-farah-wins-10000-2017-worlds-earns-10th-global-gold-toughest-race-life/.

265 **"I'm leaving Alberto":** Chris McClung, Kara and Adam Goucher, "Episode #14: Kara and Adam Goucher on the 4-Year Bans for Salazar/Brown for Doping Violations," *Clean Sport Collective* podcast, October 7, 2019, https://podcasts.apple.com/us/podcast/episode-14-kara-adam-goucher-on-4-year-bans-for-salazar/id1466187704?i=1000452633967.

266 **assumed she was a doper:** Sally Bergesen, "State of the Sport, Kara + Sally,

Part 2 podcast," Oiselle.com, December 4, 2019, https://www.oiselle.com/blogs/oiselle-blog/state-of-sport-kara.

266 "'that's the guy'": Sally Bergesen, "State of the Sport, Kara + Sally, Part 2 podcast," Oiselle.com, December 4, 2019, https://www.oiselle.com/blogs/oiselle-blog/state-of-sport-kara.

266 tripped over his dog: Abigail Lorge, "Chris Solinsky's Long Road Back," *Runner's World*, September 9, 2014.

267 "Alberto was happy about it": Steve Magness. Interview by author. Tape recording. Houston, Texas, November 11, 2018.

267 "our mortal enemies": Sara Germano, "Inside a Nike Family Feud," *Wall Street Journal*, March 7, 2014, https://www.wsj.com/articles/inside-a-nike-family-feud-alberto-salazar-and-jerry-schumacher-1394234890.

267 "Galen is going to beat Solinsky": Steve Magness. Interview by author. Tape recording. Houston, Texas, November 11, 2018.

267 twentieth place: No attribution, Athlinks.com, https://www.athlinks.com/event/34567/results/Event/162165/Course/249240/Bib/12.

267 retiring from professional running: Mario Fraioli, "Adam Goucher Officially Retires from Running," PodiumRunner.com, November 9, 2011, https://www.podiumrunner.com/adam-goucher-officially-retires-from-running_41956.

268 "I am a runner": Mario Fraioli, "Adam Goucher Officially Retires from Running," PodiumRunner.com, November 9, 2011, https://www.podiumrunner.com/adam-goucher-officially-retires-from-running_41956.

268 "your athletes will still gain an advantage": "INTERIM REPORT OF THE U.S. ANTI-DOPING AGENCY," USADA, March 17, 2016, page 124.

268 "make sure there's nothing bad in it": AMERICAN ARBITRATION ASSOCIATION ("AAA") COMMERCIAL ARBITRATION TRIBUNAL AAA Case No. 01-17-0004-0880, paragraph 83, https://www.usada.org/wp-content/uploads/Salazar-AAA-Decision-1.pdf.

270 allergic reaction to the stitches: Matt McCue, "Whatever It Takes," Runners World.com, October 4, 2012, https://www.runnersworld.com/advanced/a20789075/whatever-it-takes.

270 "he started freaking out": Steve Magness. Interview by author. Tape recording. Houston, Texas, November 12, 2018.

271 "the guy to figure things out": Steve Magness. Interview by author. Tape recording. Houston, Texas, November 11, 2018.

271 ongoing thyroid issue: AMERICAN ARBITRATION ASSOCIATION ("AAA") COMMERCIAL ARBITRATION TRIBUNAL AAA Case No. 01-17-0004-0880, page 30, paragraph 92, https://www.usada.org/wp-content/uploads/Salazar-AAA-Decision-1.pdf.

271 Magness was registered: AMERICAN ARBITRATION ASSOCIATION ("AAA") COMMERCIAL ARBITRATION TRIBUNAL AAA Case No. 01-17-0004-0880, paragraph 103, https://www.usada.org/wp-content/uploads/Salazar-AAA-Decision-1.pdf.

271 pay pacers somewhere between $100 and $500 per session: AMERICAN ARBITRATION ASSOCIATION ("AAA") COMMERCIAL ARBITRATION TRIBUNAL AAA Case No. 01-17-0004-0880, page 33, paragraph 104, https://www.usada.org/wp-content/uploads/Salazar-AAA-Decision-1.pdf.

272 **Corner Compounding Pharmacy:** AMERICAN ARBITRATION ASSOCIATION ("AAA") COMMERCIAL ARBITRATION TRIBUNAL AAA Case No. 01-17-0004-0880, paragraph 97, https://www.usada.org/wp-content/uploads/Salazar-AAA-Decision-1.pdf.

272 **"And he said it was safe":** Steve Magness. Interview by author. Tape recording. Houston, Texas, November 12, 2018.

272 **"very significant performance enhancement":** "INTERIM REPORT OF THE U.S. ANTI-DOPING AGENCY," USADA, March 17, 2016, page 155.

273 **alongside on his bicycle:** "INTERIM REPORT OF THE U.S. ANTI-DOPING AGENCY," USADA, March 17, 2016, page 165.

273 **six days after:** "INTERIM REPORT OF THE U.S. ANTI-DOPING AGENCY," USADA, March 17, 2016, page 164.

274 **"Galen was quite upset":** "INTERIM REPORT OF THE U.S. ANTI-DOPING AGENCY," USADA, March 17, 2016, page 165.

274 **$1,000 per person for a six-month supply:** "INTERIM REPORT OF THE U.S. ANTI-DOPING AGENCY," USADA, March 17, 2016, page 161.

274 **administration of the treadmill tests:** Steve Magness. Interview by author. Tape recording. Houston, Texas, November 12, 2018.

274 **"This doesn't sound legal":** "INTERIM REPORT OF THE U.S. ANTI-DOPING AGENCY," USADA, March 17, 2016, page 167.

275 **"*Clinical trial?*":** Steve Magness. Interview by author. Tape recording. Houston, Texas, November 12, 2018.

275 **"does not exceed 50 mL per 6-hour period":** "INTERIM REPORT OF THE U.S. ANTI-DOPING AGENCY," USADA, March 17, 2016, page 261.

17: LOOPHOLE SALAZAR

276 ***I can't coach you*:** Mark Daly, "Alberto Salazar: The inside story of Nike Oregon Project founder's downfall," BBC Panorama, October 1, 2019, https://www.bbc.com/sport/athletics/49853029.

276 **wasn't the L-carnitine:** "INTERIM REPORT OF THE U.S. ANTI-DOPING AGENCY," USADA, March 17, 2016, page 80.

277 **"'I'm not on Advair'":** Kara Goucher. Interview by author. Tape recording. Boulder, Colorado, September 28, 2017.

277 **"Alberto was intentionally obscuring":** Steve Magness. Interview by author. Tape recording. Houston, Texas, November 12, 2018.

277 **business manager for the Oregon Project:** "INTERIM REPORT OF THE U.S. ANTI-DOPING AGENCY," USADA, March 17, 2016, page 15.

277 **"like a million dollars":** Steve Magness. Interview by author. Tape recording. Houston, Texas, November 11, 2018.

278 **Phil Knight's bastard son:** Steve Magness. Interview by author. Tape recording. Houston, Texas, November 12, 2018.

278 **crazy-dad stories:** Steve Magness. Interview by author. Tape recording. Houston, Texas, November 12, 2018.

278 **Webb later confirmed this:** Alan Webb. Telephone interview by Parker Henry. April 16, 2020.

279 **down to $100,000:** "INTERIM REPORT OF THE U.S. ANTI-DOPING AGENCY," USADA, March 17, 2016, page 20.

279 **"We've had a great journey together":** David Monti, "Shalane Flanagan Wins 2012 Olympic Marathon Trials," LetsRun.com, January 14, 2012, https://www.letsrun.com/2012/shalane-wins-0114.php.

280 **"I commend them for reaching it":** Terry Frieden, "Prosecutors drop Lance Armstrong doping investigation," CNN, February 3, 2012, https://www.cnn.com/2012/02/03/sport/lance-armstrong/index.html?hpt=hp_t3.

281 **"it's all about, winning":** Ryan from FloTrack, "Galen Rupp mentally attacking 2012," March 6, 2012, https://www.flotrack.org/video/5369028-galen-looks-back-on-ncaa-experience.

281 **"one hundred percent trust":** Ryan from FloTrack, "Relationship of Galen Rupp and Alberto Salazar," March 6, 2012, https://www.flotrack.org/video/5369028-galen-looks-back-on-ncaa-experience.

281 **"that's the end of the world":** Steve Magness. Interview by author. Tape recording. Houston, Texas, November 12, 2018.

281 **"Alberto can get anything he wanted":** Steve Magness. Interview by author. Tape recording. Houston, Texas, November 12, 2018.

282 **shuttle him back:** Jackie Areson. Interview by author. Phone call. October 24, 2019.

282 **"I knew he had the testosterone cream":** Jackie Areson. Interview by author. Phone call. October 24, 2019.

283 **"Bizzaro world":** Steve Magness. Interview by author. Tape recording. Houston, Texas, November 11, 2018.

283 **"Not until four o'clock, Alberto":** Steve Magness. Interview by author. Tape recording. Houston, Texas, November 12, 2018.

283 **"this is where the group is":** Steve Magness. Interview by author. Tape recording. Houston, Texas, November 12, 2018.

283 **the two men argued:** Steve Magness. Interview by author. Tape recording. Houston, Texas, November 12, 2018.

284 **"he could ruin my career":** Steve Magness. Interview by author. Tape recording. Houston, Texas, November 12, 2018.

284 **formally accused of doping and drug trafficking:** No attribution, "USADA Notice Letter Against Lance Armstrong," *Wall Street Journal*, June 14, 2012, https://www.wsj.com/articles/SB10001424052702303734204577464954262704154.

284 **failed to bring charges:** Terry Frieden, "Prosecutors drop Lance Armstrong doping investigation," CNN, February 3, 2012, https://www.cnn.com/2012/02/03/sport/lance-armstrong/index.html?hpt=hp_t3.

284 **"Maybe we were wrong":** No attribution, "Why Was a Man with a 10-Year Doping Ban Enjoying the 2012 US Olympic Track & Field Trials from the Nike Sky Box?" LetsRun.com, July 3, 2012, https://www.letsrun.com/2012/block-0701.php.

285 **escorted out of the stadium:** Mario Fraioli, "Podcast: Episode 82 with Jesse Williams," *Morning Shakeout*, October 21, 2019, https://themorningshakeout.com/tag/jesse-williams/.

285 **"I was overcome with emotion":** Matt McCue, "Whatever It Takes," Runners World.com, October 4, 2012, https://www.runnersworld.com/advanced/a20789075/whatever-it-takes.

285 "You're wearing the Oregon Project uniform": Jackie Areson. Interview by author. Phone call. October 24, 2019.

285 "my best shot": No attribution, "RW Interviews: Mo Farah," RunnersWorld .com, April 8, 2011, https://www.runnersworld.com/uk/training/motivation /a767081/rw-interviews-mo-farah.

286 just trying to catch up: Matt McCue, "Whatever It Takes," RunnersWorld .com, October 4, 2012, https://www.runnersworld.com/advanced/a20789075 /whatever-it-takes.

286 According to Nielsen: No attribution, "London Olympics 2012 Ratings: Most Watched Event in TV History," *Huffington Post*, August 13, 2012, https:// www.huffpost.com/entry/london-olympics-2012-ratings-most-watched -ever_n_1774032.

286 219.4 million Americans watched: London 2012 Summer Olympics (July 27, 2012–August 12, 2012).

286 watched in utter disbelief: Sean Ingle, "John Rohatinsky backs drug allegations made against Alberto Salazar," *Guardian*, June 14, 2015, https://www .theguardian.com/sport/2015/jun/14/john-rohatinsky-backs-drug-allegations -alberto-salazar.

287 "I loved my team": Pete Julian. Interview by author. Tape recording. Boulder, Colorado, November 7, 2019.

287 Big Ben, Parliament, then Buckingham Palace: Olympic, "Athletics—Women Marathon—Day 9 | London 2012 Olympic Games," August 5, 2012, YouTube video, https://www.youtube.com/watch?v=-tjUHasesSU.

287 distorted the finishing photo: Simon Hart, "Olympics hero Mo Farah aiming to break Steve Ovett's 34-year-old British record at Aviva Birmingham Grand Prix," *Daily Telegraph*, August 25, 2012.

288 Al Sal A Drug Czar: Roy B. Thompson, "Alberto Salazar," email message to Weldon and Robert Johnson, August 24, 2012.

288 not liable for the posts therein: Weldon Johnson, "Re: Alberto Salazar," email message to Roy B. Thompson, August 30, 2012.

288 Thompson thanked Weldon: Roy B. Thompson, "Re: Alberto Salazar," email message to Weldon Johnson, August 30, 2012.

289 beginning to bear fruit: "Julian brought the University to their first NCAA Championship berth since 2006," Nike Oregon Project, https://nikeoregonproject .com/pages/pete-julian.

289 "It was terrible": Pete Julian. Interview by author. Tape recording. Boulder, Colorado, November 7, 2019.

289 One of the first athletes he met: Pete Julian. Interview by author. Tape recording. Boulder, Colorado, November 7, 2019.

290 "almost 11,000%": "INTERIM REPORT OF THE U.S. ANTI-DOPING AGENCY," USADA, March 17, 2016, page 221 of 269.

290 Welling would be the last: Pete Julian. Interview by author. Tape recording. Boulder, Colorado, November 7, 2019.

290 "Reasoned Decision": "U.S. Postal Service Pro Cycling Team Investigation," USADA.org, http://cyclinginvestigation.usada.org/.

290 "successful doping program that sport has ever seen": "U.S. Postal Service Pro Cycling Team Investigation," USADA.org, http://cyclinginvestigation.usada.org/.

290 **way back to 1998:** USADA, "U.S. Postal Service Pro Cycling Team Investigation," USADA.org, http://cyclinginvestigation.usada.org/.

290 **ninety-eight products:** Matt Williams, "Nike drops deal with Lance Armstrong after he 'misled us for a decade,'" *Guardian*, October 17, 2012, https://www.theguardian.com/sport/2012/oct/17/nike-lance-armstrong-misled-decade.

291 **Kobe Bryant was charged with felony sexual assault in 2003:** Ken Belson and Mary Pilon, "Armstrong Is Dropped by Nike and Steps Down as Foundation Chairman," *New York Times*, October 17, 2012.

291 **he lost half a dozen other sponsors:** Kurt Badenhausen, "Tiger Woods, Nike And Winning," *Forbes* magazine, March 27, 2013, https://www.forbes.com/sites/kurtbadenhausen/2013/03/27/tiger-woods-nike-and-winning.

291 **highest-paid athlete:** Barbara Smit, *Sneaker Wars* (New York: HarperCollins Perennial Edition, 2009), 331.

291 **"a bad word spoken about Tiger":** Phil Knight, *Shoe Dog* (New York: Simon & Schuster, 2016), 368.

291 **"I will conclude my chairmanship":** Houston Mitchell, "Lance Armstrong steps down as chairman of Livestrong charity," October 17, 2012, https://www.latimes.com/sports/la-xpm-2012-oct-17-la-sp-sn-lance-armstrong-livestrong-20121017-story.html.

292 **"seemingly insurmountable evidence":** Nike, "Nike Statement on Lance Armstrong," October 17, 2012, https://news.nike.com/news/nike-statement-on-lance-armstrong.

292 **"continue support of the Livestrong":** Darren Heitner, "Nike's Disassociation from Lance Armstrong Makes Nike a Stronger Brand," *Forbes*, October 17, 2012, https://www.forbes.com/sites/darrenheitner/2012/10/17/nikes-disassociation-from-lance-armstrong-makes-nike-a-stronger-brand/#331c5696df49.

292 **abandoned the Texan:** Ken Belson and Mary Pilon, "Armstrong Is Dropped by Nike and Steps Down as Foundation Chairman," *New York Times*, October 17, 2012.

292 **$75 million:** Juliet Macur, "The Eternal Martyrdom Of Lance," Deadspin.com, March 4, 2014, 3:34 p.m., https://deadspin.com/the-eternal-martyrdom-of-lance-armstrong-1537029078.

292 **stripped him:** Julien Pretot, "'Sickened' UCI strips Armstrong of Tour wins," Reuters, October 22, 2012, 5:11 a.m., https://www.reuters.com/article/us-cycling-armstrong/sickened-uci-strips-armstrong-of-tour-wins-idUSBRE89L0HC20121022.

292 **"one of the world's most sophisticated sports companies":** Juliet Macur, *Cycle of Lies: The Fall of Lance Armstrong* (New York: HarperCollins, 2014), Kindle ed., Location 377.

292 **"no fucking rat":** Juliet Macur, *Cycle of Lies: The Fall of Lance Armstrong* (New York: HarperCollins, 2014), Kindle ed., Location 395.

18: YOU'RE A NOBODY

293 **"make sure you are all on Vitamin D":** "INTERIM REPORT OF THE U.S. ANTI-DOPING AGENCY," USADA, March 17, 2016, page 111. And March 31, 2012, email from Alberto Salazar.

294 **"increase cancer risk":** "INTERIM REPORT OF THE U.S. ANTI-DOPING AGENCY," USADA, March 17, 2016, page 112.

294 **filled his next prescription:** "INTERIM REPORT OF THE U.S. ANTI-DOPING AGENCY," USADA, March 17, 2016, page 112.

295 **sit down with Oprah and explain himself:** "Lance Armstrong's Confession," *Oprah's Next Chapter*, OWN, January 17, 2013, https://www.youtube.com/watch?v=N_0PSZ59Aws.

295 **"I'll go to USADA":** Shanna Burnette, Kara Goucher, and Chris McClung. "Episode 1: Travis Tygart, CEO of the US Anti-Doping Agency," *Clean Sport Collective* podcast, cleansport.libsyn.com/episode-1-travis-tygart-ceo-of-usada.

296 **"we were suspects":** Chris McClung, Kara and Adam Goucher, "Episode #14: Kara and Adam Goucher on the 4-Year Bans for Salazar/Brown for Doping Violations," *Clean Sport Collective* podcast, October 7, 2019, https://podcasts.apple.com/us/podcast/episode-14-kara-adam-goucher-on-4-year-bans-for-salazar/id1466187704?i=1000452633967.

296 **supply them with his medical records:** Steve Magness. Interview by author. Tape recording. Houston, Texas, November 11, 2018.

296 **"more than the three tablespoons":** Steve Magness. Interview by author. Tape recording. Houston, Texas, November 11, 2018.

296 **judge the market:** Stuart Elliott, "Losing a Step, Nike Seeks to Regain Its Edge," *New York Times*, April 14, 2013.

296 **"a harder time standing out":** Stuart Elliott, "Losing a Step, Nike Seeks to Regain Its Edge," *New York Times*, April 14, 2013.

296 **ceded in October 2010:** Kurt Badenhausen, "Tiger Woods, Nike And Winning," *Forbes* magazine, March 27, 2013, https://www.forbes.com/sites/kurtbadenhausen/2013/03/27/tiger-woods-nike-and-winning/#1f5eaa26e340.

297 **"feigned ignorance":** Joshua Smith, @yeshuasmith, Twitter.com, March 26, 2013, https://twitter.com/yeshuasmith/status/316581095766364161.

297 **"beyond irresponsible":** No attribution, "Nike's Tiger Woods ad draws critics," ESPN.com, March 26, 2013, https://www.espn.com/golf/story/_/id/9100497/nike-winning-takes-care-everything-tiger-woods-ad-draws-critics.

297 **publicly called into question Brown's tactics:** Sara Germano and Kevin Clark, "U.S. Track's Unconventional Physician," *Wall Street Journal*, April 10, 2013.

298 **"'didn't have anything nice to say'":** Jackie Areson. Interview by author. Phone call. October 24, 2019.

298 **"the professional approach":** Sean Ingle, "Nike Oregon Project shut down after Alberto Salazar's four-year ban," *Guardian*, October 11, 2019, https://www.theguardian.com/sport/2019/oct/11/nike-oregon-project-shut-down-after-alberto-salazars-four-year-ban.

299 **"the most prominent figures":** Mark Daly, "Alberto Salazar: The inside story of Nike Oregon Project founder's downfall," BBC Panorama, October 1, 2019, https://www.bbc.com/sport/athletics/49853029.

299 **"control the sport":** Mark Daly, "Alberto Salazar: The inside story of Nike Oregon Project founder's downfall," BBC Panorama, October 1, 2019, https://www.bbc.com/sport/athletics/49853029.

300 **"I wanted to be brave like Adam":** Chris McClung, Kara and Adam Goucher,

"Episode #9: Adam Goucher, 4-time NCAA Champion and Olympian," *Clean Sport Collective* podcast, August 25, 2019, https://podcasts.apple.com/us/podcast/episode-9-adam-goucher-4-time-ncaa-champion-and-olympian/id1466187704?i=1000447757470.

300 **"He knew shit":** Adam and Kara Goucher. Interview by author. Tape recording. Boulder, Colorado, September 20, 2017.

301 **"beat me the next day":** Adam and Kara Goucher. Interview by author. Tape recording. Boulder, Colorado, September 18, 2017.

301 **sixth-fastest man ever:** Aimee Lewis, "Mo Farah breaks Steve Cram's 28-year British 1500m record," BBC *Sport*, July 19, 2013, https://www.bbc.com/sport/athletics/23385989.

301 **she set an American record:** Adam and Kara Goucher. Interview by author. Tape recording. Boulder, Colorado, October 3, 2017.

302 **"iron, vitamin D, and that's it":** Simon Hart, "Alberto Salazar takes a swipe at rumour-mongers who put Mo Farah's sudden run of success down to doping," *Telegraph*, August 19, 2013, https://www.telegraph.co.uk/sport/othersports/athletics/10253393/Alberto-Salazar-takes-a-swipe-at-rumour-mongers-who-put-Mo-Farahs-sudden-run-of-success-down-to-doping.html.

302 **new addition:** Amanda Brooks, "Cain and Salazar Accept Awards at 2013 World Athletics Gala 11/16/2013," USATF.org, November 16, 2013, https://web.archive.org/web/20151016194114/http://www.usatf.org/News/Cain-and-Salazar-Accept-Awards-at-2013-World-Athle.aspx.

302 **new American high school record:** Elizabeth Weil, "Mary Cain Is Growing Up Fast," *New York Times Magazine*, March 4, 2015, https://www.nytimes.com/2015/03/08/magazine/mary-cain-is-growing-up-fast.html.

303 **"Alberto is my coach":** @JordanHasay, Twitter.com, November 16, 2013, https://twitter.com/JordanHasay/status/401800833118924802.

303 **met Lomong in the mixed zone:** Jon Gugala and LetsRun.com, "Lopez Lomong Says He Was Verbally Berated By Alberto Salazar: 'To see the outrage in his eyes, he was very, very mad. . . . It was very hard to see how unprofessional, the (lack of) sportsmanship,'" LetsRun.com, February 25, 2014, https://www.letsrun.com/news/2014/02/lopez-lomong-happy-verbally-berated-alberto-salazar-see-outrage-eyes-mad-hard-see-unprofessional-lack-spo/.

303 **"very, very mad":** Jon Gugala and LetsRun.com, "Lopez Lomong Says He Was Verbally Berated By Alberto Salazar: 'To see the outrage in his eyes, he was very, very mad. . . . It was very hard to see how unprofessional, the (lack of) sportsmanship,'" LetsRun.com, February 25, 2014, https://www.letsrun.com/news/2014/02/lopez-lomong-happy-verbally-berated-alberto-salazar-see-outrage-eyes-mad-hard-see-unprofessional-lack-spo/.

303 **actively avoided having to spend time together:** No attribution, "Nike Coaches Alberto Salazar And Jerry Schumacher Have Heated Exchange, Have to Be Separated After Men's 3,000 at 2014 USA Indoors," LetsRun.com, February 22, 2014, Accessed July 17, 2019, https://www.letsrun.com/news/2014/02/nike-coaches-alberto-salazar-jerry-schumacher-heated-exchanged-separated-mens-3000-2014-usa-indoors/.

304 **"escorted from the area":** Gugala and LetsRun.com, "Lopez Lomong Says He Was Verbally Berated By Alberto Salazar: 'To see the outrage in his eyes, he was

very, very mad. . . . It was very hard to see how unprofessional, the (lack of) sportsmanship,'" LetsRun.com, February 25, 2014, https://www.letsrun.com/news/2014/02/lopez-lomong-happy-verbally-berated-alberto-salazar-see-outrage-eyes-mad-hard-see-unprofessional-lack-spo/.

305 **"I just didn't understand":** Mackenzie Lobby, "Gabe Grunewald fights to the finish," *Sports Illustrated*, March 6, 2014, https://www.espn.com/espnw/news-commentary/story/_/id/10560658/espnw-cancer-survivor-gabe-grunewald-set-run-3000-meters-world-championships-mysterious-disqualification-us-indoor-championships.

305 **at the Marriott, the event's host hotel:** Jon Gugala, "Alberto Salazar to Gabe Grunewald's Husband: "Get the f— out of my face"; Grunewald to Salazar, "She's had cancer twice. This is bullshit. How do you sleep at night? You've got a defibrillator in your chest," LetsRun.com, February 23, 2014, https://www.letsrun.com/news/2014/02/alberto-salazar-gabe-grunewalds-husband-get-f-face/.

306 **Accounts differ:** Jon Gugala, "Alberto Salazar to Gabe Grunewald's Husband: 'Get the f— out of my face'; Grunewald to Salazar, 'She's had cancer twice. This is bullshit. How do you sleep at night? You've got a defibrillator in your chest,'" LetsRun.com, February 23, 2014, https://www.letsrun.com/news/2014/02/alberto-salazar-gabe-grunewalds-husband-get-f-face/.

306 **meet record time:** No attribution, "Womens 1500m Final—USA Indoor Track and Field Championships 2014," RunnerSpace.com, https://gobeavs.runnerspace.com/gprofile.php?mgroup_id=45365&do=videos&video_id=105717.

306 **verbally assaulted:** Jon Gugala. Interview by author by telephone. February 19, 2020.

306 **operating outside of their own rules:** Sara Germano and Kevin Helliker, "USA Track Reinstates Disqualified Champ Grunewald," *Wall Street Journal*, February 24, 2014, https://www.wsj.com/articles/usa-track-reinstates-disqualified-champ-grunewald-1393291523?tesla=y.

306 **until the year 2040:** Darren Rovell, "Nike extends deal with Team USA," ESPN.com, April 16, 2014, https://www.espn.com/olympics/story/_/id/10791755/nike-extends-sponsorship-deal-usa-track-field-team.

307 **$25.3 billion in total sales:** No attribution, "NIKE, Inc. Reports FY2013 Q4 and Full Year Results," Nike.com, June 27, 2013, https://news.nike.com/news/nike-inc-reports-fy2013-q4-and-full-year-results.

307 **"Kara will not run for Nike again":** Adam and Kara Goucher. Interview by author. Tape recording. Boulder, Colorado, October 3, 2017.

308 **"get her on board":** Charles Butler, "Kara Goucher Signs with Skechers," *Runner's World*, May 5, 2014, https://www.runnersworld.com/races-places/a20783927/kara-goucher-signs-with-skechers/.

308 **improving her "marathon shuffle":** Brian Metzler, "Don't Call It A Comeback: Kara Goucher Has Her Eyes on the Prize," PodiumRunner.com, February 2, 2016, https://www.podiumrunner.com/dont-call-comeback-kara-goucher-eyes-prize_144209.

309 **in eighteen months:** Brian Metzler, "Don't Call It A Comeback: Kara Goucher Has Her Eyes on the Prize," PodiumRunner.com, February 2, 2016, https://www.podiumrunner.com/dont-call-comeback-kara-goucher-eyes-prize_144209.

309 **"I just want you guys to tell me":** Brian Metzler, "Don't Call It A Comeback: Kara Goucher Has Her Eyes on the Prize," PodiumRunner.com, February 2, 2016, https://www.podiumrunner.com/dont-call-comeback-kara-goucher-eyes-prize_144209.

309 **"I really needed to hear that":** Brian Metzler, "Don't Call It A Comeback: Kara Goucher Has Her Eyes on the Prize," PodiumRunner.com, February 2, 2016, https://www.podiumrunner.com/dont-call-comeback-kara-goucher-eyes-prize_144209.

19: OFF TRACK

311 **"I'm really angry":** Mark Daly, "Alberto Salazar: The inside story of Nike Oregon Project founder's downfall," BBC, October 1, 2019, https://www.bbc.com/sport/athletics/49853029.

311 **"the people who made you who you are":** Bonnie D. Ford, "Athletes, others who raise doping concerns in sports often left whistling into the wind," ESPN.com, https://www.espn.com/espn/otl/story/_/id/16209580/doping-whistleblowers-such-stepanovs-kara-goucher-often-left-dangling-taking-sports-bodies-governing-bodies.

311 **"good friend":** Press Association, "Lord Sebastian Coe calls on friend Alberto Salazar to defend himself," *Guardian*, June 7, 2015, https://www.theguardian.com/sport/2015/jun/07/lord-sebastian-coe-alberto-salazar-drug-allegations-mo-farah.

312 **"outrageous and baseless":** Ken Goe, "Galen Rupp's parents call allegations 'outragegous' and 'baseless'," *Oregonian*, June 24, 2015, https://www.oregonlive.com/trackandfield/2015/06/galen_rupps_parents_call_accus.html.

312 **"they are lying":** Alberto Salazar, NikeOregonProject.com, https://web.archive.org/web/20150627071828/https://nikeoregonproject.com/blogs/news/35522561-alberto-open-letter-part-1.

313 **"the dynamic of our relationship changed":** Mark Daly, "Alberto Salazar and the Nike Oregon Project," BBC, June 3, 2015, https://www.bbc.com/news/uk-scotland-32883946.

314 **"to cover up Boston":** David Epstein, "Alberto Salazar Disputes Allegations—Some of Which Were Never Made," ProPublica.org, June 24, 2015, https://www.propublica.org/article/alberto-salazar-disputes-allegations-some-of-which-were-never-made.

314 **two other independent sources:** A Nike employee and a USATF medical staff: David Epstein, "Alberto Salazar Disputes Allegations—Some of Which Were Never Made," ProPublica.org, June 24, 2015, https://www.propublica.org/article/alberto-salazar-disputes-allegations-some-of-which-were-never-made.

314 **"My relationship with Adam never recovered":** No attribution, Nike Oregon Project, https://web.archive.org/web/20150627071828/https://nikeoregonproject.com/blogs/news/35522561-alberto-open-letter-part-1.

315 **"she's as clean as a whistle":** Scott Douglas, "Former Nike Oregon Project Coach 'Not Surprised' by Doping Allegations," *Runner's World*, June 8, 2015, https://www.runnersworld.com/news/a20806570/former-nike-oregon-project-coach-not-surprised-by-doping-allegations/.

315 **"Alberto has threatened him":** Jonathan Gault, "John Capriotti, Nike Global Director of Athletics, Threatened to Kill Brooks Beasts Head Coach and Former Nike Employee Danny Mackey at 2015 USAs According to Police Report," LetsRun.com, August 13, 2015, Accessed June 30, 2019, https://www.letsrun.com/news/2015/08/police-report-nike-global-director-of-athletics-john-capriotti-threatened-to-kill-brooks-beasts-head-coach-and-former-nike-employee-danny-mackey-at-2015-usas/.

317 **"You're fuckin' dead":** Danny Mackey. Interview by author. Phone call. September 6, 2019.

318 **denied making any of the statements:** Mark Daly, "Alberto Salazar's spectacular fall from grace," BBC Panorama, February 24, 2020, https://www.bbc.com/sport/athletics/51599747.

318 **"institutional Nike aggression":** Sarah Barker, "Nike Global Director of Athletics Threatens to Kill a Brooks Coach," DeadSpin.com, August 13, 2015, Accessed August 10 2019, https://deadspin.com/nike-global-director-of-athletics-threatens-to-kill-a-b-1724003701.

318 **"people won't talk":** Jonathan Gault, "John Capriotti, Nike Global Director of Athletics, Threatened to Kill Brooks Beasts Head Coach and Former Nike Employee Danny Mackey at 2015 USAs According to Police Report," LetsRun.com, August 13, 2015, https://www.letsrun.com/news/2015/08/police-report-nike-global-director-of-athletics-john-capriotti-threatened-to-kill-brooks-beasts-head-coach-and-former-nike-employee-danny-mackey-at-2015-usas/.

318 **"I don't walk with my headphones on anymore":** Danny Mackey. Telephone interview by author. September 6, 2019.

318 **"Why else would you do it":** Ed Odeven, "Conte expects Salazar to be banned," *Japan Times*, August 4, 2015, Accessed August 27, 2019, https://www.japantimes.co.jp/sports/2015/08/04/more-sports/conte-expects-salazar-banned.

318 **denied obstructing:** Mark Daly, "Alberto Salazar's spectacular fall from grace," BBC Panorama, February 24, 2020, https://www.bbc.com/sport/athletics/51599747.

319 **"cooperating with whatever officials":** Ross Running, "Galen Rupp responds to Kara Goucher interview and doping allegations," February 17, 2016, YouTube video, https://www.youtube.com/watch?v=jl-RmOyT1Lg.

319 **"largely refused":** "INTERIM REPORT OF THE U.S. ANTI-DOPING AGENCY," USADA, March 17, 2016, page 33 of 269.

319 **"It's unprecedented":** Travis Tygart. Interview by author. Phone call. September 30, 2019.

319 **"medical records had been altered":** Mark Daly, "Alberto Salazar: The inside story of Nike Oregon Project founder's downfall," BBC Panorama, October 1, 2019, https://www.bbc.com/sport/athletics/49853029.

320 **"It's a winning bet":** Sean Ingle, "Alberto Salazar and Mo Farah still have many questions to answer," *Guardian*, February 26 2018, https://www.theguardian.com/sport/blog/2017/feb/26/alberto-salazar-mo-farah-allegations.

321 **"at the time I didn't remember":** Mark Daly, "Fresh questions over Mo Farah's relationship with Alberto Salazar," BBC, February 24, 2020, https://www.bbc.com/sport/athletics/51591701.

321 renewed emphasis on recovery: Brian Metzler, "Don't Call It a Comeback: Kara Goucher Has Her Eyes on the Prize," PodiumRunner.com, February 2, 2016, https://www.podiumrunner.com/dont-call-comeback-kara-goucher-eyes-prize_144209.

322 "nobody would have given Kara much of a chance": Brian Metzler, "Don't Call It a Comeback: Kara Goucher Has Her Eyes on the Prize," PodiumRunner.com, February 2, 2016, https://www.podiumrunner.com/dont-call-comeback-kara-goucher-eyes-prize_144209.

322 were prototypes: "The Shoes Athletes Say Will Change the Future of Running," Nike.com, May 3, 2017, https://news.nike.com/news/nike-vaporfly-4-review.

322 disguised to look like Nike's existing racing flats: Jonathan Gault, "Turning Down a Contract to Race in the Vaporflys? Vaporfly > Alphafly? How Shoes Will Affect the US Olympic Marathon Trials," LetsRun.com, February 25, 2020, https://www.letsrun.com/news/2020/02/turning-down-a-contract-to-race-in-the-vaporflys-vaporfly-alphafly-how-shoes-will-affect-the-us-olympic-marathon-trials/.

322 unfair assistance or advantage: Jeré Longman, "Do Nike's New Shoes Give Runners an Unfair Advantage?" *New York Times*, March 8, 2017, https://www.nytimes.com/2017/03/08/sports/nikes-vivid-shoes-and-the-gray-area-of-performance-enhancement.html.

323 "This shoe is a game changer": Roger Pielke, "On Par With Doping: The First Person To Miss The Olympics For Wearing The Wrong Shoes," Forbes.com, January 21, 2020, https://www.forbes.com/sites/rogerpielke/2020/01/21/on-par-with-doping-the-first-person-to-miss-the-olympics-for-wearing-the-wrong-shoes/#37757ac94a08.

323 elbow-length cooling gloves: Chris Raulli, "2016 US Marathon Trials," April 4, 2017, YouTube video, https://www.youtube.com/watch?v=ASKZpQnI2q8.

323 at mile sixteen: No attribution, "Dathan Ritzenhein drops out of U.S. Olympic Marathon Trials," MLive.com, February 13, 2016, https://www.mlive.com/sports/grand-rapids/2016/02/us_olympic_marathon_trials_liv.html.

324 an astonishing 4-minute-and-47-second pace: John Brant, "Can Galen Rupp Win an Olympic Marathon Medal?" *Runner's World*, August 8, 2016, https://www.runnersworld.com/races-places/a20814407/can-galen-rupp-win-an-olympic-marathon-medal/.

325 "Always great to be in your presence, Phil": *Good Morning America*, "Phil Knight Discusses His New Book *Shoe Dog*," April 26, 2016, https://www.youtube.com/watch?v=cSe-qR4f1Ng.

325 the arrest soon appeared online: Ben Bloom, "Mo Farah distanced from Jama Aden after controversial coach is arrested in doping raid," *Telegraph*, June 20, 2016, https://www.telegraph.co.uk/athletics/2016/06/20/mo-farah-distanced-from-jama-aden-after-controversial-coach-is-a/.

326 were Nike athletes: Bill Saporito, "How Phil Knight Built Nike Into a $100 Billion Global Empire," *Maxim*, September 29, 2016, https://www.maxim.com/entertainment/how-phil-knight-became-sultan-of-swoosh-2016-9.

326 "anti-doping rule violations": AMERICAN ARBITRATION ASSOCIATION ("AAA") COMMERCIAL ARBITRATION TRIBUNAL AAA Case No. 01-17-

0004-0880, page 4, paragraph 15, https://www.usada.org/wp-content/uploads/Salazar-AAA-Decision-1.pdf.

327 **Salazar had wooed Hasay:** Katherine Laidlaw, "Jordan Hasay Will Outrun You. While Smiling.," *Outside* magazine, April 2018, https://www.outsideonline.com/2284496/jordan-hasay-will-outrun-you-while-smiling.

328 **"I felt nothing":** Adam and Kara Goucher. Interview by author. Tape recording. Boulder, Colorado, September 18, 2017.

328 **was finally released to the public:** "How to Get the Nike Zoom Vaporfly 4%," Nike.com, October 31, 2018, https://news.nike.com/news/how-to-get-the-nike-zoom-vaporfly-4.

328 **less of a burden on the body:** Lisa Marshall, "What makes the world's fastest shoe so fast? New study provides insight," *CU Boulder Today*, November 20, 2018, https://www.colorado.edu/today/2018/11/20/what-makes-worlds-fastest-shoe-so-fast-new-study-provides-insight.

328 **They flew off the shelves:** Sean Ingle, "Nike's lightning shoes hint at power of technology to skew elite competition," *Guardian*, July 22, 2018, https://www.theguardian.com/sport/2018/jul/22/nike-shoes-vaporfly-sport.

328 **"hard to truly and genuinely get excited":** Chris Chavez, "Ahead of the 2017 New York City Marathon, Shalane Flanagan Isn't Done . . . Yet," October 25, 2017, https://www.si.com/edge/2017/10/25/new-york-city-marathon-shalane-flanagan-career.

329 **"every runner used less energy with the prototype shoe,":** Lisa Marshall, "New shoe makes running 4 percent easier, 2-hour marathon possible, study shows," University of Colorado Boulder, November 16, 2017, https://www.colorado.edu/today/2017/11/16/new-shoe-makes-running-4-percent-easier-2-hour-marathon-possible-study-shows.

329 **"We need evidence":** Kevin Quealy and Josh Katz, "Nike Says Its $250 Running Shoes Will Make You Run Much Faster. What If That's Actually True?" *New York Times*, July 18, 2018, https://www.nytimes.com/interactive/2018/07/18/upshot/nike-vaporfly-shoe-strava.html.

20: BANNED IN DOHA

330 **hit a record $86.30:** No attribution, *Yahoo Finance*, https://finance.yahoo.com/quote/NKE/history/?guccounter=1&guce_referrer=aHR0cHM6Ly93d3cuZ29vZ2xlLmNvbS8&guce_referrer_sig=AQAAAKqGsCrFTvJxVALZkRn4KWFhqYql-RpIS5kwhV0tLGDFHsQxooyRbQ6Y2DdcUwRSdpkq0nVyF7KmcY6PSuxo5khxA5HSoYfazoptEhc4Hr26BIxDDVjJjEovBko6ukgssVmQwBBYhraUjma90b3oU3uDpN0W9lrZU9AKZn9IzKEH.

331 **"an insular group of high-level managers":** Julie Creswell, Kevin Draper, and Rachel Abrams, "At Nike, Revolt Led by Women Leads to Exodus of Male Executives," *New York Times*, April 28, 2018, https://www.nytimes.com/2018/04/28/business/nike-women.html.

331 **"We went to make it better":** Julie Creswell, Kevin Draper, and Rachel Abrams, "At Nike, Revolt Led by Women Leads to Exodus of Male Executives," *New York Times*, April 28, 2018, https://www.nytimes.com/2018/04/28/business/nike-women.html.

331 **"restore trust":** Kevin Draper and Julie Creswell, "Nike's C.E.O. Vows

Changes After Claims of Workplace Harassment and Bias Image," *New York Times*, May 5, 2018.

333 **she used foul language:** LetsRun.com, "An emotional Kara Goucher talks after finishing 4th at 2016 US Olympic Marathon Trials," February 13, 2016, YouTube video, https://www.youtube.com/watch?v=S5E-jxhkKw0.

333 **"Alberto Salazar, Coach of the Nike Oregon Project, Gets a 4-Year Doping Ban":** Jeré Longman and Matt Hart, "Alberto Salazar, Coach of the Nike Oregon Project, Gets a 4-Year Doping Ban," *New York Times*, September 30, 2019.

334 **"going to be thrown out":** Pete Julian. Interview by author. Tape recording. Boulder, Colorado, November 7, 2019.

334 **his athletes were told to sever ties with Salazar:** Tariq Panja, Matthew Futterman, and Scott Cacciola, "World Track Moves to Cut Off Alberto Salazar After Doping Allegation," *New York Times*, October 1, 2019.

334 **made each of Salazar's athletes sign an agreement:** Pete Julian. Interview by author. Tape recording. Boulder, Colorado, November 7, 2019.

335 **"Yeah, if Pete [Julian] takes control":** Joshua Robinson and Rachel Bachman, "After Nike Coach's Suspension, His Star Pupil Faces Scrutiny," *Wall Street Journal*, October 3, 2019.

335 **Murphy finished dead last:** International Association of Athletics Federations, "Report: men's 800m—IAAF World Athletics Championships Doha 2019," Accessed October 1, 2019, https://www.iaaf.org/news/report/world-championships-doha-2019-men-800m-report.

335 **walked away mid–follow-up question:** LetsRun.com, "Clayton Murphy addresses Alberto Salazar's 4-year ban," October 1, 2019, YouTube video, https://www.youtube.com/watch?v=VNHWxbtQsVI.

336 **"Trafficking of testosterone":** United States Anti-Doping Agency, "AAA Panel Imposes 4-Year Sanctions on Alberto Salazar and Dr. Jeffrey Brown for Multiple Anti-Doping Rule Violations," Accessed September 30, 2019, https://www.usada.org/sanction/aaa-panel-4-year-sanctions-alberto-salazar-jeffrey-brown.

336 **found Dr. Brown guilty:** AMERICAN ARBITRATION ASSOCIATION Commercial Arbitration Panel. AAA: USADA V. DR. JEFFREY BROWN, AAA CASE NO. 01-17-0003-6197. Section 1.3 (page 7 of 108, iii).

337 **"I will not be commenting further":** Nike Oregon Project, "Alberto Statement," Accessed October 1, 2019, https://nikeoregonproject.com/blogs/news/alberto-statement.

337 **"clouded his judgment":** AMERICAN ARBITRATION ASSOCIATION ("AAA") COMMERCIAL ARBITRATION TRIBUNAL AAA Case No. 01-17-0004-0880. Paragraph 532. Page 133 of 134.

337 **"THIS BADGE MUST BE WITHDRAWN":** Image from LetsRun.com, https://www.letsrun.com/forum/flat_read.php?thread=9621470.

EPILOGUE

338 **"legal medication for the wrong reason":** *Koersduivel*, "David Walsh interview 2018 about Lance Armstrong & Team Sky," April 3, 2018, YouTube video, https://www.youtube.com/watch?v=79DB5YWV1ts.

338 **all-time legend status:** No attribution, "Not A Typo: Sifan Hassan Wins 10,000 World Title With A 3:59 Final 1500," LetsRun.com, September 28, 2019, https://www.letsrun.com/news/2019/09/not-a-typo-sifan-hassan-wins-10000-world-title-with-a-359-final-1500/.

339 **"wind down the Oregon Project":** Mark Parker, Nike Internal Memo, *Wall Street Journal*, https://s.wsj.net/public/resources/documents/Nike%20Memo.pdf?mod=article_inline.

340 **"the most talented athlete":** Mary Cain, "I Was the Fastest Girl in America, Until I Joined Nike," *New York Times*, November 7, 2019, https://www.nytimes.com/2019/11/07/opinion/nike-running-mary-cain.html.

341 **"We Believe Mary":** Jeff Manning, "'We believe Mary': Nike workers protest as company unveils renovated 'Alberto Salazar' building," *Oregonian*, December 9, 2019, https://www.oregonlive.com/business/2019/12/i-believe-mary-nike-workers-protest-as-company-unveils-renovated-alberto-salazar-building.html.

341 **"Do the Right Thing":** Sophie Peel, "Hundreds of Nike Employees Stage Walk to Protest Alberto Salazar on Beaverton Campus," *Willamette Week*, December 9, 2019, https://www.wweek.com/news/business/2019/12/08/hundreds-of-nike-employees-stage-walk-to-protest-alberto-salazar-on-beaverton-campus/.

341 **in a statement:** Chris Chavez, "Inside the Toxic Culture of the Nike Oregon Project 'Cult,'" *Sports Illustrated*, November 13, 2019, https://www.si.com/track-and-field/2019/11/13/mary-cain-nike-oregon-project-toxic-culture-alberto-salazar-abuse-investigation.

341 **"glad that it happened":** Chris McClung and Kara Goucher, "Episode #38: Colleen Quigley, Olympic Steeplechaser," *Clean Sport Collective* podcast, 43m 50s, March 1, 2020, https://podcasts.apple.com/us/podcast/episode-38-colleen-quigley-olympic-steeplechaser/id1466187704?i=1000467131465.

342 **calling for a full investigation:** Rob Draper, "Talk to us now! Jo Pavey and Co want their say on Alberto Salazar and use of thyroxine as British athletes call on authorities to investigate UK Athletics," *Daily Mail*, March 7, 2020, https://www.dailymail.co.uk/sport/sportsnews/article-8086893/Jo-Pavey-want-say-Alberto-Salazar-use-thyroxine.html.

342 **upgraded Osaka World Championship silver medal:** Kara Goucher, Twitter.com, https://twitter.com/karagoucher/status/1241182857269137409.

342 **"I don't even know what place I got anymore":** Adam and Kara Goucher. Interview by author. Tape recording. Boulder, Colorado, September 26, 2017.

INDEX

ABOUT THE AUTHOR

MATT HART IS A JOURNALIST WHOSE WRITING COVERS SPORTS SCIENCE, HUMAN-powered adventure and exploration, performance-enhancing drugs, nutrition, and evolution. His work has appeared in the *Atlantic*, the *New York Times*, *National Geographic Adventure*, *Outside*, and *Men's Journal* magazine, among others. His reporting on the investigations into Salazar appeared on the front page, above the fold, of the *New York Times* in May 2017. An avid runner himself, Matt was once a professional ultrarunner who won trail-races from 50 kilometers to 100 miles before turning his energy and attention to journalism. He lives in the Front Range of Colorado. This is his first book.